# BETATRON UND TELEKOBALTTHERAPIE

INTERNATIONALES SYMPOSION

AM CZERNY-KRANKENHAUS FÜR STRAHLENBEHANDLUNG

DER UNIVERSITÄT HEIDELBERG

VOM 1. BIS 3. JULI 1957

HERAUSGEGEBEN VON

## J. BECKER UND K. E. SCHEER

MIT 151 ABBILDUNGEN

SPRINGER-VERLAG

BERLIN HEIDELBERG GMBH

1958

ISBN 978-3-540-02251-0          ISBN 978-3-642-85752-2 (eBook)
DOI 10.1007/ 978-3-642-85752-2





Brühlsche Universitätsdruckerei Gießen

# Inhaltsverzeichnis

# Einleitung

### Von
### J. Becker und K. E. Scheer

Da in der Röntgentiefentherapie die erzielbare relative Tiefendosis unter anderem von der Höhe der Spannung der Röntgenröhre abhängt, bestand schon immer das Bestreben zur Therapie möglichst hohe Spannungen an die Röntgenröhre zu legen. Aus praktischen Gründen ergab sich die Grenze nach oben bei etwa 200—250 kV. In den dreißiger Jahren wurden verschiedentlich Versuche unternommen, eine Hochvolttherapie zu betreiben mit Röhrenspannungen von 300 bis etwa 600 kV, doch konnten sich solche Geräte nie eine größere Verbreitung erobern, da sie einmal wesentlich komplizierter waren als 200 kV-Apparate und der Gewinn an Tiefendosis bescheiden war und in keinem rechten Verhältnis zum Aufwand stand.

Die unterschiedliche Strahlenabsorption in den verschiedenen Geweben, insbesondere die sehr viel stärkere Absorption in der Knochensubstanz, ist für die Strahlentherapie immer nachteilig. Bei Übergang zu Strahlen höherer Energie vermindern sich diese Unterschiede, doch zeigte es sich, daß gerade bei einer Steigerung von 200 auf 400 kV oder 600 kV dieser Unterschied sich nur geringfügig verkleinert. Auch aus diesem Grund vermag die Hochvolttherapie im Bereich einiger hundert kV dem Strahlentherapeuten keine besonders attraktiven Eigenschaften anzubieten.

Diese Verhältnisse ändern sich allerdings grundlegend, wenn die 1 Million-Voltgrenze überschritten wird. Bei Photonenstrahlungen in diesem Energiebereich ist die Absorption der Strahlung in den verschiedenen Körpergeweben nahezu gleich geworden. Die erzielbare relative Tiefendosis wächst natürlich mit der verminderten Absorption ebenfalls. Steigert man die Photonenenergie noch auf einige 10 Millionen Volt, so kommt es zu einem merklichen Dosisanstieg von der Oberfläche nach der Tiefe des Gewebes hin, der durch die den Primärstrahl begleitenden Compton-Elektronen langer Reichweite bedingt ist. Zwischen 15 und 30 Millionen Volt kann dieser Dosiszuwachs unter der Oberfläche einen ganz entscheidenden Effekt auf die Dosisverteilung zwischen Herd und Umgebung haben und damit ebenso wie die Angleichung der Absorption innerhalb der verschiedenen Gewebsarten zu Bedingungen führen, die sich grundsätzlich von denen der klassischen percutanen Röntgentherapie unterscheiden.

Aus diesen Gründen erscheint es gerechtfertigt, für die Strahlenbehandlung mit Photonen von Energien von mehr als einer Million Volt eine neue Gruppe mit der Bezeichnung Supervolttherapie zu bilden, der dann die konventionelle Therapie mit Spannungen bis 250 kV und die heute kaum mehr bedeutungsvolle Hochvolttherapie zwischen 250 kV und 1 MeV gegenübersteht.

Daß der Beginn der Supervolttherapie mit dem gewaltigen Aufschwung der Kernphysik in den letzten beiden Jahrzehnten eng verknüpft ist, ergibt sich einfach daraus, daß die Erzeugung solcher energiereicher Photonen nur mit

Geräten möglich ist, wie sie von der modernen Kernphysik als Teilchenbeschleuniger entwickelt wurden oder mit radioaktiven Isotopen als Strahlenquellen, die ja ihrerseits ihre Erzeugung den Errungenschaften der modernen Physik verdanken.

Sieht man von einigen van der Graff-Generatoren ab, die nach dem Prinzip der klassischen Röntgenröhre arbeitend Maximalspannungen bis 2 Millionen Volt erreichen, so bedienen sich alle anderen Verfahren der Mehrfachbeschleunigung von Elektronen. Als Geräte kommen hierfür das Betatron, der Linearbeschleuniger und das Synchrozyklotron in Betracht.

Das Prinzip des Betatrons, die Beschleunigung von Elektronen in einem magnetischen Wirbelfeld, ist erstmals in einer Patentanmeldung des Amerikaners SLEPIAN aus dem Jahre 1921 enthalten. Der erste Schritt zur praktischen Anwendung des Prinzips erfolgte durch den norwegischen Physiker WIDEROE, der an der Technischen Hochschule in Aachen theoretisch die Bedingungen für die Stabilisierung der Elektronen auf einer Kreisbahn entwickelte. Seine praktischen Versuche verliefen jedoch negativ, da er die Stabilisierung der Elektronen nur in der Ebene der Kreisbahn durchführte und durch Ausbrechen von Elektronen senkrecht zu dieser Ebene so große Verluste entstanden, daß er am Ende keine beschleunigten Elektronen nachweisen konnte.

1933 gelang dem Deutschen STEENBECK ein weiterer Schritt zur Verbesserung des Betatronprinzips. Durch eine geeignete Ausbildung des Magnetfeldes, das die Elektronen auf der Kreisbahn hält, konnte er auch die Stabilitätsbedingungen senkrecht zur Kreisebene schaffen und damit Elektronen in solcher Zahl auf der Bahn festhalten, daß Röntgenstrahlen von 1,8 MeV erzeugt und für Experimente verwendet werden konnten.

Den letzten Schritt zur Erzielung praktisch verwertbarer Ausbeuten an Strahlen unternahm der Amerikaner KERST, der mit seinem Injektor die Raumladungsschwierigkeiten bei Beginn der Beschleunigung überwinden konnte und damit in der Lage war, Elektronen in solcher Zahl auf der Kreisbahn zu halten und zu beschleunigen, daß sie nachher für eine praktische Anwendung zur Verfügung standen.

Während im Laufe der weiteren Entwicklung sich KERST in den USA vorwiegend dem Bau großer Geräte zur Erzeugung von Photonen höchster Energie zuwand, haben in Europa GUND und WIDEROE die Entwicklung mittelgroßer Geräte betrieben, die besonders für die medizinische Anwendung gedacht waren. Hierbei gelang dem leider zu früh verstorbenen GUND, die Lösung des Problems mit dem gleichen Gerät wahlweise die energiereichen Elektronen auftreten zu lassen oder sie auf eine Antikathode auftreffen zu lassen und die entstehende Bremsstrahlung zur therapeutischen Nutzung nach außen zu führen.

Die erste für die medizinische Anwendung geeignete Betatronkonstruktion ist das 24 MeV-Gerät von ALLIS-CHALMERS von 24 MeV. Es ist ein schweres ortsfestes Gerät mit waagerecht austretendem Strahl, daß mit 180 Perioden pro sec arbeitet. Die Umstellung auf Elektronentherapie erfordert das Auswechseln des Gefäßes, so daß bisher nur an vereinzelten Stellen mit diesem Gerät mit Elektronen gearbeitet wurde. Mit diesem Gerät arbeiteten HAAS, HARVEY u. Mitarb. in Chikago, WATSON, BURKELL u. Mitarb. in Saskatoon (Canada), SEAMAN und CORDONNIER in St. Louis, sowie die Gruppe um TUBIANA in Paris.

Als nächstes Gerät ist das 31 MeV-Betatron von Brown-Boveri zu nennen, das von WIDEROE konstruiert wurde. Es ist ebenfalls ein schweres Gerät, das mit 50 Hertz arbeitet und zwei Strahlenkegel in um 180° versetzter Richtung austreten läßt. Beide Strahlenkegel können unabhängig ein- und ausgeschaltet werden. In jüngster Zeit gibt es dieses Gerät auch in einer kardanischen Aufhängung, die die Durchführung von Bewegungsbestrahlung ermöglicht.

Mit diesem Gerät haben SCHINZ u. Mitarb. in Zürich und später ZUPPINGER in Bern, POPPE in Oslo und BELLION in Turin Pionierarbeit geleistet. Seit kurzer Zeit kann mit diesem Gerät auch Elektronentherapie betrieben werden.

Schließlich ist das 15 MeV-Betatron der Siemens-Reiniger-Werke, eine Konstruktion von GUND, zu nennen. Dieses Gerät hat einen Vorläufer in dem 6 MeV-Modell, das als Versuchsgerät in Göttingen aufgestellt war und nur zur Erzeugung schneller Elektronen diente. An diesem Gerät haben der heutige Heidelberger Ordinarius für Physik KOPFERMANN und die medizinisch-biologische Arbeitsgruppe SCHUBERT, PAUL, KEPP, BODE u. Mitarb. entscheidende, grundlegende, physikalische und biologische Forschungsarbeit geleistet.

Auf Grund der damaligen Arbeiten entstand das heutige 15 MeV-Gerät als Weiterentwicklung. Entsprechend der schon geschaffenen Tradition in der Elektronentherapie ist auch das 15 MeV-Gerät besonders für diese Sonderform der Supervolttherapie geeignet, da es als einziges der beschriebenen Betatronkonstruktionen sehr leicht beweglich und schwenkbar ist, so daß seine Handhabung völlig der eines 200 kV-Gerätes entspricht.

Mit dem 15 MeV-Betatron konnten SCHUBERT u. Mitarb. die mit 6 MeV begonnenen strahlenbiologischen Untersuchungen fortsetzen und wertvolle Kenntnisse für die klinische Anwendung dieser Strahlen gewinnen.

In England führte die Entwicklung der letzten Jahre zur Konstruktion von Linearbeschleunigern von zunächst 4, dann 8 und schließlich 16 MeV. Als Pioniere der klinischen Anwendung dieser Geräte sind SMITHERS u. Mitarb., sowie MITCHEL u. Mitarb. zu nennen. In jüngster Zeit kommt noch ein Synchrozyklotron hinzu, das ebenfalls von der Smithers-Gruppe klinisch angewendet wurde. Im Vergleich zum Betatron sind die Erfahrungen mit diesen Geräten noch gering.

Zu den Geräten für Supervolttherapie müssen auch die Kobaltbomben und mit anderen Isotopen geladenen Fernbestrahlungseinheiten gerechnet werden. Diese Apparate setzen unmittelbar die Entwicklung der früheren Radiumkanonen fort und zwar an der Stelle, an der wegen der geringen Menge des verfügbaren Radiums die Erhöhung der Focushautdistanz und der Dosisleistung ihr Ende gefunden hatte.

Es mag in diesem Zusammenhang erlaubt sein, darauf hinzuweisen, daß das erste Radiumfernbestrahlungsgerät im Jahre 1906 von WERNER in Heidelberg gebaut und praktisch erprobt wurde. Diese wahrhafte Pionierleistung ist weitgehend unbeachtet geblieben und heute kaum mehr bekannt.

Zu großer Bedeutung kam die Fernbestrahlung erst mit den künstlich radioaktiven Isotopen, unter denen ganz besonders das Kobalt$^{60}$ für diese Form der Therapie geeignet ist. Die ersten Entwicklungen von Kobalteinheiten mit großen Quellen erfolgten in Canada, das damals über sehr leistungsfähige Reaktoren verfügte, die zur Erzeugung von Kobaltquellen hoher spezifischer Aktivität eingesetzt werden konnten. So konnten in Canada durch die Gruppe von WATSON

u. Mitarb. die ersten klinischen Erfahrungen gesammelt werden und konstruktive Verbesserungen erfolgen.

Neben den canadischen Geräten sind eine ganze Reihe amerikanischer und europäischer Geräte inzwischen entwickelt worden und die Verbreitung der Kobaltfernbestrahlung ist heute schon so groß, daß es nicht mehr möglich ist, die einzelnen Stellen, die damit arbeiten, aufzuführen.

In Zusammenhang mit der ersten klinischen Anwendung der Supervoltbestrahlung ergaben sich wichtige Fragen zur Dosimetrie dieser Strahlen und zur biologischen Wirkung. Unser bisheriger Begriff der Strahlendosis gibt die unterschiedlichen Bedingungen bei der Supervoltbestrahlung nicht richtig wieder und der Meinungsstreit über die zweckmäßige Kennzeichnung des Dosisbegriffs auch für Supervoltstrahlen ist mit der Einführung des rad noch nicht beendet. Auch bezüglich der Meßtechnik ergeben sich eine Reihe von Schwierigkeiten gegenüber der Messung von 200 kV-Röntgenstrahlen.

In der auf Empirie fußenden Strahlentherapie kommt der Strahlenbiologie als experimenteller Basis eine wichtige Stellung zu. In einigen klinischen Beobachtungen hat man den Eindruck, daß die Supervoltstrahlung eine größere Elektivität gegenüber Tumorgewebe hat. Sollte sich dieser Eindruck bestätigen, so bleibt noch immer die Frage, auf welche der vielen, von den konventionellen Strahlen abweichenden, Eigenschaften eine solche höhere Elektivität zurückzuführen ist. Hier kann nicht die Klinik, sondern nur das biologische Experiment weiterführen.

Es wird eine ganze Anzahl von Jahren dauern, bis die biologischen Grundlagen für die Supervolttherapie erarbeitet sind. Der Kliniker aber steht unter dem Zwang, seinem Patienten sofort die besten verfügbaren Möglichkeiten für eine Heilung oder wenigstens für einen palliativen Erfolg anzubieten. Wenn die Geräte vorhanden sind, kann der Kliniker nicht warten, bis die theoretischen und praktischen Probleme der Dosimetrie endgültig geklärt und die biologischen Grundlagen der Supervolttherapie erarbeitet sind. Er muß mit der Therapie beginnen und seine Technik aus eigener Erfahrung heraus verbessern.

Gerade deswegen ist es außerordentlich wichtig, daß Physiker, Strahlenbiologen und Kliniker sich immer wieder gegenseitig über ihre Beobachtungen und Erfahrungen unterrichten und daß die Ergebnisse der Therapie und der Forschung innerhalb der einzelnen Zentren, die sie gewonnen haben, ausgetauscht werden. Aus dieser Vorstellung heraus entstand der Gedanke, die Vertreter der Klinik, Physik und Biologie der wichtigsten mit Supervoltstrahlen arbeitenden Zentren zu diesem Symposion nach Heidelberg zusammenzurufen, um ihnen im Austausch ihrer Erfahrungen und Ansichten neue Anregungen zu geben.

Die Anwendung von Supervoltstrahlen, sei es als ultraharte Röntgenstrahlen des Betatron, sei es als Gammastrahlen der Kobaltgeräte, bringen im Vergleich zu den konventionellen Röntgenstrahlen dem Radiologen und dem Patienten wesentliche Vorteile. Es kann heute kein Zweifel mehr darüber bestehen, daß in einigen Jahren die Supervolttherapie die klassische Röntgentiefentherapie mit 200 oder 250 kV-Strahlen in breiter Front ablösen wird.

# Klinische Erfahrungen
# mit energiereichen Photonen

---

## Clinical Experiences with Supervoltage Radiation

By

**E. D. Jones**, London

In considering the value of any form of supervoltage therapy it is useful to consider how it differs physically and biologically from conventional therapy. We have had a 2-million volt Van de Graaff generator in use as a clinical machine since 1950 and have used a 30 MeV synchroton on small experimental groups of patients for almost as long, and in November last a telecaesium unit went into clinical use. Each of these, to a greater or lesser extent, differ in the same ways from more conventional therapy.

| *Physical:* | *Biological:* |
|---|---|
| 1. Increased depth dose | 1. Better survival rate |
| 2. Decreased bone absorption | 2. Greatly improved palliation |
| 3. Diminished ion density | 3. Less severe tissue reaction |
| 4. Displaced maximum dose | 4. Better normal tissue recovery |
| 5. Improved estimation of dose distribution | 5. Less fibrosis and bone necrosis |
| 6. Decreased total energy absorbed | 6. Less radiation sickness |

*Increased depth dose* is very desirable in many ways, but it can have its disadvantages, because with increasing penetration, so there is increasing dose at the exit point.

There is obvious improvement in depth dose at 2 MeV and again at 24 MeV, but even so the maximum is too superficial to reach many tumours by a single direct field. By using moving beam techniques the increase at 2 MeV can be used to the best advantage.

Our telecaesium unit works at a S. S. D. of 40 cms. and the central axis depth closely resembles that of an X-ray beam of H. V. L. 3 mm. cu. at 50 cm. F. S. D. Our Unit has a cylindrical source 2.9 cm. in diameter and 3.9 cm. long. The adjustable diaphragms are so mounted that a line passing along their edge will always hit the circumference of the front end of the source. Due to the penumbra a problem arises as to what size any given field should be called and the convention that is used is that the light beam and the 50% penumbra correspond and it is these dimensions that are used as the nominal field size. This works well in clinical practice and where there are two fields adjacent, a gap of about 1 cm. is left between them and the penumbra evens out the dose at the junction.

*Displacement of maximum energy absorption.* The displacement of the maximum dose at 24 MeV is obvious and is deep enough at 2 MeV to be well below the erythema producing layer with consequent great diminution in skin reaction and epilation. Using a telecaesium unit, by calculation the displacement of the maximum should just be sufficient to cause useful reduction in skin reaction and this is found in practice to be so. With this unit any build-up in the form of a jig will bring the maximum into the skin, as also may happen if the beam passes very obliquely through the skin.

*Decreased differential bone absorption.* Decreased differential bone absorption is probably the most important single advantage of supervoltage therapy. At

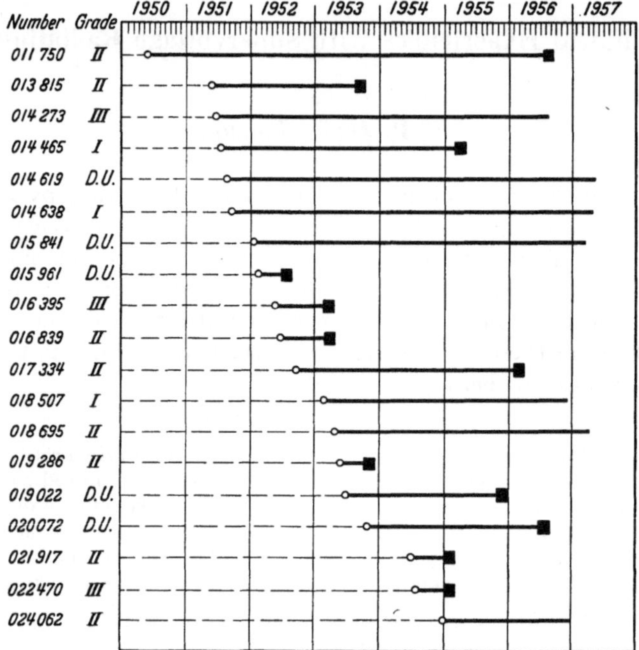


Fig. 1. Patients with cerebral gliomas treated on the 24 MeV-Synchrotron
at the Royal Marsden Hospital 1950—1954

low voltages bone absorption may be 4—5 times that of muscle, which means that the bone will suffer undue damage and will shield a tumour lying deep to it. The advantages of using a high energy beam to overcome this is not new; short S. S. D. teleradium units have been used with success for many years on the head and neck where the beam has to pass through bone, but great penetration is not needed.

The range from about 2—20 million volt X-rays has the advantage of a good depth dose, minimal difference in relative bone absorption without too great an exit dose. This range is already well known as supervoltage. If voltages higher than these are used new problems will arise and a new name for this type of therapy would be useful.

In considering the relative bone absorption, in practice it must not be forgotten that effective hardness of a beam of radiation falls off at a depth due to scatter and that the spectral distribution of the radiation is of significance.

Using a telecaesium unit, the primary radiation is monochromatic 660 kV radiation and the energy absorbed by bone will be much less than that from higher kV peak voltage of continuous spectrum ray, which contains a proportion of comparatively soft radiation.

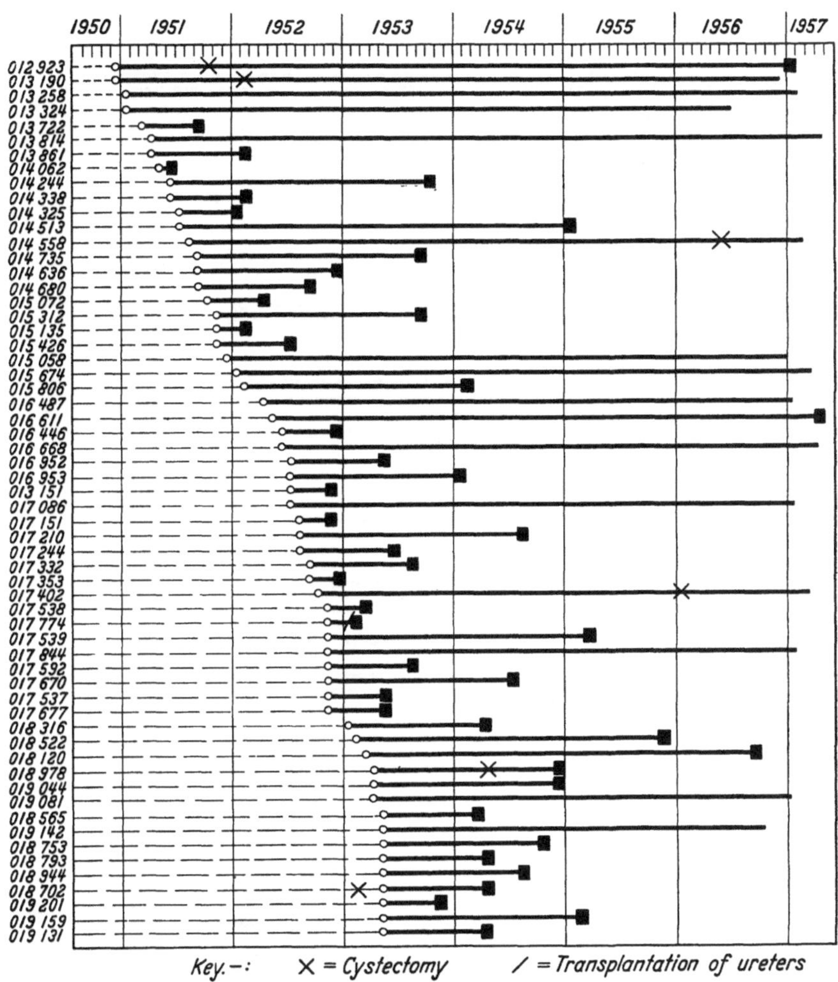


Fig. 2. Patients with advanced carcinoma of the bladder treated with 2 MeV rotationtherapy
at the Royal Marsden Hospital 1950—1953

Looking at it in the simplest terms a low voltage radiograph shows bone well but a supervoltage radiograph will not. With the telecaesium unit the effect is the same, but due to the large source size there is much unsharpness.

*Improved assessment of dose distribution.* From the fact previously mentioned that bone may shield a tumour lying deep to it, it follows that estimations of dose taken from measurements in a water phantom will cause errors usually on the high side. This is particularly true in the 250 kV range but in the supervoltage range and with telecaesium this difference is too small to be of great clinical significance.

*Diminished ion density.* There is a physical difference in the energy absorption by tissue with increasing hardness of the radiation. The primary disturbance produced by irradiation only occurs in a few hundred molecules for each cubic micron of tissue but its effect is profound. Experimental evidence shows that

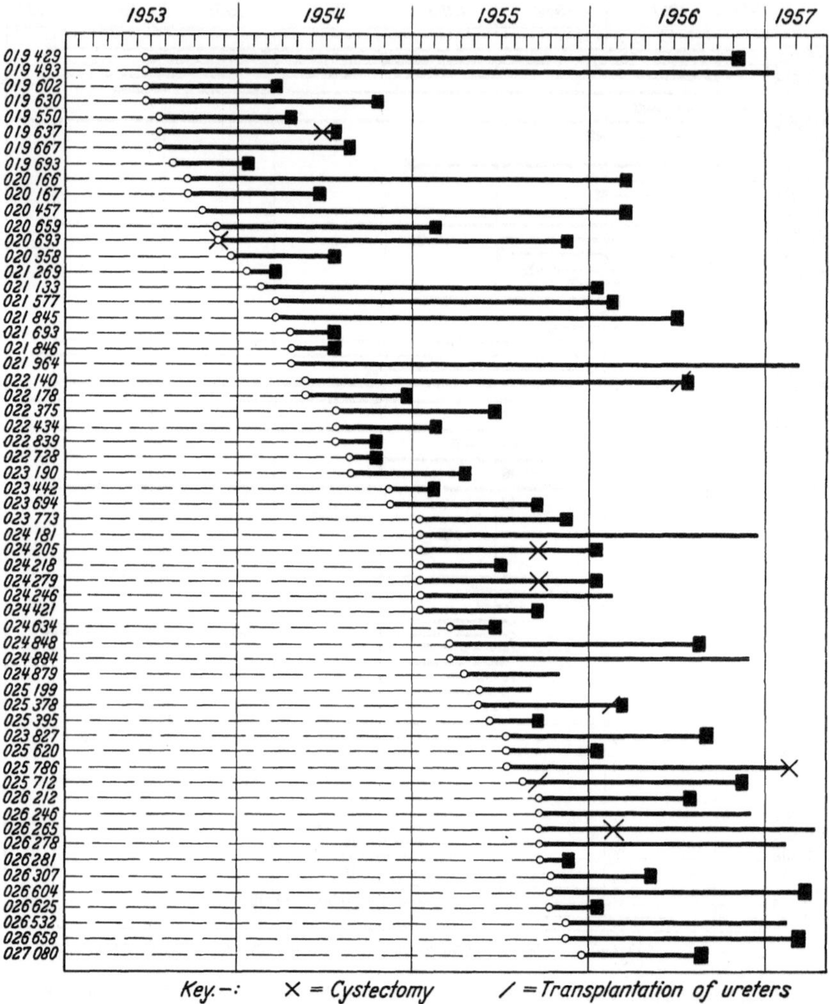



Fig. 3. Patients with advanced carcinoma of the bladder treated with 2 MeV rotationtherapy at the Royal Marsden Hospital 1953—1955

the linear ion density or linear energy transfer has great influence on the changes that occur. The linear energy transfer changes by a factor of about 8 as the hardness of the radiation increases from 250 kV to 2 MeV and with further increases changes slowly.

What the direct influence of this on clinical radiotherapy should be is difficult to assess, but it must always be considered when the effect on tissue of irradiation at differing voltage ranges are being compared.

*Lower integral dose.* As the energy of the radiation is increased, so the percentage of scatter is decreased.

When fixed physical factors and the thickness of the patient remain constant and the energy of the radiation increases, so the integral dose decreases, up to the million volt range, with slight further decrease with very high voltages. Beams directed nearly parallel to the long axis of the trunk would have a larger integral dose.

**Biological Differences.** *Less bone necrosis.* The use of supervoltage has to a very large extent eliminated bone necrosis, and enables treatment to be carried out through bone without the hazard of necrosis, which was always present when only lower energy radiation was available.

*Less severe tissue reaction.* There can be little doubt that supervoltage therapy causes less tissue reaction. This is partly due to physical factors, such as displacement of the maximum below the surface. There is a slight but definite diminution in skin and tissue reaction with a telecaesium unit, as compared to 250 kV X-rays and a further diminution with 2 MeV therapy. This makes possible useful palliation in sites where it would be hazardous to use other than supervoltage.

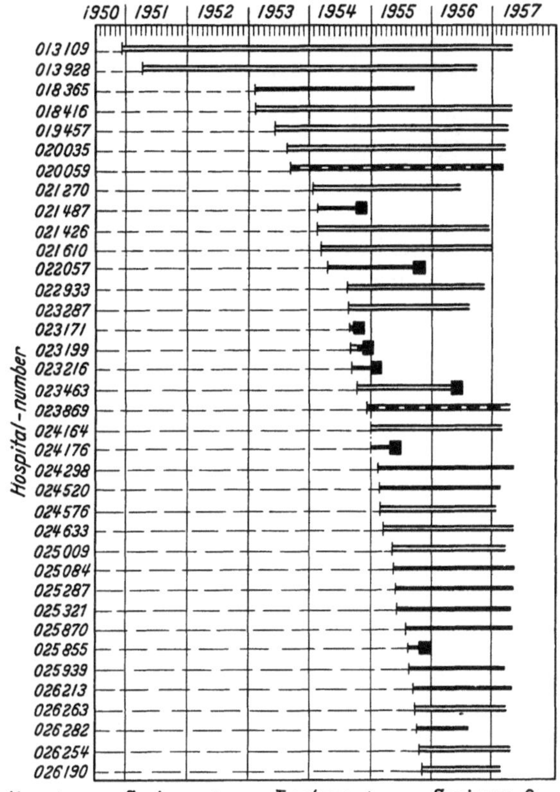


Fig. 4. Patients with tumours of the testicle treated postoperatively on the 2 MeV machine at the Royal Marsden Hospital 1950—1955

*Normal tissue repair* seems to take place more quickly. It is all part and parcel of the tissue damage which has just been mentioned and is probably closely related to differences in ion density.

*Greatly improved palliation.* With supervoltage greatly improved palliation can be achieved, both in regard to the results in any individual case and also in the number of cases in which it is possible to undertake treatment at all. This is closely tied up with biological advantages previously mentioned and also with the fact that the treatment is less likely to cause severe local or general reaction and is therefore more readily undertaken. To a lesser extent this is probably true of telecaesium.

*Survival rates.* The usual way to express results is by survival rates and supervoltage therapy has already shown that it is capable of improving these. A few cases are now appearing that are surviving for a time that would have been most unusual a few years ago when supervoltage was not available.

Table 1. *Rough Estimate of Cost of Providing Good Quality Radiation at a Depth*

| Apparatus | Initial cost written off in 10 years £ | Protection of staff written off in 20 years £ | Maintenance per annum. £ | No. of patients treated per annum | Cost per patient treated £ |
|---|---|---|---|---|---|
| 250 kV | 4,500 | 1,000 | 500 | 400 | 2. 10. 0. |
| Cs 137 | 5,000 | 4,000 | 400 | 400 | 2. 15. 0. |
| Co 60 | 12,000 | 6,000 | 700 | 400 | 5. 10. 0. |
| 2 MeV | 20,000 | 6,000 | 1500 | 600 | 6.  6. 8. |
| 4 MeV | 45,000 | 8,000 | 2000 | 1100 | 6.  5. 5. |
| 30 MeV | 50,000 | 10,000 | 2000 | 500 | 15.  0. 0. |

Although it is interesting to consider both what might be achieved with very high energy machines and what is available, the cost of each unit must be considered.

A telecaesium unit has several claims to be classed in the supervoltage category but though it has many advantages over conventional therapy, it must not be regarded as having advantages other than cost over a true supervoltage machine.

# The 4 MeV Linear Accelerator for X-ray Therapy

By

F. T. Farmer, Newcastle upon Tyne

It is well known that above about 500 kV, the acceleration of electrons by steady d. c. voltages becomes extremely difficult, and the generation of electron beams or X-rays at higher energies has depended largely on the use of multiple acceleration techniques. Of these the Betatron provides the best known example and this has been used very successfully in Germany and elsewhere as a therapeutic instrument. Voltages as high as 20—30 MeV are now readily obtainable.

In the Betatron the electrons travel in a circular orbit making many thousands of revolutions before they finally emerge or strike a target, producing X-rays. The idea of accelerating electrons in a straight line to energies of a comparable order, by an alternating electric field, was not thought to be feasible until the high power Magnetron valve was developed for Radar[1], which opened up the possibility of generating extremely high intensity wireless waves of 10 cm. wavelength or less. Such waves may be caused to travel inside a copper tube or "wave-guide" which has a diameter of the order of 1 wavelength, and under these conditions a new and important feature of the wave emerges, namely that it acquires a *forward* component of electric field — a field in the direction in which the wave is travelling. Thus, if an electron is moving along the axis of the wave-guide the possibility arises that it will remain in step with the electric field and so gain in energy continuously as it travels along. The total energy received will then be equal to the electric field of the wave multiplied by the distance travelled. With modern Radar valves generating 2 million watts or more of power, this electric field may be extremely strong, and hence a high energy may be acquired by the electron in a short distance of travel.

The linear accelerator depends on this principle. The acceleration can only be achieved if the electrons can be kept in step with the travelling wave over a distance of many wavelengths. To bring this about it is clear that the velocity of the wave must be carefully adjusted so that it is always exactly equal to that of the moving electrons: it must be slow where these enter the guide and fast where they approach the target. The electrons are injected with an initial energy of 50 kV, which means that at the start the wave velocity must be about half that of light, and it must increase gradually towards that of light as the electrons gain energy in their travel.

This control of wave speed is obtained by means of iris diaphragms or "corrugations" closely spaced along the inside of the guide (Fig. 1). Where these corrugations are deep, i. e. the aperture through the centre of each diaphragm is

---

[1] Pulsed radio transmissions of very short wavelength for navigation

small, the velocity is low; where the corrugations are shallow and the aperture correspondingly large, the velocity is greater. In practice the aperture increases from about 2—2.6 cms. as the energy of the electrons increases from 50 kV to 4 MeV.

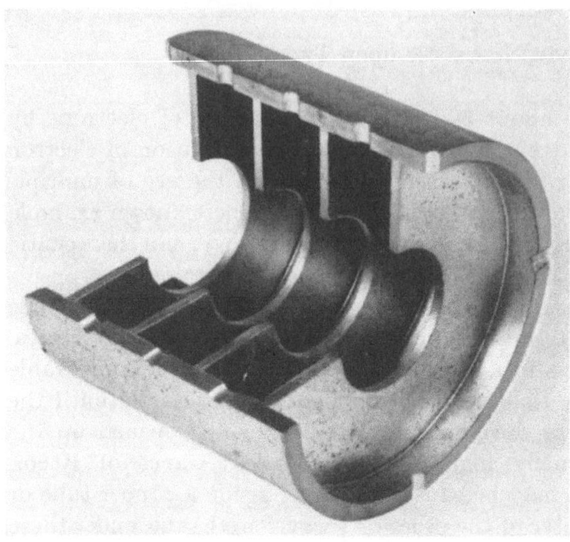

Fig. 1. Portion of corrugated wave-guide. The size of the "iris" diaphragms determines the velocity of the wave, which is made to increase exactly in step with the accelerating electrons

For successful operation the dimensions of each section of waveguide must be extremely precise so that the phase of the wave is preserved within a few degrees relative to the accelerating electrons over the entire length of the guide, 10 wavelengths in the 4 MeV machine. This imposes severe design and constructional problems and the success with which it has been achieved is a credit to the original designers, D. W. Fry and his colleagues at Malvern [1] and to the manufacturers[1], Messrs. Philips Electrical Ltd., London in the present instance.

The lay-out of the 4 MeV linear accelerator, including the magnetron and wave guide system, is shown in Fig. 2. The electron gun is seen at the top, from

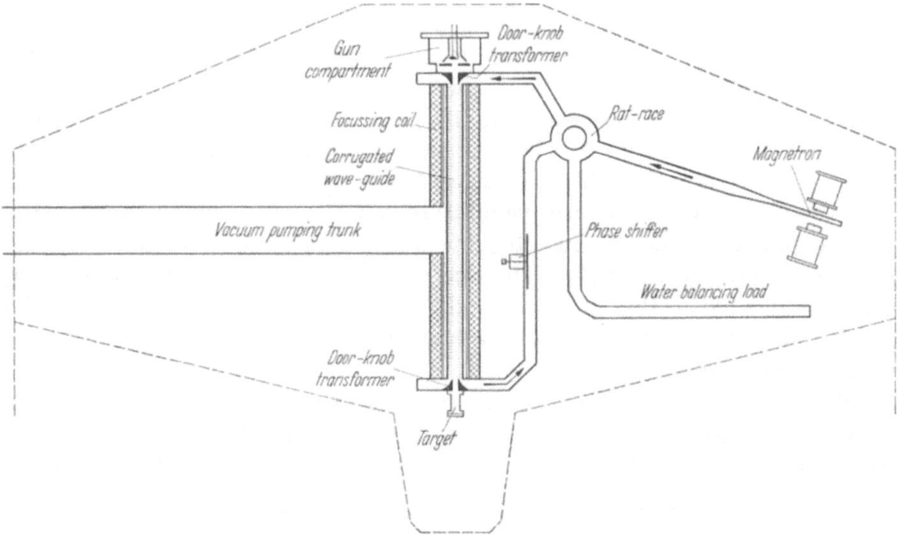


Fig. 2. Lay-out of 4 MeV machine. The magnetron, corrugated wave-guide and feed-back loop are shown. The whole system is evacuated through the duct on the left

<hr>

[1] At Mullard Research Laboratories, Salfords, Surrey

which the electrons enter the main corrugated guide and through this travel towards the platinum target at the bottom. This guide is 100 cms. long. The extremely rapid acceleration achieved in this distance is facilitated by the addition of a "feed back" system [2] carrying radio wave power from the lower end of the guide through the non-return transformer or "rat race" seen on the right, and so back to the start, instead of its being wasted in a water load. The effect of this is to increase the 2 MW of power from the magnetron to 6 MW flowing in the main wave guide, and so gain correspondingly in accelerating voltage. The accelerating voltage in this guide is approximately 40 kV per cm.

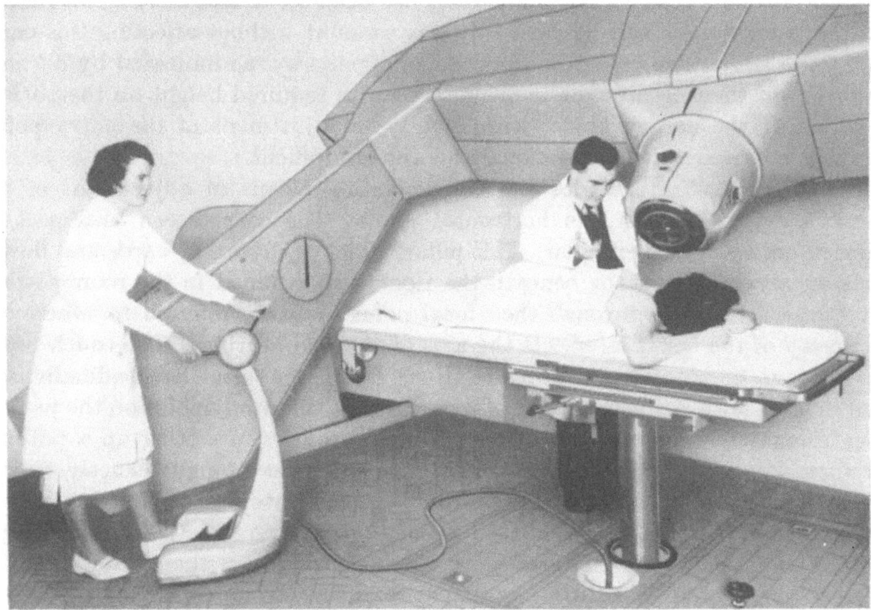

Fig. 3. The gantry mounted 4 MeV linear accelerator and treatment couch

The electron energy obtainable from a linear accelerator depends on the length of the wave guide and also on the power of the magnetron. The length of guide may be increased to several metres without electrical difficulties and this has been the basis of design of the 8 MeV machine at Hammersmith Hospital, London [3], and of the 15 MeV machine at St. Bartholomews Hospital, London [4].

Higher energies still can be obtained by the use of Klystron valves in place of Magnetrons, and machines giving energies up to 50 MeV are already being produced. These valves are still, however, in an experimental stage.

**Gantry Mounting.** An important feature of the present machine is the fact that the whole accelerating system is mounted on a moving gantry which is pivotted on bearings at the two ends of the treatment room, as can be seen in Fig. 3. By this means the source of X-rays can be moved in a circle about a fixed centre or "focal point" in the treatment room. The beam can thus be directed at any angle to the patient, always from a fixed distance, in this case 100 cms. This mounting adds to its size, but gives greatly increased freedom for setting up patients and has been found to be well worth the elaboration involved.

It may be used for pendulum therapy in the same manner as the 15 MeV Betatron or other types of moving field apparatus. For this purpose the tumour is brought to the centre of rotation, and the machine is driven through its required range of angle by a variable speed electric motor.

Alternatively it may be used in conjunction with its specially designed couch to facilitate quick and accurate setting up of patients for fixed field treatment. In this case the centre of rotation is made to coincide, not with the tumour, but with the skin entry point for any given field. The focal-skin distance is now constant (100 cms.), and when the patient has been set to this distance, it follows that the gantry angle can be varied by any amount without affecting this entry point. The advantage of this is that the emergent ray, as indicated by a "back pointer", can then be brought immediately to its required height on the patient by adjusting the gantry angle, without any re-adjustment of the entry point, such as is required with conventional therapy equipment.

The couch design provides the corresponding means of adjustment of the emergent ray pointer in the horizontal plane. The couch, seen in Fig. 3, is mounted on a single steel pillar. This pillar, which is driven upwards and downwards by an electric motor beneath the floor, is positioned in the room so that its vertical axis passes through the "focal point" already referred to, where the central ray of the beam intersects the axis of the two bearings. The couch top is free to move horizontally in two directions at right angles (longitudinally and transversely), with sufficient range of movement to allow any point on the patient to be brought to the fixed "focal point" in the room. In setting up a patient, therefore, the required entry point on the skin is first brought exactly to this "focal point" as shown by light beams from the accelerator head. This is achieved by variation of couch height and by adjustment of the two directions of movement of the couch top. When this has been done, not only can the gantry angle be varied as already explained to bring the emergent ray to its correct height on the patient, but the couch can be rotated about its pillar to bring this ray to its correct position horizontally on the patient — both without affecting in any way the entry point of the beam.

This system allows rapid and accurate positioning of fixed fields. It avoids the "trial and error" process that is required with most X-ray apparatus and reduces considerably the time taken for setting-up of precisely directed beams.

It should be noted that this type of mounting can be used only if the wave guide does not exceed about 100 cms. in length. Any greater length would not allow the guide to be mounted at right angles to the main span of the gantry; therefore, the energy attainable with this form of construction is limited to approximately 4 MeV. Similar movements of the machine relative to the patient have been achieved in the 8 MeV machine, though in this case by using servo couplings to a rising floor and moving couch instead of allowing the accelerator itself to move.

**Physical Performance.** The X-ray output of the machine can be set to any level up to 250 r/min. This is with the beam "flattening filter", required to correct for the forward direction of transmission of X-rays from the target, in use. This filter absorbs about 30% of the radiation as measured on the central ray. Normally it is operated at about 130 r/min., for which treatment times are of the order of two minutes. The depth of maximum X-ray dose in the body is about 1 cm. and the electron range about 2 cms. The first of these results in an almost complete sparing of the skin from radiation damage, and the usual advantages of supervoltage therapy are achieved. The limited electron range means that electron therapy is of doubtful value at this energy, and it has not in fact been used with the present machine.

A sharp cut-off at the edges of the X-ray beam is achieved by mounting the beam-defining diaphragms (which are of lead 10 cms. thick) on arms pivotted at target level, instead of arranging them as is usual to move in a plane perpendicular to the axis of the beam. This is illustrated in fig. 4. The inner surfaces of the diaphragms are thus kept always tangential to the edge of the beam, and the penumbra is thereby reduced to a very small value. Measurements show that at the surface of the patient the penumbra is about 5 mm. wide and at a depth of 20 cms. it is 9 mm. This sharp boundary to the beam assists in keeping the integral dose down to as low a level as possible.

**Treatment Planning Unit.** Since time occupied on the linear accelerator itself is valuable, it was considered uneconomical to use this machine for the planning of treatments. It was decided, therefore, to build a full size model of the accelerator and its mounting in which a diagnostic X-ray tube takes the place of

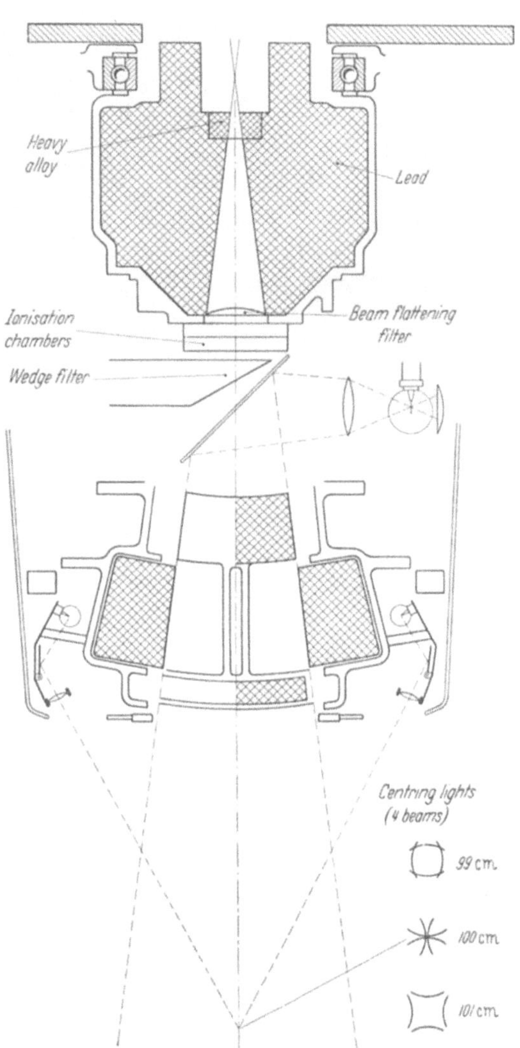


Fig. 4. Section of X-ray head, showing diaphragms pivotted at target level moving with their inner surfaces tangential to the beam. The "range finder" lights centring at 100 cms target distance are seen

the 4 MeV wave guide and target. An outline of this is given in Fig. 6. The X-ray tube is mounted on one arm of a U-shaped support, the opposite arm carrying an image intensifier for observation of the exit beam. Owing to the serious problem of protection of the operator with the wide range of beam angles and large focal distances employed it has been arranged that the screen of the image intensifier is

 F. T. Farmer:

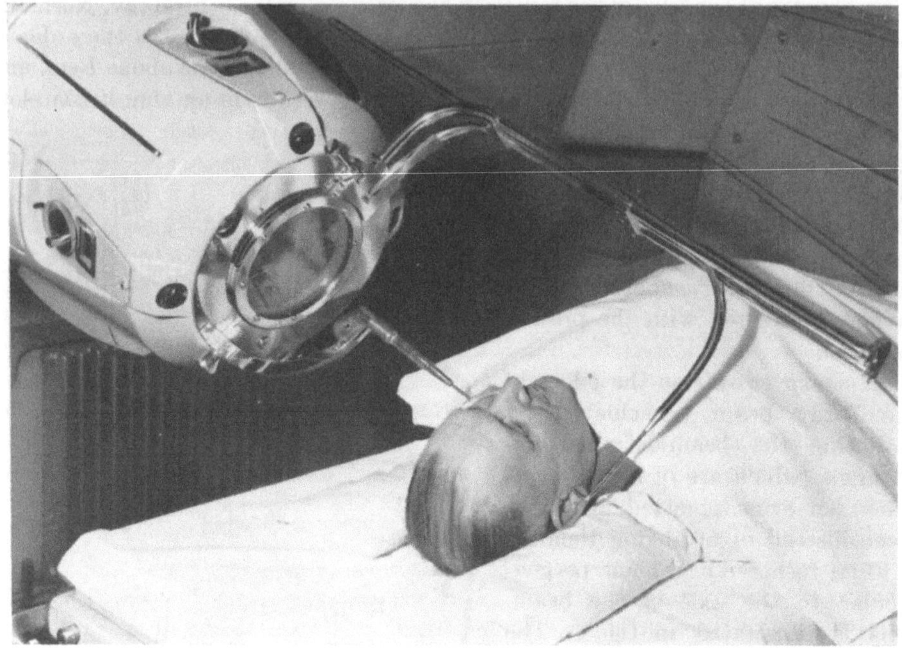

Fig. 5. The mounting facilitates rapid and accurate setting-up of patients. The forward pointer seen in this figure is normally substituted by the four range-finder lights

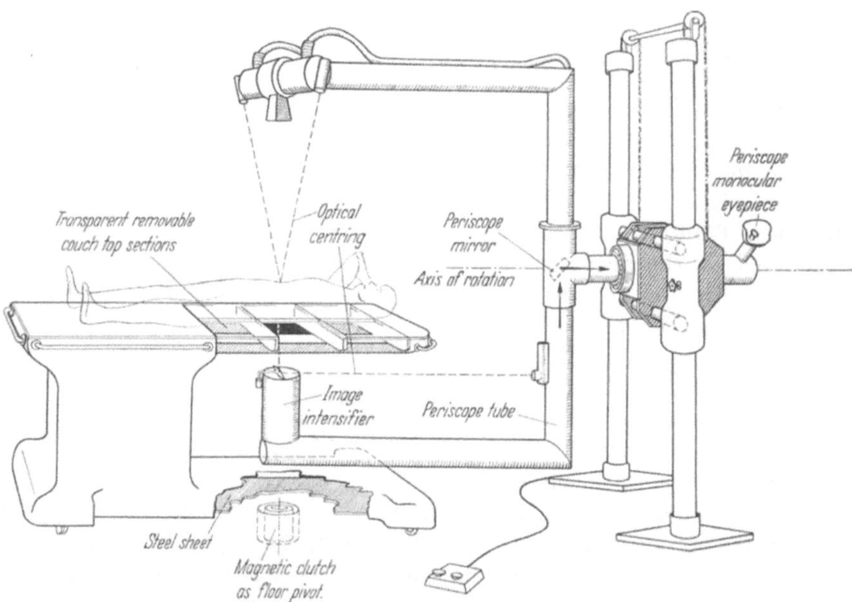


Fig. 6. Treatment planning unit, with diagnostic X-ray tube. The dimensions and relative movements are identical to those of the linear accelerator

viewed through a periscope which passes through the main bearing, leading to an eye-piece on the opposite side of the main stand. A protective screen can thus be interposed between the patient and operator without difficulty. The arrangement gives very satisfactory viewing and is much more convenient than direct observation of a fluorescent screen in a darkened room. It means, of course, that the image in the eye-piece rotates about its axis as the U-shaped arm is turned. The addition of a directional marker to the intensifier screen however, overcomes this difficulty satisfactorily.

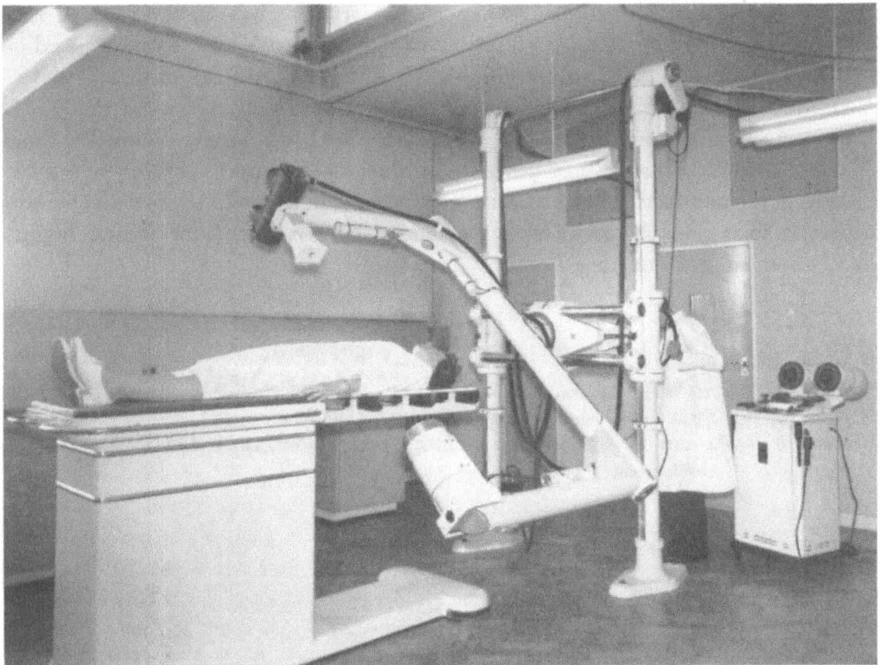

Fig. 7. Treatment planning unit. The operator can be protected by a lead-rubber screen as he views the image intensifier through the periscope system

The vertical movement of the treatment couch of the linear accelerator is replaced by a vertical movement of the whole tube and image intensifier assembly on the planning unit, so that a fixed height of couch may be used. This enables the top of the couch to be built in an overhanging manner as seen in Fig. 6 and 7 and the pivoting centre in the floor provided by the treatment couch pillar is now replaced by a magnetic clutch in the floor which can be switched on when it is desired to rotate the patient about this point. The geometrical relationships of couch, axis of gantry and direction of central ray from the tube are all identical to those of the treatment unit.

Casettes holding films can be placed in grooves under the Perspex panels of the couch top, and any one of these panels can also be removed at will to give access to a part of the patient. Optical beams for the forward and emergent ray centring points are provided and an optical "range finder" indicates the 100 cms. distance from the tube target.

The treatment planning unit is used to study generally the best approach to treatment of any given tumour. Once a plan has been tentatively prepared this unit serves to check the entry and exit points of oblique or vertical beams against anatomical landmarks in the body or against the tumour itself if this is visible. Normally it is employed only at the commencement of a course of treatment. Experience shows that the setting up of any given field can be repeated with the necessary degree of accuracy is successive daily treatments, though the beam directions can, of course, be checked by this unit at any stage.

**Conclusions.** As a means of generating high power X-rays of energy 4—20 MeV, the linear accelerator has come to take an important place in the field of radio-therapeutic equipment. With its high output of 200 r per min. or more, and ease of operation and maintenance, it allows an intensive clinical programme to be carried out with the minimum of interruption for servicing.

The 4 MeV machine installed at Newcastle, with its gantry mounting, facilitates the accurate setting up of patients and allows as many as 60 patients to be treated in a normal day. This 4 million volt radiation appears to give all the major advantages of supervoltage therapy without the complications which begin to arise at much higher energies.

## References


1. Fry, D. W.: Philips Techn. Rev. 14, 1 (1952).
2. Shersbie-Harvie, R. B. R., and L. B. Mullett: Proc. Phys. Soc. **62 B**, 270 (1949).
3. Miller, C. W.: Proc. Inst. Electr. Engrs. **101**, 207 (1954).
4. Bareford, C. F., and M. G. Kelliher: Philips Techn. Rev. **15**, 1 (1953).

# Clinical Experience with the 4 MeV Linear Accelerator*

By

C. J. L. Thurgar, Newcastle upon Tyne

Although the linear accelerator is a relative newcomer to the field of mega-voltage therapy it is, in our judgement, an important machine and possesses several features which are of particular value in clinical use. The 4 MeV set at Newcastle has been in operation for $3^1/_2$ years and although some 1700 cases of malignant disease have now been treated, it is not possible yet to give any statistical record of the results, as we have no figures relating to more than three years survival.

I would summarise the main features of this machine as first, its high X-ray output of penetrating radiation and second, the special form of mounting which allows rapid and accurate setting up of treatment fields. Like all megavoltage equipment, it is expensive but the cost is justified by the large number of patients that can be treated as a result of the two features mentioned, neither of which can be used to maximum advantage without the other. Working with a dose rate of 100—150 r/min., an average of 16 fields can be treated in one hour. As a result, we can deal with 60 patients in a normal day's programme without undue stress and during the year 1956 we were able to treat 565 patients with the linear accelerator out of a total of 2,250 new cases of malignant disease referred to the Radiotherapy Centre in Newcastle.

After an initial period of varying the total dose and treatment time, we have come to the conclusion that a dose rate of 1,000—1,200 rads a week is an optimum figure for the majority of cases. The tumour dose varies from 4500—7000 rads according to the site and the volume of tissue treated, while the duration is normally between four and six weeks. We use the accelerator for the irradiation of deep seated tumours anywhere in the body but our main experience has been gained in cancer of the bladder, the cervix, the breast, the lung and epithelial cancers in the head and neck region.

Bladder carcinoma is treated by giving the whole bladder a dose of 6,500 rads in six weeks through three fields, one anterior and two posterior oblique. This avoids giving a high dose either to the skin or to the rectum. Cancer of the cervix is first treated by local radium and then the lymph nodes on the pelvic wall are given a dose of 4,000 rads in four weeks by two large fields to the pelvis, front and back, with the central portions of each field blocked out by a lead screen to prevent overdosage of the tissues already treated by the radium. This lead screen is supported on a tray of Lucite which is attached below the X-ray head of

---

* Der Beitrag kann aus technischen Gründen nur als Zusammenfassung wiedergegeben werden.

2*

the accelerator and is sufficiently far from the skin to prevent secondary radiation reaching the patient. Breast cancer and lung cancer are both treated postoperatively in the majority of cases. After a radical mastectomy, a single anterior field is applied to the mediastinum and the root of the neck, the lung being protected by a lead screen, and a dose of 5,000 rads is given in four weeks. The same dose is delivered to the mediastinum for lung cancer after pneumonectomy or lobectomy. For epithelioma in the head and neck region, treatment is given through one, two or three fields to a dose of 6,000—7,000 rads in six weeks and much use is made of wedge filters in order to avoid opposing fields when the emergent beam is apt to produce too high a dose in the subcutaneous tissues.

The radiation quality of the 4 MeV machine brings it into a class between $Co^{60}$ telecurie apparatus and the betatron. The maximum dose is delivered at a distance of 10 mms. under the skin and as would be expected, skin reactions are normally non existent, though moist peeling can occur with fields tangiential to the skin or in folds such as the groin and between the buttocks. Mucosal reactions appear to develop rather more rapidly than with X-ray therapy at conventional voltages but they are less severe and they heal more quickly. Soft tissue reactions adjacent to the tumour are less marked and for an equal tumour dose, the late effects of radiation are correspondingly less severe. Such sequels as a permanent dry mouth or a contracted bladder have only rarely been observed and our surgical colleagues state that the difficulties of surgical excision following radiotherapy are much less when 4 MeV radiation has been used than after treatment at 250 kV. Systemic effects during treatment, notably radiation sickness, are also less troublesome than at the lower voltages.

In conclusion I would like to say that although the 4 MeV linear accelerator is large in size, it is a precision instrument and on grounds of speed, ease of operation and lack of discomfort to the patient, I believe it to be the best all round X-ray therapy set that I have yet seen.

## Diskussionsbemerkungen

R. WIDERÖE (Baden/Schweiz):

Werden in England die hochenergetischen Elektronen auch extrahiert? Bei einer Energie von 24 MeV wäre Elektronentherapie von großem Interesse.

H. R. SCHINZ (Zürich):

1. Wenn wir relativ wenig Patienten mit dem Züricher Betatron von 31 MeV bestrahlt haben im Vergleich zu den Herren Engländern, so rührt das davon her, daß wir im Anfang sehr vorsichtig und rückhaltend waren, denn wir hatten das 1. Betatron in Europa und konnten uns nicht auf die Erfahrungen anderer stützen, auch nicht auf die Schubertsche Schule mit dem Betatron von 4—6 MeV, denn das war keine $\gamma$-Therapie wegen der zu geringen Intensität, sondern eine Elektronentherapie. Begonnen haben wir in Zürich vor 6 Jahren 3 Monaten.

2. Wie groß waren bei den Herren aus England die Maximaldosen und in welcher Zeit wurden sie appliziert?

3. Wie ist das Intensitätsspektrum des Linearaccelerators? Das Züricher Betatron hat seinen Schwerpunkt bei etwa 11 MeV; dies ist der Durchschnittswert.

O. DAHL (Stockholm):

1. Wenn man eine genaue Vorkontrolle der Einstellung bei jeder einzelnen Behandlung macht, was wir am Radiumhemmet für wichtig halten, wie ist es dann möglich, 60 Patienten pro 6 Std. zu behandeln?

2. Haben Sie in der klinischen Arbeit Anhaltspunkte für die Möglichkeit gefunden, höhere Tumordosen mit energiereichen Strahlen als mit den konventionellen Röntgenstrahlen in den Fällen zu geben, in denen man mit den letztgenannten Strahlen eine ausreichende Herddosis nicht geben kann?

3. Verwenden Sie das Durchleuchtungsgerät jeden Tag bei jedem Patienten?

M. TUBIANA (Paris):

1. Gibt es eine Verschiebung der Lage beim Blasentumor, wenn der Patient auf dem Rücken oder auf dem Bauch liegt?

2. Können Sekundär-Elektronen von der Blende den Patienten erreichen?

E. SCHERER (Marburg):

Wir durchleuchten seit einiger Zeit die Blasencarcinome auf dem Müller-Gerät UGX, d.h. also im Liegen, in der gleichen Lage wie sie bestrahlt werden im frontalen Strahlengang, machen gleichzeitig eine Cystographie und sehen, daß die Blasen ganz anders liegen, als man es sich vorher anhand der Hohlfelderschen, an der fixierten Leiche gewonnen, Situationsskizzen vorstellt. Beim Mann rutscht die Blase wesentlich tiefer in die Kreuzbeinhöhle hinein. Die durchschnittlichen Herdtiefen liegen zwischen 10, 12 bis zu 16 cm von der Oberfläche.

B. RAJEWSKY (Frankfurt a. M.):

Ich möchte die Herren JONES, THURGAR, FARMER und SCHINZ fragen, ob sie Neutronen im Strahl mit einem Gerät gemessen haben, wie sie diese Neutronen gemessen haben und wie groß ihr Anteil war.

P. PALEANI (Roma):

Ich erlaube mir anzufragen, ob Herr Dr. THURGAR bei seinem Fall einer mediastinalen Bestrahlung Veränderungen am Herzen beobachtet hat. Wir selbst sahen bei einem Fall eines Lungentumors eine extraventrikuläre Arrythmie mit letalem Ausgang nach Bestrahlung unter Einbeziehung der Herzregion.

H. J. MAURER (Bern):

Demonstration von Isodosen, die an einem Beckenphantom (Parafin-Magnesiumoxyd) mit eingebautem knöchernen Becken bei Rotationsbestrahlung mit 15 MeV Röntgenstrahlen (Betatron: SRW, Erlangen) gewonnen wurden; die Untersuchung wurde durch das liebenswürdige Entgegenkommen von Herrn Prof. Dr. S. SCHUBERT, Hamburg-Eppendorf, ermöglicht und mit Unterstützung des S-Labors, (Leiter: Dr. MALSCH) der SRW, Erlangen, vorgenommen; die Feldgröße betrug $4 \times 10$ cm der Rotationswinkel $\pm 90°$.

Es ergibt sich bei diesem relativ großen Feld ein Maximum im Bereich der Parametrien; der Dosisabfall gegenüber der Beckenmitte zum Schenkelhals ist steil. Die maximale Belastung der Haut beträgt etwa 20%.

Bei Wahl schmalerer Felder kann die übliche kombinierte Radium-Röntgen-Bestrahlung, bei breiteren Feldern (bis 6 cm) u. a. eine alleinige Röntgenbestrahlung durchgeführt werden (15 MeV).

K. E. SCHEER (Heidelberg):

Nach unserer Erfahrung ist für die Einstellung eines Patienten und die Durchführung der Bestrahlung ein guter Lagerungstisch fast ebenso wichtig wie ein zuverlässiges und gut bewegliches Strahlungsgerät. Ich möchte daher Herrn Dr. THURGAR fragen, welche Eigenschaften der Lagerungstisch nach seiner Meinung haben sollte und ihn bitten, uns noch einige Einzelheiten über die Konstruktion des von ihm verwendeten Tisches zu sagen.

E. D. JONES (London):

Das 24 MeV-Synchrotron, mit dem wir gearbeitet haben, war ein physikalisches Experimentiergerät mit niedriger Dosisleistung. Es war nicht geeignet, Elektronenstrahlen austreten zu lassen.

Neutronenmessungen wurden nicht vorgenommen.

F. T. FARMER (Newcastle):

Zur Frage Einstellung des Patienten und der Behandlung von 60 Patienten innerhalb von 6 Std. möchte ich sagen, daß sie nur mit einem schwenkbar aufgehängten Linearbeschleuniger möglich erscheint und nicht mit irgendeinem anderen Gerät, das ich jemals gesehen habe.

Mit Hilfe der Lichtzeiger dauert es etwa 30—40 sec, um ein Einzelfeld einzustellen, vielleicht 1—2 min, wenn es sich um ein kompliziertes Feld handelt. Die Behandlung selbst nimmt 1—2 min in Anspruch. Im Mittel ist es also möglich, alle 5 min einen neuen Patienten bzw. ein neues Feld zu bestrahlen.

Das Diagnostikgerät wird nicht jeden Tag für jeden Patienten angewendet.

Zur Frage der Möglichkeit, höhere Tumordosen mit Hochvoltstrahlen zu erzielen, möchte ich sagen, daß wir den Eindruck haben, daß dies möglich ist. Bisher sind wir in der Steigerung der Dosis sehr langsam vorgegangen, und wir haben bisher keine Herddosen von mehr als 7500 rad gegeben. Wir sind jedoch der Ansicht, daß dies keine absolute Grenze ist und daß es im Einzelfall sicher möglich sein wird, höhere Herddosen zu geben, wenn dies wünschenswert erscheint.

Eine Messung von Neutronen wurde nicht durchgeführt.

C. J. L. THURGAR (Newcastle):

Zur Frage der Dosierung möchte ich sagen, daß wir 7000 rad in nicht weniger als 6 Wochen verabreichen. In 4 Wochen würden wir höchstens 5000—5500 rad geben und wir sind nie über eine Dosis von 7500 rad hinausgegangen.

Wir sind auch der Ansicht, daß der Behandlungstisch genauso wichtig ist wie der Apparat zur richtigen Behandlung von tiefgelegenen Tumoren. Er sollte in jeder Richtung getrennt beweglich sein und getrennt zu bedienen sein. Wir haben es nicht als notwendig angesehen, den Strahlenkegel durch den Tisch zu schicken, da dies durch eine Öffnung am oberen Ende des Tisches gemacht werden kann, wenn der Holzkasten mit einem Austrittsstrahlanzeiger versehen ist.

Veränderungen oder Beeinflussungen des Herzens haben wir bei allen Bestrahlungen des Thorax niemals beobachtet.

Die Blase bewegt sich natürlich, wenn der Patient auf das Gesicht gelegt wird. Wenn man jedoch große Felder anwendet, um die gesamte Blase zu bestrahlen, dann gelangt diese nicht außerhalb des Bereichs des Dosismaximum. Wenn schmale Felder angewendet werden, um nur die Hälfte der Blase zu erreichen, dann muß ein Spielraum vorgesehen werden für eine Bewegung der Blase und Kontroll-Röntgen-Aufnahmen sind in der supinierten und pronierten Stellung erforderlich.

Wenn Bleiblenden verwendet werden, um den Röntgenstrahl-Kegel zu begrenzen, so müssen diese genügend weit vom Patienten angeordnet sein, um Sekundär-Elektronen zu vermeiden, die die Dosisverteilung beeinflussen können. Der Abstand beträgt bei uns 13 bis 15 cm.

H. R. SCHINZ (Zürich):

Im Strahlenkegel des Betatrons in Zürich gibt es Neutronen, doch tragen diese nur zu weniger als 1% zur Dosis bei. Nach der Bestrahlung ist der Patient an der bestrahlten Stelle für ganz kurze Zeit radioaktiv infolge von Kernprozessen im Körper. Für die therapeutische Dosis spielt diese Radioaktivität keine Rolle.

Das Intensitätsspektrum unseres Betatrons wurde durch Spaltung von schwerem Wasserstoff bei 11 Atm. Druck bestimmt mit Hilfe der Messung der Reichweite der entstandenen Protonen in Spezialfilmen.

K. E. SCHEER (Heidelberg):

Ein bemerkenswerter Unterschied in der Therapie mit einem 15 MeV- und einem 30 MeV-Betatron besteht darin, daß nach Bestrahlung mit dem 15 MeV-Betatron keine Radioaktivität über dem Bestrahlungsfeld des Patienten nachgewiesen werden kann. Diese Radioaktivität entsteht vorwiegend durch $\gamma$-n-Prozesse, für die die Schwelle über 18 MeV liegt. Erst bei merklich höheren Photonenenergien als 18 MeV kommt es daher zur Bildung radioaktiver Isotope im Bestrahlungsgebiet.

# Le traitement du cancer bronchique au moyen du bétatron.
## Technique et résultats préliminaires*


Par

**B. L. Pierquin**, Paris


Depuis le mois de mars 1954, nous traitons les cancers bronchiques par radio-thérapie de haute-énergie (photonthérapie) au moyen d'un bétatron d'une énergie maximum de 24 MeV (Allis-Chalmers).

**Technique**[1]. En nous inspirant de la «méthode de repérage et de centrage» de l'Institut G. Roussy, notre technique, qui aboutit à une irradiation fixe en position assise, fait appel aux tomographies transversales du thorax pour le système de référence anatomique et à un corset plastique avec repères en plomb pour le système de référence extérieure.

1. Après une détermination approximative de la position et du volume tumoraux grâce aux clichés standards et à la bronchoscopie.

2. Le malade subit, dans la position exacte où il sera traité (position assise, avec bras relevés), un moulage thoracique en matière plastique (laine de verre imprégnée de métacrylate de méthyle solubilisé dans l'acétone).

3. Equipé de son corset en matière plastique, sur les faces antérieure et postérieure duquel 2 fils de plomb dessinant des lettres N sont appliqués (Fig. 1), le malade subit une série de tomographies transversales étagées de 2 en 2 cm sur toute la hauteur de la zone tumorale (5 à 7 coupes au total).

4. Après lecture de ces différentes coupes, celle qui passe par le centre du volume-cible est choisie pour plan de centrage. A l'aide d'un pantographe, cette coupe est réduite sur calque aux dimensions réelles du thorax.

5. Les dimensions et la forme du volume-cible sont dessinées directement sur ce calque réduit. A partir de la forme standard d'un parallèlépipède de section losangique, nous établissons pour chaque cas la forme la plus satisfaisante en nous efforçant, grâce à l'angle interne du losange, de pénétrer largement dans la zone lymphatique du médiastin tout en respectant le coeur et la moelle épinière. Les dimensions de ce parallèlépipède sont de l'ordre de 12 cm de hauteur sur 8 à 12 cm de côtés. A partir de ce volume-cible nous pouvons mettre en place 3 faisceaux d'irradiation, avec 2 faisceaux en vis-à-vis et un faisceau sensiblement perpendiculaire (Fig. 2).

6. Les points d'entrée et de sortie des axes de ces 3 faisceaux sont marqués sur le corset plastique. La détermination de ces points est établie à partir des

---

* Travail de l'Institut Gustave Roussy (Villejuif).

[1] Pierquin, B., Mme. A. Dutreix, J. Dutreix et M. Tubiana: Conditions techniques de l'irradiation des cancers bronchiques par le Bétatron (22 MeV). J. de Radiol. **38**, 504—509 (1957).

images en coupe des fils de plomb. Leur écartement ou leur rapprochement continu permet, à l'aide d'un compas, de retrouver le niveau exact de la coupe de centrage sur le corset plastique. Ce niveau établi, il suffit de mesurer sur le calque la distance entre chaque point d'entrée ou de sortie et le fil le plomb le plus proche et de reporter cette distance sur le corset. En ajoutant enfin une ligne verticale sur la paroi du corset, on est assuré d'obtenir une position reproductible dans les 3 dimensions de l'espace.

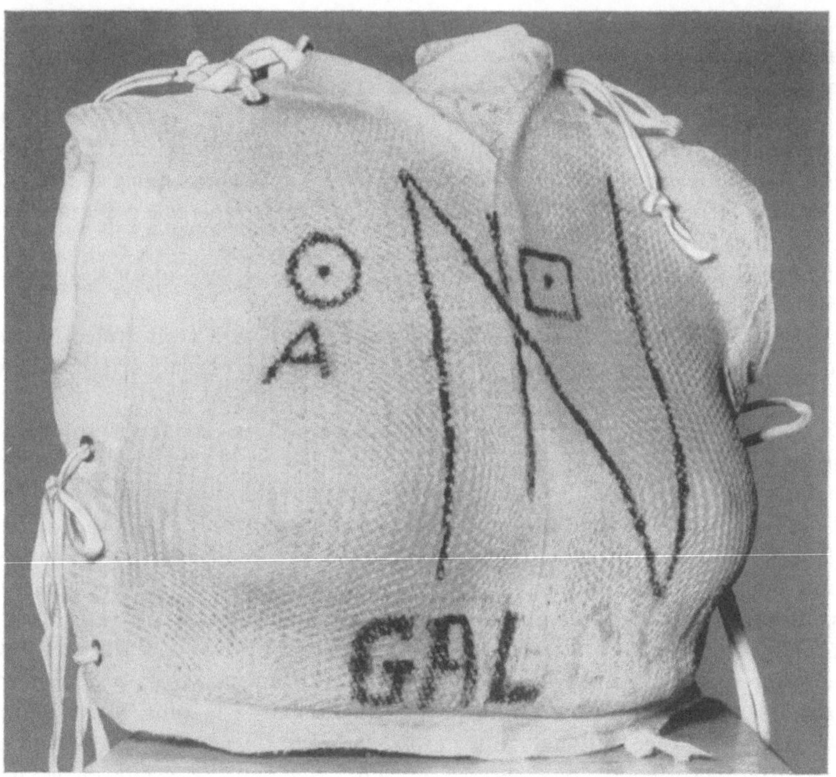


Fig. 1. Corset en matière plastique avec points de centrage des faisceaux d'irradiation

7. Après un contrôle dosimétrique de ce plan d'irradiation par le service de physique, le traitement est mis en route à raison de 5 séances hebdomadaires de 200 rad-tumeur par séance. Au total 1000 rad par semaine. La dose totale atteignant 7500 rad à raison de 2500 rad par faisceau. La durée totale du traitement atteint alors près de 8 semaines.

8. Pour chaque séance, la mise en place du malade devant le bétatron s'effectue de la façon suivante: fixé dans son corset, le malade est assis sur un tabouret avec appui-tête. Une sangle abdominale le maintient dans une position fixe. Le tabouret est avancé en position rapprochée devant le faisceau du bétatron. Les points d'entrée et de sortie du faisceau d'irradiation choisis pour la séance sont alors mis en coïncidence exacte avec les réticules du centreur optique antérieur et

du rétro-centreur grâce à une mobilisation du bétatron. Après vérification de la distance focus-corset, l'irradiation peut être mise en route. Une cellule photo-électrique éclairée par le rétro-centreur est placée sur le corset à la sortie du faisceau. La moindre déviation du malade en cours de traitement peut être ainsi enregistrée sur un cadron placé dans la salle de commande. Chaque séance d'irradiation dure, selon les cas, entre 5 et 10 minutes.

**Resultats.** De mars 1954 à juillet 1957, nous avons irradié 183 malades par cette méthode. Sur ce total, 85 malades ont plus de 1 an d'observation radio-

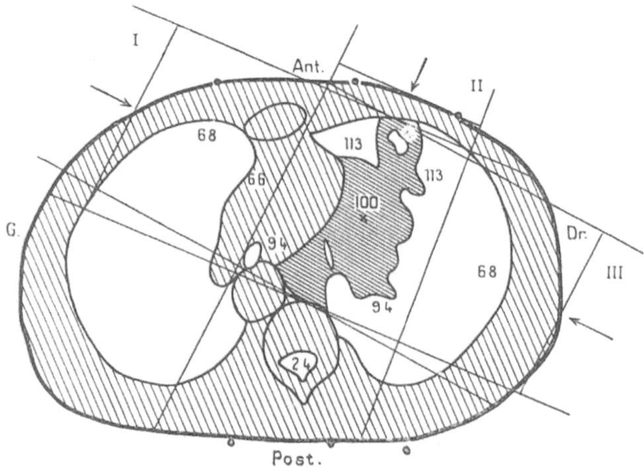


Fig. 2. Calque reduit de la coupe de centrage. Mise en place des 3 faisceaux avec dosimétrie

thérapique. Etant donné la médiocre durée de survie des cancers bronchiques, nous pouvons sur ce groupe de 85 malades, essayer de dégager quelques faits.

Sur ces 85 malades une première série (jusqu'en fin 1955) n'a reçu que 5500 rad; une seconde série fin 1955 — Juillet 1956 a reçu 7500 rad. En voici les données numériques:

*Tableau 1*

| Série à 5500 rad | Série à 7500 rad |
|---|---|
| 54 | 31 |
| Ayant terminé leur traitement | Ayant terminé leur traitement |
| Vivants:  13 ⎫ 46 | Vivants:   9 ⎫ 22 |
| Décédés: 33 ⎭ | Décédés: 13 ⎭ |
| Ayant interrompu leur traitement | Ayant interrompu leur traitement |
| Vivants:  0 ⎫ 8 | Vivants:  0 ⎫ 9 |
| Décédés:  8 ⎭ | Décédés:  9 ⎭ |

Dans la série des 5500 rad, la survie globale à 1 an est de l'ordre de 25% (13 survivants sur 54). Sur les 13 survivants, 5 sont en pleine récidive. Ce pourcentage est en concordance avec les autres statistiques radiothérapiques publiées dans le monde, qu'il s'agise de haute-énergie ou de radiothérapie conventionnelle.

Dans la série des 7500 rad, la survie globale de 1 an paraît un peu plus élevée, aux environs de 30% (9 survivants sur 31). En fait ces chiffres sont trop faibles pour tirer une conclusion statistiquement précise. Par ailleurs cette série est plus récente que celle des 5500 rad. On peut donc, sans risquer de se tromper, affirmer que les résultats de cette irradiation à dose tumorale plus élevée ne semblent pas meilleurs que ceux à dose tumorale plus faible.

Nous poursuivons actuellement cette étude comparative entre 7500 et 5500 rad. Elle prendra d'ici un an une signification statistique plus nette. Nous pensons cependant dès maintenant, à la lumière des résultats actuels, que la radiothérapie à dose très élevée (7500 rad) n'apporte pas, au niveau du cancer bronchique, de meilleurs résultats que celle à dose moyenne (5500 rad).

Devant ces résultats assez décévants, nous nous sommes retournés vers les chirurgiens pour leur demander de s'associer avec nous dans une tentative radio-chirurgicale systématique[1]. Les conditions d'une exerèse chirurgicale post-radiothérapique nous paraissaient très favorables avec le bétatron. La plupart de nos malades termine en effet leur irradiation dans un bon état fonctionnel et général: pas de «mal des rayons», état sthénique avec prise de poids, état respiratoire amélioré avec bonnes épreuves fonctionnelles. Devant ces bons résultats immédiats, plusieurs équipes chirurgicales se sont associées à notre effort. Actuellement nous pouvons citer le cas de l'équipe DAUMET qui depuis la fin de l'année 1956 est intervenu sur 19 malades. Avec cette équipe il n'est plus question de faire une distinction préalable entre cas chirurgicaux (les formes les moins étendues) et malades radiothérapiques (cas inopérables). Tous ces malades proviennent d'un pétrutement uniforme (Hôpital St. Joseph) et subissent sauf con-treind-ication absolue à tout traitement curateur, une irradiation préalable (7500 rad) suivie, si possible 10 à 15 jours plus tard d'une intervention chirurgicale à visée radicale (pneumonectomie). Ce court délai entre la fin de la radiothérapie et l'acte chirurgical a été fixé pour des raisons essentiellement psychologiques, le malade restant ainsi en permanence sous contrôle médical. Quelques cas antérieurs en effet nous avaient appris qu'un plus long délai, de l'ordre de 5 à 6 semaines, entrainant fréquemment la défection chirurgicale du malade, celui-ci redoutant après quelques semaines de repos d'affronter le bistouri.

Les résultats immédiats de cette série de 19 cas se présentent de la façon suivante:

Série DAUMET-DELARUE

| | |
|---|---|
| Nombre de malades traités par Bétatron | : 19 |
| *Malades considérée comme opérables d'emblée* | : 15 |
|    *Opérés: 11,* se répartissant en: | |
|       Exérèses sans complications | : 8 |
|       Exérèses suivies d'exitus | |
|       per ou post-chirurgie | : 2 |
|       Contre-indication peropératoire | |
|       devant métastase | : 1 |
|    *Non opérés: 4* | |
|       Par refus du malade | : 2 |
|       Par complications cliniques | : 2 |
| *Malades considérés comme inopérables d'emblée* | : 4 |
|    *Opérés* (bonnes suites immédiates) | 2 |
|    *Non opérés* (mauvais état général ou métastases) | 2 |

---

[1] Cette association radio-chirurgicale reste dans l'esprit de la doctrine de l'Institut G. ROUSSY vis-à-vis du cancer bronchique (HUGUENIN, REDON, FAUVET et LEMOINE).

Cette expérience radio-chirurgicale est encore trop récente pour que nous puissions en tirer une conclusion optimiste. Ce que l'on peut affirmer cependant, c'est que la radiothérapie pré-opératoire en haute-énergie a permis d'effectuer sur des cas non sélectionnés une exérèse chirurgicale avec succès immédiat (10 cas sur 19) dans une forte proportion.

**En conclusion,** il se dégage trois faits de notre expérience actuelle d'irradiation des cancers bronchiques par le bétatron:

1. La radiothérapie est bien supportée par le malade tant sur le plan général que fonctionnel, avec un minimum d'incidents ou d'accidents immédiats. Les résultats de la photonthérapie de haute-énergie sont donc immédiatement bons.

2. Avec un délai d'observation minimum de 1 an, les résultats médiats sont médiocres, et ceci quelle que soit la dose tumorale (5500 ou 7500 rad). La radiothérapie de haute-énergie n'améliore donc pas sensiblement à elle seule le pronostic des cancers bronchiques.

3. L'association systématique radio-chirurgicale nous a donné des résultats immédiats satisfaisants, la majorité des malades pouvant supporter avec succès ce traitement combiné. L'irradiation par le bétatron semble donc permettre aux chirurgiens d'opérer de façon radicale une forte proportion de cancers bronchiques.

# Klinische Erfahrungen mit dem 31 MeV-Betatron

## Züricher Bestrahlungsergebnisse 1951 — 1957[*]

Von

**U. Cocchi, Zürich**

Die Bestrahlungen mit dem 31 MeV-Brown-Boveri-Betatron wurden in Zürich im April 1951 begonnen. Bis heute (Juni 1957) sind insgesamt etwas mehr als 660 Patienten bestrahlt worden, und zwar hauptsächlich solche mit intrakraniellen Tumoren, mit Tumoren der Speiseröhre, des weiblichen Genitalapparates, der Bronchien, Harnblase, des Colon und Rectum, in etwas geringerem Maße auch Tumoren des Mediastinum, der Nieren, des Retroperitoneum, des Knochensystems und des Magens, sowie vereinzelte Tumoren anderer Lokalisationen.

Es handelt sich also vornehmlich um in der Körpertiefe gelegene Tumoren, bei denen die Verwendung ultraharter Strahlung — im Vergleich zur konventionellen Therapie — besonders vorteilhaft erscheint. Wichtig ist hierbei auch die geringe Hautbelastung. Bei rund 11 % der Patienten ist daher auch eine Betatronbestrahlung in rezidivierenden Fällen durchgeführt worden, bei denen der Zustand der Haut infolge früher verabfolgter konventioneller Strahlentherapie eine weitere Behandlung mit dieser Methode nicht mehr zuließ.

Die weiteren Vorteile, welche die ultraharte Röntgenbestrahlung bietet, sind die durch die in gleichem Maße stattfindende Absorption der Strahlung im Knochen wie in den Weichteilen bedingte homogene Verteilung der Dosen im bestrahlten Raum, wodurch besonders eine bessere Wirksamkeit auf die hinter Knochen befindlichen Geschwulstpartien erzielt wird, und ferner die scharfe seitliche Begrenzung des Strahles.

Die biologische Wirksamkeit ultraharter Strahlung ist jedoch mit einigen wenigen Ausnahmen qualitativ die gleiche wie bei den übrigen Strahlenqualitäten. Gegenüber konventioneller Therapie wenig reagierende Tumorarten zeigen auch gegenüber ultraharter Bestrahlung keine wesentlich günstigere Reaktion. Eine Ausnahme scheinen in dieser Hinsicht die Tumoren der Harnblase darzustellen.

Betrachten wir die in den ersten 5 Jahren mit dem Betatron bestrahlten Patienten, so haben wir 532 Patienten, die mindestens 1 Jahr nach Abschluß der Strahlenbehandlung beobachtet werden konnten. Von diesen Patienten hatten

73 Tumoren des Gehirnes und der Hypophyse,
73 Tumoren der Speiseröhre,
69 Bronchialtumoren,
50 Tumoren der Harnblase,

[*] Aus der Radiotherapeutischen Klinik der Universität Zürich (Direktor: Prof. Dr. H. R. Schinz).

45 Portiocarcinome,
42 Rectum- und Coloncarcinome,
39 Carcinome corporis uteri,
21 Ovarialgeschwülste,
11 Nierengeschwülste (Hypernephrome, Sarkome),
24 Mediastinaltumoren (Lymphogranulomatose, Sarkome, Metastasen),
23 Lymphknotenmetastasen der Hals- und Abdominalregion,
14 primäre und sekundäre Geschwülste des Knochensystems,
 7 Magengeschwülste,
18 Tumoren der oberen Luft- und Speisewege,
 4 Rückenmarktumoren (Metastasen, Ependymom),
 3 Tumoren der Orbita (Cylindrom, Carcinom, Sarkom),
 3 Hauttumoren,
je 2 Urethra-, Prostata-, Gallenblasen- und Parotisgeschwülste,
 2 solitäre Lungenmetastasen,
 1 Morbus Paget,
je 1 Tumor des Pankreaskopfes und der Glandula submandibularis.

Was die *Bestrahlungstechnik* anbetrifft, so wurden in den ersten Monaten nach Beginn der Betatronbestrahlungen täglich Einzeldosen von 300 r (in 5 cm Tiefe) in 1 m FH-Abstand verabreicht. Die Gesamtdosen betragen 4000—8000 r (in 5 cm Tiefe), bei den Oesophagustumoren anfänglich sogar bis zu 12000 r in 45—76 Tagen. Diese hohen Dosen bewirkten jedoch schwere gangränescierende Veränderungen, die rasch zum Exitus führten, worauf Einzel- und Gesamtdosen bald herabgesetzt wurden. Heutzutage werden täglich 2mal 100 r auf ein vorderes und hinteres Feld verabreicht, bei Hirntumoren auf 2 seitliche und je nach Lage des Tumors mitunter auch noch auf ein frontales und occipitales Feld, bei Gesamtdosen von 5000—6000 r (in 5 cm Tiefe) in 30 Tagen und evtl. Zusatzbestrahlung nach einem Intervall von 2—3 Monaten (früher von 2—8 Wochen) mit einer Gesamtdosis von 3000—4000 r. Die Tumordosis betrug hierbei 4900 bis 6000 r pro Serie, also je nach der Lage des Tumors 81—100%.

Die mit dem 31 MeV-Betatron im Gewebe erzielten Dosiseffekte sind allerdings, bei Berücksichtigung des Density-Effektes, nicht gleich den mit 200 kV-Energie erzielten. Zur Erlangung des gleichen Dosiseffektes müßten mit Betatron die Dosen um 15% erhöht werden, bzw.. müßten die oben angegebenen Betatrondosen also, um mit den entsprechenden mit konventioneller Bestrahlung erzielten Dosiseffekten im Gewebe verglichen werden zu können, um 13% herabgesetzt werden.

Fast allgemein handelte es sich bei unseren Patienten um solche mit ausgedehnten, inoperablen Tumoren oder um solche, bei denen der schlechte Allgemeinzustand oder Herzbefund einen chirurgischen Eingriff nicht zuließ. Nur in seltenen Fällen wurden uns die Patienten zu einer postoperativen prophylaktischen Nachbestrahlung zugewiesen. Zumeist waren es Patienten mit operierten Hirntumoren oder Ovarialtumoren, ganz selten Patienten zur Vorbestrahlung vor einem chirurgischen Eingriff. Grosso modo unterscheidet sich das zur Strahlenbehandlung zugewiesene Patientengut nicht wesentlich von demjenigen früherer Jahre, in denen uns nur konventionelle Therapie zur Verfügung stand. Bei gewissen Tumoren, wie hauptsächlich beim Oesophagustumor, hat sich das Patientengut gegenüber früher allerdings wesentlich verschlechtert, da auf Grund der Fortschritte, welche die Operationstechnik seit dem letzten Weltkrieg gemacht hat, ein großer Teil der Patienten, die früher ausschließlich strahlenbehandelt wurden, heute einem operativen Eingriff unterzogen werden.

In Tab. 1 sind die Behandlungsresultate der Betatrontherapie bei einigen größeren Patientengruppen wiedergegeben. An erster Stelle stehen die 3 Gruppen mit Geschwülsten der weiblichen Genitalorgane mit zusammen 105 Patienten. 45 Patienten hatten ein *Portiocarcinom.* Von diesen wurden 18 einer prophylaktischen Nachbestrahlung nach Radikaloperation (Stad. I und II) unterzogen, während die übrigen Patienten wegen nicht operierten, inoperablen oder rezidivierenden (Stad. III und IV, einmal Stad. II), schon vorher mit konventioneller Therapie bestrahlten Tumoren behandelt wurden. Zusatzbehandlung mit intravaginaler Ra-Bestrahlung wurde bei 2 prophylaktisch und bei 11 kurativ bestrahlten Patienten verabfolgt. Von allen Patienten leben mindestens 1 Jahr über

Tabelle 1. *Überlebensdauer nach Betatron-Bestrahlung*

| mit Patienten Anzahl | 1 Jahr | 2 Jahren | 3 Jahren | 4 Jahren | 5 Jahren |
|---|---|---|---|---|---|
| Portio-Ca. 45 | 25 von 45 | 16 von 34 | 11 von 26 | 5 von 18 | 1 von 11 |
| symptomfrei | 13 von 45 | 10 von 34 | 6 von 26 | 5 von 18 | 1 von 11 |
| Uterus-Ca. 39 | 31 von 39 | 19 von 28 | 15 von 22 | 10 von 18 | 4 von 11 |
| symptomfrei | 23 von 39 | 17 von 28 | 15 von 22 | 10 von 18 | 4 von 11 |
| Ovarial-Tu. 21 | 11 von 21 | 9 von 20 | 7 von 19 | 3 von 9 | 1 von 7 |
| symptomfrei | 7 von 21 | 6 von 20 | 4 von 19 | 2 von 9 | 0 von 7 |
| Astrocytom 16 | 11 von 16 | 6 von 11 | 0 von 2 | — | — |
| symptomfrei u. gebessert | 8 von 16 | 4 von 11 | 0 von 2 | — | — |
| Renale Tu. 11 | 9 von 11 | 4 von 9 | 3 von 9 | 2 von 6 | 1 von 4 |
| symptomfrei | 3 von 11 | 3 von 9 | 3 von 9 | 2 von 6 | 1 von 4 |
| Harnblasen-Tu. 50 | 9 von 50 | 4 von 38 | 3 von 31 | 1 von 20 | 0 von 10 |
| symptomfrei | 4 von 50 | 2 von 38 | 1 von 31 | 1 von 20 | — |
| Bronchus-Ca. 69 | 8 von 69 | 1 von 47 | 1 von 28 | 1 von 23 | 1 von 18 |
| symptomfrei | 2 von 69 | 1 von 47 | 1 von 28 | 1 von 23 | 1 von 18 |
| Oesophagus 72 | 6 von 73 | 1 von 66 | 0 von 53 | 0 von 32 | 0 von 19 |
| symptomfrei | — | — | — | — | — |
| Colon-Rectum 42 | 12 von 42 | 7 von 41 | 3 von 37 | 2 von 29 | 2 von 17 |
| symptomfrei | 3 von 42 | 3 von 41 | 3 von 37 | 2 von 29 | 2 von 17 |

die Hälfte (56%), symptomfrei rund 29%, mindestens 2 Jahre leben rund 47%, symptomfrei prozentual gleich viele wie 1 Jahr, 3 Jahre leben rund 42%, symptomfrei rund 23%, während 5 Jahre nur 1 Patient symptomfrei von 11 lebt. Von den 27 inoperablen Fällen leben heute 5, und zwar 1 mit Rezidiv 12 Monate, 2 in stationärem Zustand 32 bzw. 38 Monate, 1 symptomfrei 48 Monate und der 5. in zur Zeit nicht sicher beurteilbarem Zustand 44 Monate. Ein 6. Patient lebte mit Rezidiv 44 Monate und starb an seinem Tumor. Die übrigen dieser Patientengruppe starben noch innerhalb der ersten 12 Monate nach Abschluß der Bestrahlung (Tab. 2).

Bei 8 Patienten trat nach der Betatronbestrahlung ein Rezidiv auf, und zwar 2mal nach Radikaloperation und prophylaktischer Nachbestrahlung und 5mal bei nur bestrahlten Tumoren. Die Rezidive traten wenige Monate nach Abschluß der Strahlenbehandlung auf, nur in einem Fall erst nach $2^{1}/_{2}$ Jahren. In 5 Fällen wurde erneut mit Betatron bestrahlt; 1 Patient wurde nach Abschluß der Bestrahlung symptomfrei, die übrigen starben nach 7—26 Monaten. 1 Patient wurde mit konventioneller Therapie behandelt (Exitus nach 20 Monaten), 2 Patienten mit Radiumeinlagen. Die letzten beiden Patienten leben 3 Jahre 5 Monate symptomfrei bzw. 3 Jahre 2 Monate in stationärem Zustand. Die durchschnittliche Überlebensdauer sämtlicher 22 gestorbener Patienten beträgt 11 Monate (maximal 44 Monate). 1 Patient ist nach 1jähriger Symptomfreiheit verschollen.

Von den 39 Patienten mit *Corpus uteri-Carcinom* wurden 25 einer prophylaktischen Nachbestrahlung unterzogen, während es sich bei den restlichen 14 um Bestrahlung inoperabler, nicht radikal operierter oder rezidivierender Tumoren handelte. Die Behandlungsresultate stellen die besten unserer Patientengruppen dar. Rund 79% leben mindestens 1 Jahr, davon rund 59% symptomfrei, rund 68% leben mindestens 2 Jahre (symptomfrei rund 61%), 3 Jahre symptomfrei rund 68%, 5 Jahre leben 4 von 11 symptomfrei. Von den 12 inoperablen und

Tabelle 2. *Überlebensdauer der Patienten mit Portio- und Corpuscarcinom nach A. Radikaloperation + Nachbestrahlung und B. alleiniger Betatronbestrahlung*

| Patienten mit | 1 Jahr | 2 Jahren | 3 Jahren | 4 Jahren | 5 Jahren |
|---|---|---|---|---|---|
| | | | A. | | |
| Portio-Ca. | 14 von 18 | 9 von 11 | 6 von 10 | 4 von 6 | 1 von 1 |
| symptomfrei | 9 von 18 | 7 von 11 | 4 von 10 | 4 von 6 | 1 von 1 |
| | | | B. | | |
| Portio-Ca. | 11 von 27 | 7 von 23 | 5 von 16 | 1 von 12 | 0 von 10 |
| symptomfrei | 4 von 27 | 3 von 23 | 2 von 16 | 1 von 12 | — |
| | | | A. | | |
| Corpus-Ca. | 24 von 25 | 17 von 17 | 14 von 14 | 8 von 12 | 3 von 6 |
| symptomfrei | 19 von 25 | 15 von 17 | 14 von 14 | 8 von 12 | 3 von 6 |
| | | | B. | | |
| Corpus-Ca. | 7 von 14 | 2 von 11 | 2 von 8 | 2 von 6 | 1 von 5 |
| symptomfrei | 4 von 14 | 2 von 11 | 2 von 8 | 2 von 6 | 1 von 5 |

2 nicht radikal operierten Fällen wurden durch die Betatronbestrahlung und in 4 Fällen mit zusätzlicher intravaginaler Ra-Behandlung 4 symptomfrei mit einer Überlebensdauer von 1 Jahr, 1 Jahr 5 Monaten, 4 Jahre 7 Monate und 5 Jahre 4 Monate (s. Tab. 2). 2 weitere, unterdessen verstorbene Patienten lebten in ungebessertem Zustand 18 bzw. 19 Monate und ein 3. Patient 21 Monate symptomfrei; er starb aber nach weiteren 3 Monaten infolge Hirnmetastase.

Die durchschnittliche Überlebensdauer der 14 gestorbenen Patienten betrug 13 Monate (maximal 24 Monate). Rezidive traten nur 3mal auf, und zwar 2mal nach Radikaloperation und Nachbestrahlung und 1mal nach alleiniger Betatronbestrahlung. Durch erneute Betatronbestrahlung wurden 2 derselben symptomfrei mit Überlebensdauer von 4 Jahren 7 Monaten bzw. 5 Jahren bis heute, bei dem 3. Patienten trat nur vorübergehende Besserung auf und der Zustand ist heute progredient.

Von den 21 Patienten mit *Ovarialtumoren* hatten 18 Carcinome, je 1 ein Chorionepitheliom, Teratom und Spindelzellsarkom. Bei 15 Patienten handelte es sich um prophylaktische Nachbestrahlung, darunter befanden sich die Patienten mit Teratom und Spindelzellsarkom, und nur 6mal um Bestrahlung von Rezidiven oder inoperablen Tumoren. Die Überlebensdauer und Symptomfreiheit in den einzelnen Jahren nach der Bestrahlung unterscheidet sich nicht wesentlich von derjenigen, die beim Portiocarcinom beobachtet wurde. Von den nur bestrahlten Fällen wurde keiner symptomfrei; sämtliche Patienten starben an ihrem Tumor oder an ausgedehnter Metastasierung. Die durchschnittliche Überlebensdauer aller 13 gestorbenen Patienten beträgt 13 Monate (maximal 48 Monate). Rezidive wurden 3 mal beobachtet. In 1 Fall trat das Rezidiv nach einer symptomfreien Periode von $3^1/_2$ Jahren auf; eine erneute Behandlung war nicht mehr möglich und Patient kam ad exitum. Bei den beiden anderen Patienten wurde das Rezidiv erneut bestrahlt, einmal mit 240 kV-Therapie (Exitus nach vorübergehender Besserung nach weiteren 8 Monaten) und das andere Mal nach 8 Monaten mit Betatron. Ein 2. Rezidiv wurde nach 2 Jahren 2 Monaten wiederum mit Betatron bestrahlt. Nach vorübergehender Besserung starb Patient 4 Jahre nach Abschluß der 1. Bestrahlung an seinem Tumor. Die total verabreichte Dosis betrug 14000 r.

Unter den Patienten mit *intrakraniellen Tumoren* fanden sich 63 mit Hirntumoren und 9 mit Hypophysentumoren. Ein weiterer Patient, der wegen eines klinisch diagnostizierten inoperablen Ponstumors bestrahlt wurde, starb 1 Monat nach Bestrahlungsabschluß. Bei der Sektion stellte es sich jedoch heraus, daß kein Tumor vorlag, sondern ein Hirnabsceß. Hirn-Bestrahlungen mit Betatron wurden erst seit 1953 ausgeführt.

Die größte Gruppe unter diesen Patienten stellt diejenige mit *Astrocytom* dar. Von diesen sind 6 radikal operiert worden. 1 ist verschollen, die übrigen leben beschwerdefrei oder gebessert. 8 wurden nur subtotal operiert und dann einer kurativen Bestrahlung unterzogen. Von diesen sind 3 innerhalb der ersten 11 Monate gestorben, die übrigen 5 leben: 1 symptomfrei 16 Monate, 2 gebessert 24 bzw. 30 Monate, 1 in stationärem Zustand 23 Monate und der 5. mit Rezidiv nach 11 Monate andauerndem stationären Zustand. 2 Patienten wurden nicht operiert: 1 starb nach 6 Monaten, der 2. nach 25 Monaten nach vorübergehender Besserung bzw. Beschwerdefreiheit beim 2. Patienten. 11 von 16 Patienten lebten mindestens 1 Jahr, davon 8 beschwerdefrei oder gebessert, und 6 von 11 Patienten 2 Jahre. Von den beiden vor 3 Jahren bestrahlten Patienten lebt keiner. Von den 1935—1955 mit 200, 240 und 400 kV und ziemlich gleich hohen Herddosen bestrahlten 35 Patienten lebten dagegen mindestens 20 von 35 ein Jahr, beschwerdefrei und gebessert 15 von 35, mindestens 2 Jahre 17 von 33 (10 von 33) und mindestens 3 Jahre 13 von 32 (5 von 32). Die Erfolge erscheinen in der letzteren Gruppe etwas geringer zu sein, doch erlaubt die kleine Zahl der mit Betatron behandelten Patienten keine sichere vergleichende Beurteilung.

Tabelle 3. *Überlebensdauer und Behandlungsfolge der Patienten mit Astrocytom nach konventioneller und Betatron-Bestrahlung*

|  | 1 Jahr | 2 Jahren | 3 Jahren |
|---|---|---|---|
| *A. Konventionelle Therapie* | | | |
| Lebende . . . . . . . . . . . . . . . . . . . . | 20 von 35 | 17 von 33 | 13 von 32 |
| symptomfrei und gebessert . . . . . . . . . | 15 von 35 | 10 von 33 | 5 von 32 |
| *B. Betatron-Therapie* | | | |
| Lebende . . . . . . . . . . . . . . . . . . . . | 11 von 16 | 6 von 11 | 0 von 2 |
| symptomfrei und gebessert . . . . . . . . . | 8 von 16 | 4 von 11 | 0 von 2 |

Was die übrigen Hirntumoren anbetrifft, so wurden 2 Patienten mit *Medulloblastom* (nicht operiert) bestrahlt. Der eine kam schon kurz nach Beginn der Bestrahlung ad exitum, der andere, bei dem zuerst mit konventioneller Therapie eine Serie mit 8985 r am Herd verabfolgt wurde, erhielt wegen erneuter Verschlechterung 12 Monate später eine 2. Serie mit Betatron mit 6359 r am Herd Exitus 6 Wochen später infolge Metastasierung im Wirbelkanal.

2 Patienten mit radikal operiertem *Oligodendrogliom* leben noch beschwerdefrei 10 Monate bzw. 2 Jahre, während ein weiterer Patient 10 Monate in stationärem Zustand lebt. 2 Patienten mit radikal operiertem *Astroblastom* kamen nach 7 bzw. 10 Monate Überlebensdauer ad exitum. Von den 6 Patienten, die wegen *Glioblastoma multiforme* bestrahlt wurden, ist einer (radikal operiert) nach 6 monatiger Beschwerdefreiheit verschollen, 1 subtotal operierter starb 3 Monate nach Bestrahlungsabschluß und die übrigen 4 nichtoperierten Patienten kamen ebenfalls innerhalb 2—9 Monaten ad exitum. Außerdem wurden noch 5 Patienten mit *malignen, nicht näher klassifizierten Gliomen* bestrahlt, 3 nach Radikaloperation und 2 nach subtotaler Tumorentfernung. Einer der radikaloperierten Patienten lebt noch in gebessertem Zustand 13 Monate, die übrigen sind gestorben nach 4—15 Monaten.

Von den 3 Patienten mit *Ependymom* wurden 2 nach radikaler Tumorentfernung mit Betatron bestrahlt. Beide leben: 1 symptomfrei länger als 16 Monate, der 2. Patient war nach der ersten, mit konventioneller Bestrahlung durchgeführten Serie 2 Jahre symptomfrei (Herddosis 6390 r) und wurde wegen eines Rezidivs mit Betatron in 2 Serien: mit 6000 r und nach 1 Monat mit 3000 r bestrahlt. Wiederum Beschwerdefreiheit für 2 Jahre 2 Monate. Dann erneute Betatronbestrahlung mit 5000 r und bis jetzt 3 Monate Besserung. Der 3., subtotal operierte Patient blieb nach der Betatronbestrahlung bis jetzt 2 Jahre 3 Monate beschwerdefrei. 2 Patienten mit radikal operiertem malignem sarkomatösem und mit entdifferenziertem *Meningeom* wiesen nach Abschluß der Bestrahlung eine Überlebensdauer von nur 8 bzw. 9 Monaten auf. Von 3 Patienten mit *arteriovenösem Aneurysma* wurde einer bis jetzt 21 Monate lang beschwerdefrei, ein weiterer, der nach einer 1. Bestrahlungsserie mit 180 kV 11 Jahre lang beschwerdefrei blieb (3500 r/H) und infolge rezidivierender Blutung 8250 r/H in 2 Serien mit Betatron erhielt, lebt jetzt 21 Monate in gebessertem Zustand, während der 3. Patient (6 jährig) nach einer 1. Betatronbestrahlung mit 2400 r über $2^1/_2$ Jahre in gebessertem Zustand blieb. Infolge Verschlechterung wurde nach 3 Jahren eine 2. Serie mit 4000 r verabreicht, nach der aber keine Besserung des Zustandes auftrat (bisherige Beobachtungszeit 7 Monate).

Weiterhin kamen 2 Patienten mit radikal operierten solitären *Hirnmetastasen* (Melanom und polymorphzelliges Sarkom) zur Nachbestrahlung. Beide kamen nach 4 bzw. 9 Monaten an generalisierter Metastasierung ad exitum. Bei einem 37 jährigen Patient mit *Haemangioma cavernosum* wurden anfangs in 2 Serien 6194 r/H (konventionelle Bestrahlung) verabfolgt und nach vorübergehender Besserung nach 15 Monaten weitere 5640 r/H mit Betatron. Bei der 9 Monate später durchgeführten Autopsie fanden sich außerhalb des ausgedehnten hämangiomatösen Gebiets nekrotische Hirnpartien, die als *Röntgenspätfolgen* aufgefaßt wurden, wobei als unterstützender Faktor durch den Druck des Tumors bedingte veränderte Kreislaufverhältnisse in der Nachbarschaft des Hämangiom hinzukommen (Kombinationsschaden). Schließlich kamen noch 18 Patienten nur mit der *klinischen* Diagnose *Hirntumor* allein zur Bestrahlung. 1 von diesen ist jetzt 3 Jahre lang beschwerdefrei, 3 leben 8, 18 und 24 Monate in gebessertem Zustand,

2 leben 16 bzw. 18 Monate in stationärem Zustand, während die übrigen 12 innerhalb der ersten 13 Monate nach Asbchluß der Strahlenbehandlung starben.

Von den 9 Hypophysenpatienten wurde 1 mit solidem, nicht operiertem *Kraniopharyngeom* mit 5640 r bestrahlt; er lebt jetzt 13 Monate in gebessertem Zustand. Ferner wurden 6 Patienten mit *chromophobem Hypophysenadenom* bestrahlt. Ein Patient mit radikal entferntem Tumor lebt jetzt 3 Monate symptomfrei nach der Nachbestrahlung. Der Tumor war in 3 weiteren Fällen nur subtotal entfernt worden. 2 der Patienten sind verschollen, der 3. lebt gebessert bis jetzt 20 Monate. In den restlichen 2 Fällen wurde der Tumor nicht entfernt; beide Patienten blieben durch die Bestrahlung unbeeinflußt. Schließlich wurde noch bei 2 Patienten mit *Akromegalie* Stationärbleiben des klinischen Befundes über 12 bzw. 20 Monate nach der Betatronbestrahlung festgestellt.

Ferner wurden 11 Patienten mit *Nierentumoren* zur Bestrahlung eingewiesen. Bei 6 Patienten handelte es sich um prophylaktische Nachbestrahlung nach radikaler Tumorentfernung, bei 1 Patient um Bestrahlung nach subtotaler Entfernung und in 4 Fällen um Bestrahlung nicht operierter Tumoren. 8 Patienten hatten ein Hypernephrom und je einer ein lymphoblastäres Sarkom, ein Spindelzellsarkom und einen Birch-Hirschfeld-Tumor. 9 von diesen 11 lebten mindestens 1 Jahr, symptomfrei allerdings nur 3 von diesen, mindestens 2 Jahre lebten 4 von 9, symptomfrei 3 (s. Tab. 1). 3 Patienten leben symptomfrei über 3 Jahre 10 Monate, 4 Jahre 2 Monate und 5 Jahre 2 Monate. Bei den ersten beiden handelt es sich um Hypernephromfälle, beim 3. um den Birch-Hirschfeld-Tumor. Alle 3 Tumoren wurden radikal entfernt und nachbestrahlt. Ein 4. Patient mit nicht operiertem (rezidivierendem) Spindelzellsarkom lebt jetzt 10 Monate in progredientem Zustand. Die übrigen sind nach 4—34 Monaten, in 2 Fällen nach vorübergehender Symptomfreiheit, gestorben.

50 Patienten mit *Harnblasentumoren* (Pflasterzellcarcinome, solide und papilläre Carcinome) wurden mit Betatron bestrahlt, meistens wegen Rezidiven nach vorangegangenen Operationen und nur bei 3 Patienten prophylaktisch nach Radikaloperation. Mindestens 1 Jahr lebten nur 18%, 8% symptomfrei, mindestens 2 Jahre 10% (5%), mindestens 3 Jahre lebten ebenfalls rund 10%, doch waren nur noch 3% symptomfrei. 5 Fälle sind verschollen; vermutlich sind dieselben ad exitum gekommen. Von den 50 Patienten sind 42 gestorben; die durchschnittliche Überlebensdauer betrug 9 Monate (maximal 4 Jahre 2 Monate). Von den gestorbenen war ein Patient 4 Jahre 2 Monate beschwerdefrei; er starb angeblich an einem interkurrenten Leiden. 2 waren vorübergehend einige Monate symptomfrei und 6 einige Monate gebessert; ein 7. Patient war 3 Jahre in gebessertem Zustand und ist jetzt seit 2 Jahren verschollen. Bei einem weiteren der verstorbenen Patienten fand sich nach der in 3 Serien innerhalb von 5 Monaten durchgeführten Bestrahlung mit einer Gesamtdosis von 11 900 r autoptisch 3 Monate nach Abschluß der letzten Serie kein Tumor in der Blase; der Tod trat infolge multipler Fernmetastasen ein. Bei den wenigen übrigen zur Sektion gelangten Fällen wurde, nach allerdings etwas geringeren Dosen, noch Tumor in der Blase angetroffen.

In neuester Zeit sind in einigen Fällen präoperative Betatronbestrahlungen mit Gesamtdosen von 3000 r ausgeführt worden, über die sich der operierende Urologe äußerst günstig äußerte. Spätbeobachtungen fehlen noch.

Ein weiterer Patient mit rezidivierender gutartiger *Papillomatose* der Harnblase, die mehrere Male operativ angegangen wurde, zeigte auch nach Betatronbestrahlung mit 8400 r nach 3 Monaten wieder ein Rezidiv. Cystoskopisch wurde 5 Monate ante exitum wiederum eine Papillomatosis festgestellt ohne Anhaltspunkte für Malignität. Exitus 9 Monate nach der Betatronbehandlung (Ursache nicht feststellbar gewesen).

2 Patienten mit *Urethra-Carcinom* kamen 12 bzw. 1 Monat nach Bestrahlung ad exitum. Im 1. Fall handelte es sich um Rezidivbestrahlung nach Operation, im 2. Fall um Vorbestrahlung vor Operation.

Von den 69 Patienten mit *Bronchialcarcinom* wurde 1 Patient einer Vorbestrahlung unterzogen (Pflasterzell-Ca.) und 5 nach Pneumektomie nachbestrahlt (1 kleinzelliges Ca., 4 Pflasterzell-Ca.). Bei den übrigen Patienten wurden die Bestrahlungen wegen nicht radikal operierten oder inoperablen Tumoren vorgenommen. 2 Patienten sind verschollen und vermutlich verstorben. 12% der Patienten lebten nur mindestens 1 Jahr, symptomfrei waren nur 3%; 2% lebten mindestens 2 Jahre und ebensoviele waren symptomfrei. Mindestens 3 Jahre und länger lebte nur 1 Patient; es handelt sich hier um den vorbestrahlten und pneumektomierten Fall. Patient ist heute 5 Jahre 8 Monate symptomfrei. Die durchschnittliche Überlebensdauer der Verstorbenen beträgt 5 Monate (maximal 29 Monate). Der Exitus trat infolge Progredienz, Metastasierung oder Kreislaufversagens ein. 3 dieser Patienten waren vorübergehend gebessert worden, 5 Patienten zeigten ein vorübergehendes Stationärbleiben des Befundes.

Diese Behandlungsergebnisse entsprechen ungefähr denen der Statistik (SCHÄRER 1951) über die in unserem Institut von 1921—1951 mit konventioneller Therapie bestrahlten Bronchuscarcinome.

Noch schlechter sind die Bestrahlungsergebnisse bei den folgenden 2 Gruppen, den Patienten mit Oesophagus-, Rectum- und Colontumoren. Hier handelt es sich fast ausschließlich um ausgedehnte Tumoren, sehr oft gleichzeitig mit Vorliegen eines schlechten Allgemeinzustandes. Von den 72 Patienten mit *Oesophagustumoren* lebten mindestens 1 Jahr nur 8% und mindestens 2 Jahre nur noch 2%. Symptomfreiheit wurde bei keinem Patienten erzielt. 18 wiesen eine vorübergehende Besserung (bis zu 12 Monaten) auf. Die durchschnittliche Überlebensdauer betrug 6 Monate (maximal 31 Monate).

Betrachten wir diese Oesophagusfälle und vergleichen sie mit der Statistik der mit konventioneller Bestrahlung behandelten Oesophagusfälle des gleichen Institutes (SCHÄRER, 1954 und 1955), so fällt uns schon sofort auf, daß uns in den 5 Jahren 1951—1956 nur 87 Patienten mit Oesophagustumoren zugewiesen wurden (15 wurden mit konventioneller Therapie behandelt), während in den 8 Jahren 1943—1951 225 Patienten, in den 6 Jahren 1936—1942 gar 251 Patienten zugewiesen wurden, d. h. in früheren Jahren jährlich etwa 30—40, heutzutage dagegen nur etwa 17. Es dürften somit heute eine ansehnliche Anzahl von Patienten anderen chirurgischen Behandlungsmethoden zugeführt werden. 1943/51 lebten 24 von 225 mindestens 1 Jahr symptomfrei, 3 Jahre 12 von 225 und 5 Jahre 5 von 181, Resultate, die, wie SCHÄRER schrieb, „den chirurgischen nicht nachstehen".

Von den 42 Patienten mit *Tumoren des Colons und Rectums* haben 29% mindestens 1 Jahr gelebt, mindestens 2 Jahre lebten 17%, symptomfrei 7%

(s. Tab. 1). Von den 3 symptomfrei gewordenen Fällen hatte 1 Patient ein Carcinom des Colon transversum, er lebt jetzt 3 Jahre nach Operation und Nachbestrahlung, die beiden anderen Patienten hatten einen Tumor der Appendix (Carcinoma adenomatosum cylindrocellulare bzw. Pseudomyxoma), die ebenfalls operiert und nachbestrahlt wurden. Ein 4. Patient mit rezidivierendem und metastasierendem Sigma-Carcinom lebte über 2 Jahre nach Bestrahlung eines inoperablen Tumors über 2 Jahre in gebessertem Zustand (Gesamtdosis 16050 r in 3 Serien innerhalb 8 Monaten), bekam aber dann ein Astroblastom, das exidiert und nachbestrahlt wurde, und starb 7 Monate später infolge eines Rezidives des Hirntumors. Die durchschnittliche Überlebenszeit der 39 gestorbenen Patienten beträgt 9 Monate (maximal 35 Monate). 7 von diesen Patienten wiesen eine vorübergehende Besserung des Befundes auf.

Weitere *Abdominalgeschwülste*, die mit Betatron bestrahlt wurden, waren Rezidive von *Magencarcinomen* (7 Fälle), 1 inoperables Pankreaskopfcarcinom, Rezidiv eines Fibrosarkomes, *Metastasen* im Bereich des Netzes, des Omentum, des Douglas, lymphogranulomatöse, lymphosarkomatöse und reticulosarkomatöse Lymphknotenschwellungen, Metastasen im Bereich der Porta und inoperable Gallenblasentumoren. Zusammen mit den Magentumoren handelt es sich um 21 Patienten. Die meisten starben infolge Progredienz der Tumoren durchschnittlich nach einer Überlebensdauer von 6 Monaten. Vorübergehende Besserung von einigen Monaten wurde bei einem der Magenpatienten, bei den beiden Lymphosarkompatienten und bei 1 der 4 lymphogranulomatösen Patienten beobachtet. Von diesen 21 Patienten leben noch 3, und zwar 1 Patient mit Metastasen eines op. Seminoms des linken Hodens im Mediastinum und Pankreas. Die Pankreasmetastase wurde in 3 Serien mit total 11500 r bestrahlt und die Mediastinalmetastasen mit 8000 r; Patient ist $2^1/_2$ Jahre nach der letzten Bestrahlungsserie symptomfrei. Eine weitere Patientin mit Carcinommetastase eines unbekannten Primärtumors lebt jetzt 10 Monate in stationärem Zustand, während der 3. Patient mit reticulosarkomatösen Lymphknotenschwellungen nach den ersten beiden Bestrahlungsserien mit insgesamt 10000 r (mit Intervall von 3 Monaten) 8 Monate beschwerdefrei blieb. Eine 3. Serie mit 4400 r wegen eines Rezidives brachte nur eine vorübergehende Besserung. Patient lebt jetzt 9 Monate nach der letzten Serie, 2 Jahre nach der 1. Serie in einem sich langsam verschlechternden Zustand.

Von den 14 Patienten mit primären und sekundären *Tumoren des Skeletsystems* lebt nur einer mit Status nach Bestrahlung einer Wirbelmetastase nach Mamma-Carcinom jetzt 12 Monate symptomfrei. Ein weiterer Patient, ebenfalls mit Wirbelmetastasen, lebt jetzt in progredientem Zustand 4 Monate nach Abschluß der Bestrahlung. Die übrigen Patienten kamen nach durchschnittlich 7 Monaten Überlebensdauer ad exitum. Die längste Überlebensdauer wies dabei ein Patient mit rezidivierendem Chondrosarkom des Beckens auf, das innerhalb von 4 Monaten mit 13500 r in 3 Serien bestrahlt wurde. Der Patient kam nach 2 Jahren 5 Monate nach der 1. Serie bzw. 1 Jahr 2 Monate nach der letzten Serie ad extium.

24 Patienten mit *Mediastinaltumoren* wurden einer Betatronbestrahlung unterzogen. Von diesen hatten 12 Patienten eine *Lymphogranulomatosis Hodgkin*. 8 Patienten sind nach durchschnittlich 7 Monaten Überlebenszeit gestorben; in 2 dieser Fälle war der Mediastinaltumor verschwunden, in 6 war nur ein gewisser

Rückgang zu beobachten gewesen. 1 Patient ist verschollen. 3 leben noch symptomfrei 19, 24 und 28 Monate.

Von 3 Patienten mit *lymphoblastärem Sarkom* lebt nur einer noch über 7 Monate in progredientem Zustand. Von den 6 Fällen mit *Metastasen* anderer Tumoren haben wir schon einen mit Seminommetastasen weiter oben erwähnt. Von 4 Carcinom-Metastasen ist keiner gebessert worden; 2 sind gestorben. Von den 3 histologisch nicht verifizierten Mediastinaltumorfällen sind 2 ungebessert ad exitum gekommen, während der Tumor des 3. Patienten durch die Bestrahlung verschwunden ist; er starb 3 Jahre 2 Monate später an abdominal aufgetretenen Tumoren. Der 6., ohne Erfolg bestrahlte Patient hatte eine ausgedehnte Adenokanthommetastase.

Von den mit Betatron bestrahlten 2 Patienten mit *solitären Lungenmetastasen* starb der eine (Adenocarcinom-Metastase eines unbekannten Primärtumors) ungebessert nach 3 Monaten, während der zweite mit Metastase eines Mamma-Carcinoms heute 14 Monate in gebessertem Zustand lebt (sonst keine Anhaltspunkte für Rezidiv oder anderen Metastasen).

Von 3 Patienten mit *Rückenmarktumoren* (1 rad. op. Pflasterzellmetastase, 1 histologisch nicht verifizierter Tumor) lebt nur noch 1 Patient mit Ependymom, der schon vorher hohe Dosen mit konventioneller Therapie erhalten hatte (12400 r/o) und nach 5 Jahren Betatronbestrahlung mit insgesamt 5000 r, jetzt 5 Jahre 7 Monate in stationärem Zustand.

Keine Besserung wurde erzielt bei 2 Patienten mit *Parotistumor* (bei einem desselben Verlängerung der Überlebensdauer nach der 1. Bestrahlungsserie um über 28 Monate), bei 3 Patienten mit *Schädeltumoren* (Sympathogoniommetastase des Schädels, Cylindrom der Orbita, Spindelzellsarkom der Orbita), 3 Patienten mit ausgedehnten, vielfach und hoch bestrahlten *Basaliomen* (in einem Fall mit Infiltration in die Orbita), bei einem *Melanom* der Fußhaut mit Lymphknotenmetastasen, bei 2 Patienten mit subtotal entferntem *Prostatacarcinom* und 1 Patient mit *Ca. der Glandula submandibularis*.

Von den 18 Patienten mit *Tumoren der oberen Luft- und Speisewege* hatten 6 einen *Epipharynxtumor*, und zwar 3 ein lymphoepitheliales Carcinom (SCHMINCKE-REGAUD). 1 von diesen lebt 2 Jahre 6 Monate symptomfrei, während die beiden anderen mit rezidivierenden, schon vorher konventionell bestrahlten Tumoren nur vorübergehend gebessert wurden. 1 Patient mit Reticulo-Sarkom blieb lokal symptomfrei, starb aber an seinen ausgedehnten, sich ziemlich strahlenresistent erweisenden Halsmetastasen (bei der Sektion war der Epipharynx tumorfrei). Ein weiterer Patient mit Carcinoma solidum ist jetzt lokal 5 Jahre 9 Monate symptomfrei, während der 6. Patient mit Pflasterzellcarcinom 4 Monate nach Abschluß der Bestrahlung am Tumor starb. 9 weitere Patienten mit rezidivierenden Carcinomen (1: ausgedehntes Mundhöhlencarcinom, 3: Gaumenbogenzungenwinkelcarcinom, 1: Hypopharynx-, 1: inneres Larynx-, 3: Zungencarcinom, 1: ausgedehntes Mundbodencarcinom) wiesen keinen oder nur vorübergehenden Strahlenerfolg auf. Es leben jetzt noch 2 Patienten mit Hypopharynxcarcinom über 5 Jahre symptomfrei, ebenso war ein Patient mit Larynxcarcinom 4 Jahre 4 Monate symptomfrei, ist aber jetzt verschollen.

**Zusammenfassung.** Wie die bisherigen 6jährigen Erfahrungen an 532 von insgesamt über 660 Patienten zeigen, erweist sich die Behandlung mit ultraharten Strahlen (31 MeV-Betatron) als indiziert bei tiefliegenden Tumoren, besonders des Schädelinnern und des Stammes, infolge der Möglichkeit, relativ einfach hohe Herddosen unter Schonung der Haut und des umgebenden nicht-tumorösen Gewebes zu erzielen. Von großem Vorteil ist hierbei auch die in gleichem Maße stattfindende Absorption der Strahlung im Knochen und in den Weichteilen. Eine weitere Indikation stellt auch die Rezidivbehandlung schon mit 200 kV-Therapie vorbestrahlter Patienten dar, bei denen die Hautveränderungen keine weitere konventionelle Bestrahlung mehr erlauben. Im übrigen hängen die Bestrahlungserfolge, genau wie bei der konventionellen Strahlentherapie, von der Natur, Größe und Verbreitung des Tumors ab, sowie von dem Allgemeinzustand des Patienten. Als schlagender Beweis hierfür mögen die oben aufgeführten Ergebnisse unserer Patienten mit Oesophagustumoren dienen. Im Vergleich zur konventionellen Strahlentherapie scheint die Betatrontherapie einen besseren Effekt bei der Behandlung von Harnblasentumoren zu haben, besonders als präoperative Behandlung. Dies gilt möglicherweise auch für die Bronchus-carcinome. Sicheres läßt sich darüber aber erst anhand einer größeren Zahl von Fällen aussagen.

## Literatur

Buschke, F.: Schweiz. med. Wschr. **1953**, 641.
— Amer. J. Roentgenol. **71**, 9 (1954).
— Oncologia **6**, 1 (1953).
Schärer, K.: Oncologia **4**, 65 (1951).
— Oncologia **8**, 211 (1955).
— Schweiz. med. Wschr. **1954** 1025 (Diskuss.beitr.).
— Strahlenther. **97**, 508 (1955).
Schinz, H. R.: Fortschr. Röntgenstr. **80**, 1 (1954) (mit ausführlichem Verzeichnis der Arbeiten mit dem Zürcher Betatron bis 1953).
— Fortschr. Röntgenstr. **86**, 104 (1957).
— Fortschr. Röntgenstr. **86**, 363 (1957).

# Our Experience of Betatron Therapy*
## on Certain Clinical Aspects

By

**B. Bellion** and **G. Lovera**, Torino

I. One of the problems to which we have been paying great attention during the last few years has been the study of the characteristics of the beam of electrons of which we dispose and whose physical characteristics we will first briefly describe.

Our betatron (31 MeV B. B. C.) permits us to use both the beam of photons and the beam of electrons. As has already been pointed out, in order to extract the electrons from our apparatus, after the acceleration cycle of the electrons, these, instead of being aimed at the target, are made to emerge from the doughnut by a process of magnetic interference. Their pathway through the thin glass window situated in the wall of the doughnut

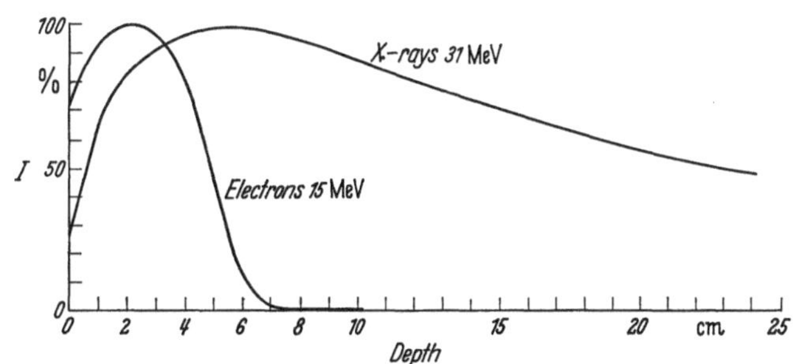



Fig. 1. Transition curves in plexiglass for γ-ray and electrons

has two results: 1. the electrons lose on the average one MeV of energy; 2. they are diffused. In our apparatus, the beam of electrons is brought to the desired dimensions by means of a plexiglass collimator: at 1 metre from the focus the field of the electrons measures $5 \times 12$ sq.cm., the greater side of the rectangle being the horizontal one. When using the electrons, we naturally always use screens and localizers made of elements with a low atomic number, in order to reduce to pratically insignificant values the effects of the "Bremsstrahlung". We attempt to reduce to a minimum the number of photons produced, in order to preserve one of the most interesting characteristics of the transition curve as regards treatment: its terminal part which drops rapidly in an evident manner. In Fig. 1 are shown the transition curves of the gamma rays corresponding to 31 MeV and of the electrons corresponding to 15 MeV (that is to say, electrons which are accelerated by an energy equivalent to 15 MeV at the moment of their expansion in the tube). At present it is possible to utilize a monoenergetic beam of electrons in the range of from 30 to 14 MeV, whose intensity can be compared with that of gamma rays. The end part of the

---


* From the Radiology Department and Centre for Physico-Biological Research of the Clinica Medica Generale of the University of Turin (Director: Prof. B. Bellion).

transition curve of the electrons has an almost linear course, which allows extrapolation. The "tail" of the curve is due to the photons produced by the "Bremsstrahlung". We believe that the ionization produced by "Bremsstrahlung" has a secondary importance: the electrons of 20 MeV energy dissipate 90% of their energy in an organic tissue by ionization and 10% by "Bremsstrahlung".

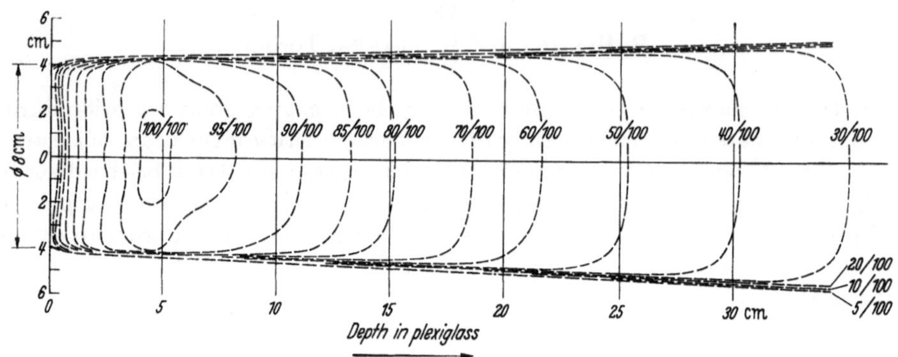


Fig. 2. Isodoses curves of 31 MeV photons

As regards high energy gamma rays, Fig. 2 shows the isodose curves of the photons at 31 MeV of energy: the collimation of our beam is satisfactory, as is shown by the percentage values of lateral diffusion.

We believe that, in practice, the photometric determination of the isodose curves and of the transition curves has the advantage of a greater precision and

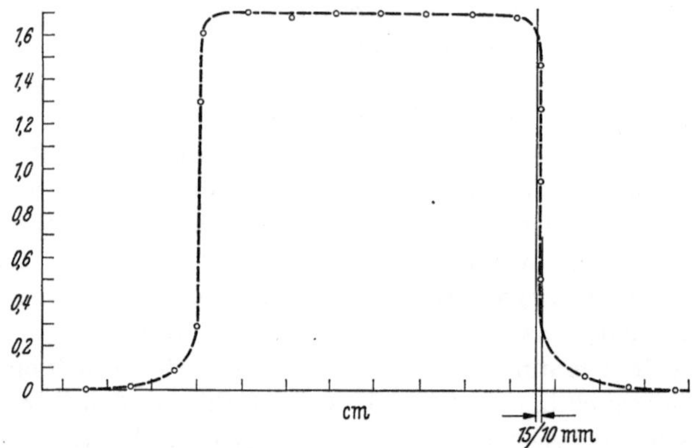


Fig. 3. Section of a beam of γ-rays determined with photometric method

of being better demonstrable as compared with determination with an ionization chamber. Thus, if a measurement of a section of the beam of γ—rays emitted by the betatron, as determined by the photometric method, is considered (Fig. 3), one is convinced of the accuracy of this method which allows the measurement of four different points in the space of 15/10 mm.; this would be in practice impossible with an ionization chamber.

With the same photometric method we have traced the isodose curves for our beam of electrons at the energy of 15 MeV (Fig. 4); these show eloquently the characteristic distribution of the dose of these corpuscular radiations.

II. Those concerned with betatron therapy or with treatment with high energy radiations must so far conclude that these new weapons represent essentially a *technical* progress. This however increases the interest of this type of application as it has been stated and demonstrated that, in our era, progress in the experimental scientific field is impossible without corresponding progress in the technical field. For this reason we also feel it is necessary to obtain a satisfactory quality in our work in order to guarantee us, as far as possible, from criticism. In this field one of the greatest difficulties in routine work is the exact topographic localization of the focus and the consequent centering of the subject. The rather high number of treatments that we perform daily has so far prevented us from applying constantly an exact radiogeometric technique, such as others apply usually; this is however one of our aims. At present we usually verify the internal disposition of the beam and the distribution of the doses in relation to the focus to be irradiated by means of the photographicmethod. In Fig. 5 is shown a practical example of our technique for the centering of a vegetating neoplasm of the middle third of the oesophagus. We use two lead windows, one coinciding with the cutaneous portal of entry and the other with the cutaneous portal of exit of the beam. The two windows must coincide and must both include all the focus to be irradiated. In Fig. 6 is shown the registration made by inserting a film in a transverse section of a dummy corresponding in size to the thorax of one of our patients with a pulmonary tumour, showing the topographic distribution and the relative dose values of two crossed beams. The values obtained from these films by densitometric determinations show a good correspondence with the data obtained by calculation and by the graphic reconstruction of the intensity of the dose (especially in the case of crossed and opposed fields). Fig. 7 represents an example of graphic and schematic representation of the focus dose in the treatment of a tumour of the oesophagus with four crossed and opposed portals of entry. In

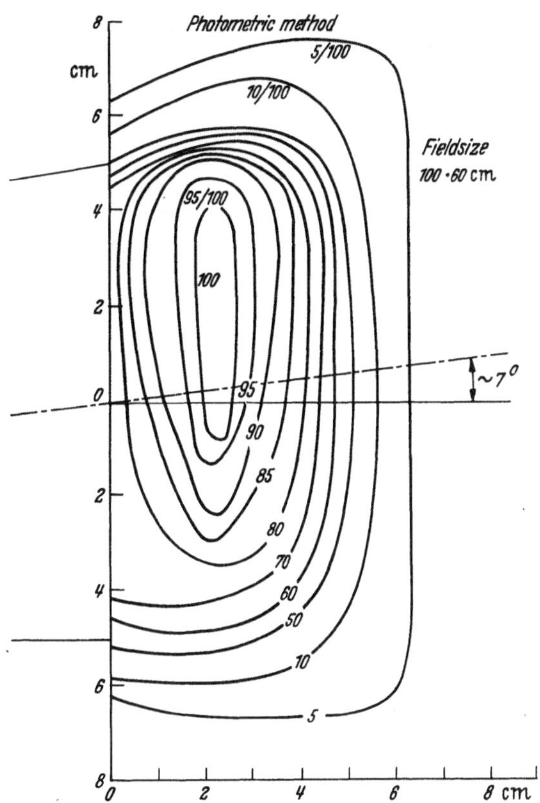



Fig. 4. Isodose curves for 15 MeV electrons

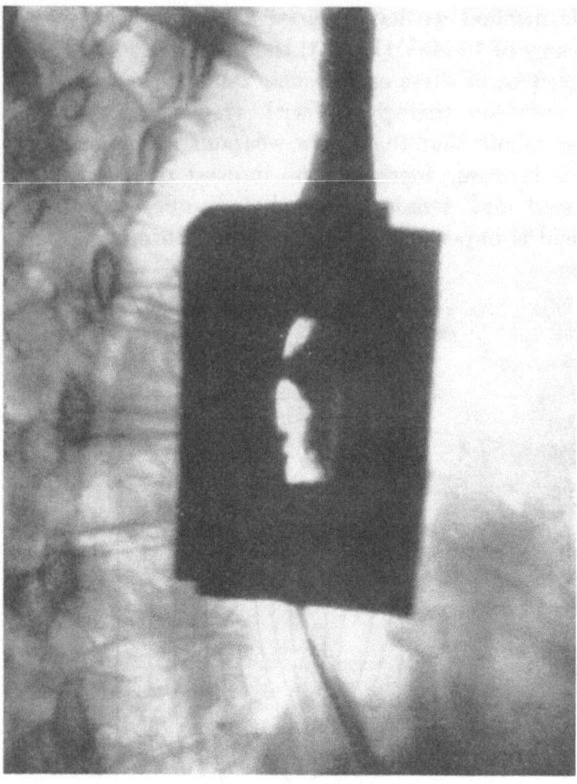

Fig. 5

our opinion, oesophageal tumours are the most easily identifiable topographically and those more easily controlled periodically; in this field the results correspond to those reported by others: see in Figs. 8 and 9 the same case of oesophageal tumour illustrated in Fig. 5, before and after 6000 r focus Victoreen.

III. The study of the *relative biological efficacy* of the various types of radiations is now undoubtedly one of the most interesting problems. We have already stated elsewhere that it seems to be difficult to establish absolute criteria for comparison, such as can be expressed numerically, of the biological effects of different types of radiations. These attempts, in which we have had the collaboration of biologists and naturalists, appear to be on the way to produce marked advances in the biological field and in the evaluation of the finer biological structures and of their functions. On the other hand, in the clinical field, which is the one that has the greater interest for us as physicians, the comparison of the effects of the various types of radiations seems to be very

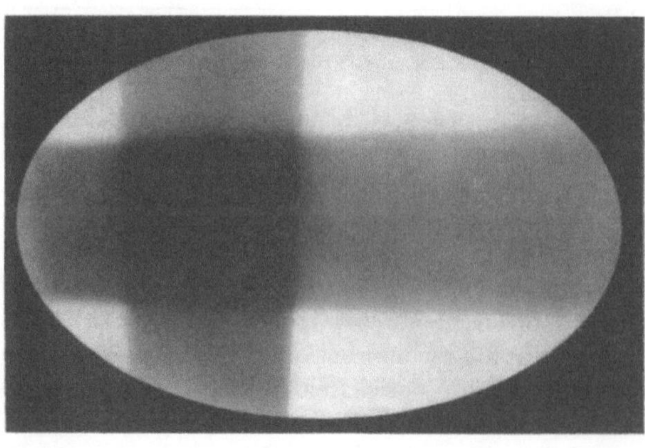

Fig. 6

difficult, as has already been stated, for these reasons: 1. it is difficult to evaluate exactly the results obtained and it especially difficult to dispose of sufficiently homogenous and numerous material to permit statistical evaluation; 2. it is impossible to obtain analogous and exactly correspondant experimental conditions. For clinical purposes, in fact, not only the number of biological events produced by a given quantity of radiations in the cells of a tissue, but also other features, such as the general reactions of the patient, changes of the endocrine system, of the blood picture, of the various enzymes, and so on, are of importance and sometimes are of prevalent significance. (We have given particular attention to these general reactions in the course of our researches on patients treated with betatron therapy.) The factors connected with the beam of radiations used must also be borne in mind,

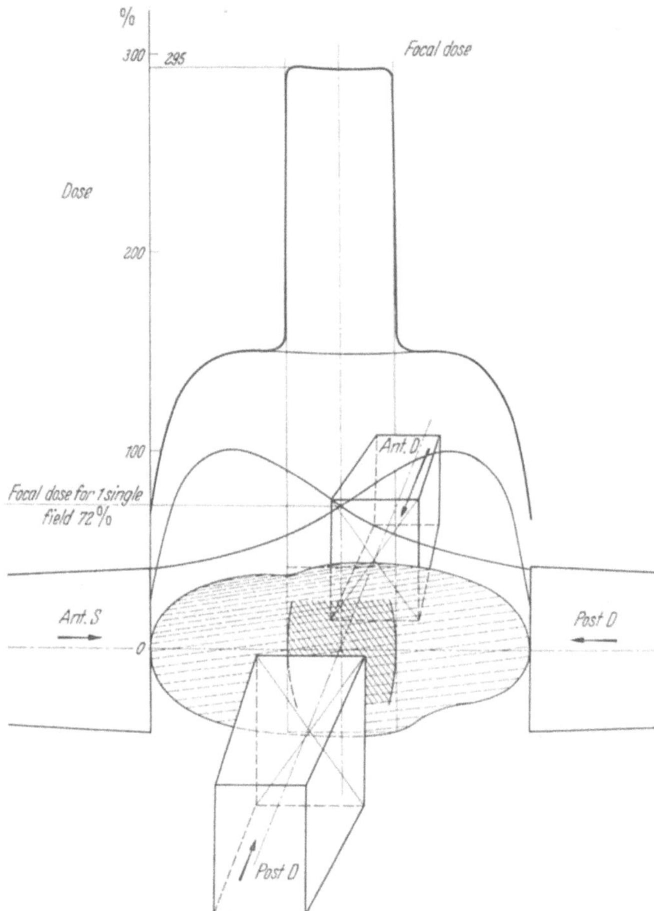



Fig. 7. Dose curves for four crossed portals of entry in the treatment of a Tumor of the oesophagus

as must the relations between the dose received by the tumour and the dose received by the healthy tissues and the possibility of resolving particular radiotherapeutic problems with a given type of radiation.

Our experience seems to indicate that the advantages of treatment with gamma or high energy beta rays, as regards the tolerance, the distribution of the energy in depth and the ratio between the dose received by the tumour and that received by the healthy tissues, are such as can amply compensate a possible minimal biological effect, in the absolute sense, of the beta or high energy gamma rays.

As regards the relative biological effects, we would like to stress, on the basis of Laughlin and Wachsmann's reports, the possibility of greater effects, in a certain sense elective, of the radiations possessing a low specific capacity of ionization, such as gamma and high energy beta rays, and the necessity, which must be again stressed, of a great number of observations of homologous lesions, possibly in the same patient, in order to obtain a comparison of the biological activity of various types of truly effective radiations.

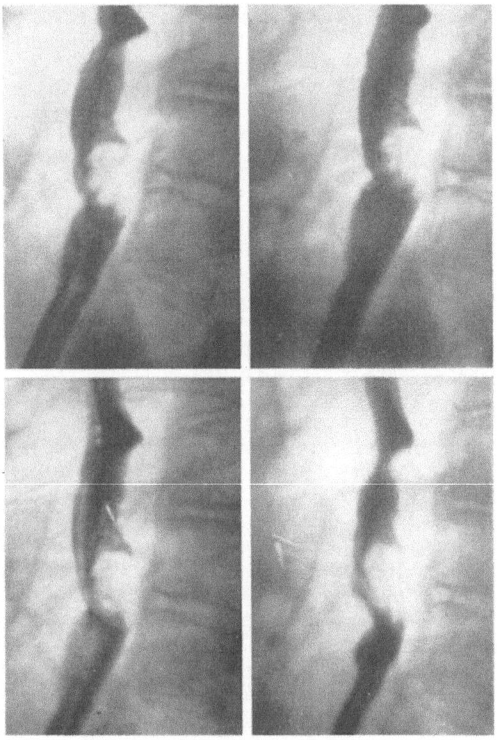

In conclusion, our impression is that the researches on the relative biological effects of the various types of radiations are only rarely of clinical significance. For example, it would be difficult to explain, on the basis of reduced relative biological effects, certain destructive changes observed by other workers and by ourselves with relatively low doses, and especially certain successes presented in our statistics and in those of others. For this reason we believe that, at present, the results of laboratory experiments differ from those of clinical experience and that the relative biological effect must be distinguished from the therapeutical effect.

IV. The disagreement on the efficacy of gamma and high energy beta rays as compared with the conventional type of X-rays is undoubtedly also connected in part with the problems of dosimetry, which assume a position of particular importance in the field of treatment with high energy radiations.

Fig. 8

In routine work, also with beta and high energy gamma rays, the unit of measurement adopted is the "roentgen", because this unit of measurement is handy and easily reproduced. In fact, the "roentgen" unit refers to the quantity of energy dissipated in the air and transferred from the ionizing radiation considered to the ionizing particles generated in the considered point; this, in the case of X rays of less than 3 MeV energy, coincides with the energy absorbed by a body in the considered point, because the ionizing particles give up their energy close to this point. With energies of more than 3 MeV the pathway of the electrons increases so much that the above described approximation is no longer valid and especially the measurements with the ionization chambers used for low tension

X rays and the measurement of the ionization as expressed in "roentgens" in a given point is no longer an exact expression.

Both for these reasons and for the desire to dispose of a unit of measurement independant of the characteristics of the radiations used the "rad" unit has been introduced (this is equivalent to the absorption of 100 ergs of energy for each gramme of tissue). Even though we believe that the use of the "rad" unit is most appropriate, we feel that its introduction cannot as yet free us from confusion in the terms used, as thus we pass from a unit of ionization to a unit of absorbed energy. The fact has not been sufficiently stressed in the past that the measurements of ionization performed with classical types of dosimeters at high energies and expressed as "roentgens" by the instruments in use in most clinics refer to units of measurement that have nothing in common with the "roentgen" of the classical definition. It would be necessary to distinguish in practice the two factors measured, ionization and absorption of energy.

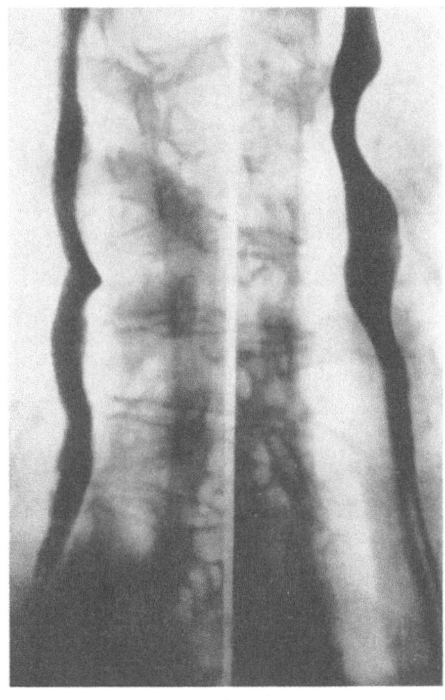

Fig 9

We feel that marked progress, especially as regards comparison of the results and methods of treatment, could be achieved if all laboratories and clinics adopted, for high energy radiations, a conventional but generally accepted unit of measurement, such as the "r victoreen" determined on a water phantom, or on one made of organic matter similar to living matter, with certain constant characteristics.

V. As regards the clinical aspects and the results of the treatment of tumours with high energy radiations, we must first of all stress that we consider our experience (536 cases treated in 3 years) too limited to be able to draw unequivocal conclusions.

In this regard, HAAS and HARVEY, in their most recent paper on the clinical aspects of betatron therapy, with which we fully agree, stress the necessity of the study of large groups of patients, with forms similar for localization and stage, and who can be controlled for a long time, in order to reach, in future, definitive conclusions as to the indications, methods of application and results of treatment with high energy radiations.

We can however present a number of observations beased on our personal experience. First of all, with this type of radiation, it is necessary to establish with great caution the results that are being aimed at. The treatment will evidently be very different in cases in which a "major remission" can be expected, together with a good survival, without side effects, and in those, unfortunately more numerous, in whom only a palliative effect can be expected. In the first, betatron therapy offers the unique possibility of reaching a presumably primary focus with very nearly cancericidal doses and in this case a certain number of risks are justifiable in order to obtain this result. In the second group, especially in patients in poor general conditions, previously operated or irradiated, betatron therapy offers the possibility of obtaining an improvement of their complaints with minimal general effects, obtaining a favourable ratio of the symptoms due to the disease to those provoked by irradiation.

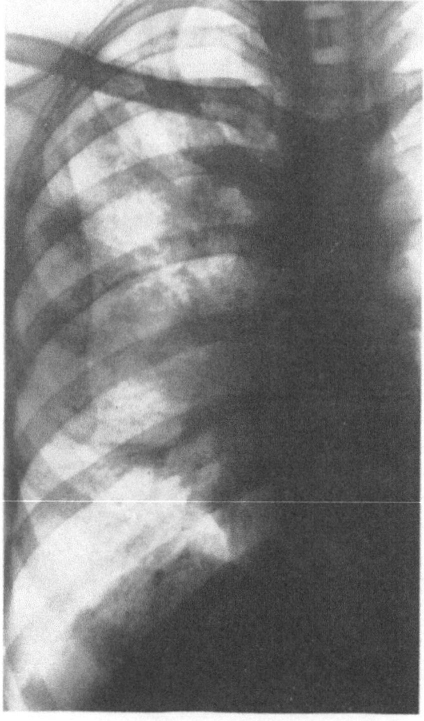

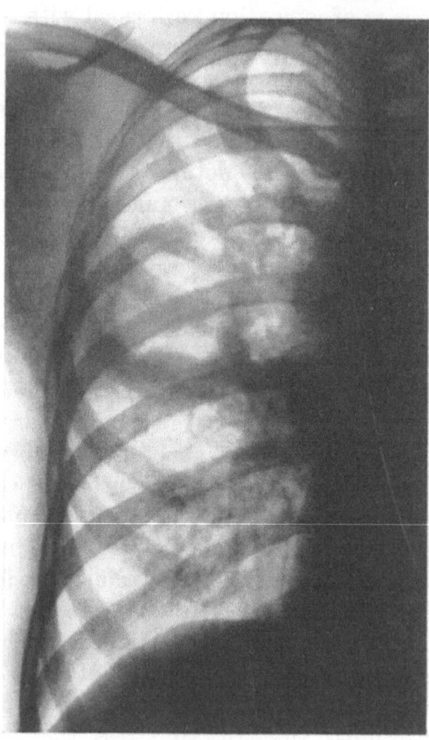

Fig. 10a                                              Fig. 10b

The second point is the method of treatment. Our experience and that of other workers has shown that betatron therapy, both on account of the physical characteristics of the beam of gamma rays and of its excellent general tolerability, allows the application of a plan of treatment with focus doses such as are not obtainable with other types of radiation therapy. To this possibility must probably be attributed almost all of the major complications reported by some Authors in the course of treatment with high energy gamma rays, such as visceral or vascular perforations, etc.; these can be at least partly attributed to the rapid regression of the neoplastic tissues.

We have adopted the principle, in accordance with Haas and Harvey, of attempting to obtain the greatest possible utilization of the factor "time" and especially of avoiding the risk of excessive doses and of adapting the plan of treatment to the clinical characteristics of each case.

In general, we believe that, also for "curative" purposes, a total focus dose of 6000 r "victoreen", given at least in 30 administrations, represents a sufficient and generally safe quantity, naturally subordinated to the localization and nature of the tumour; we feel that it is advisable not to exceed this dose.

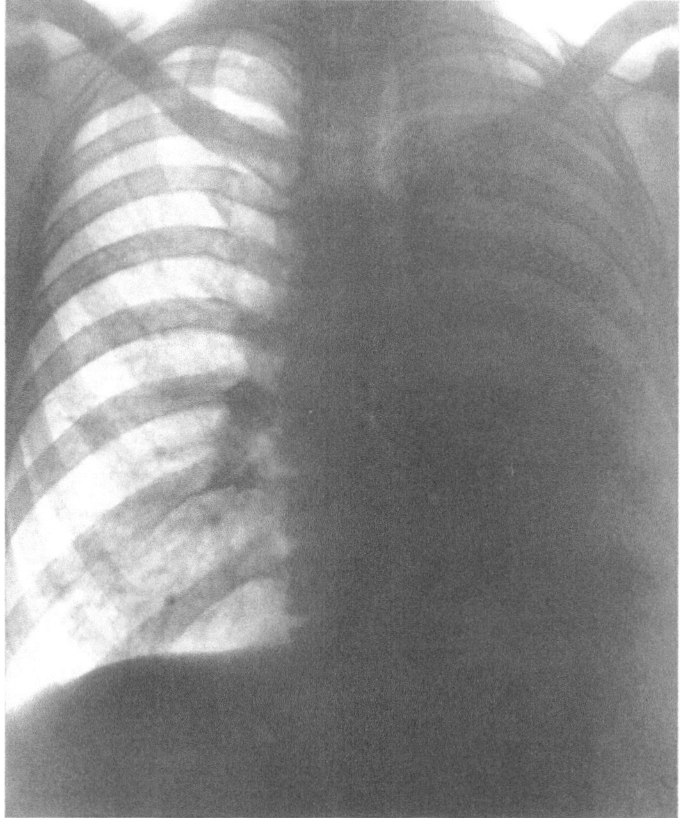

Fig. 11a

A third observation is the possibility of resolving, with high energy radiations, problems of radiotherapy which, for geometrical, biological or clinical reasons, it is impossible to resolve with conventional radiotherapy. In this regard a number of successes obtained in cases of cancer of the lung with marked pleural effusion must be remembered; it is notorious that these cases are a clear contraindication to irradiation with the conventional type of apparata.

As regards our results, we may first of all state that we have observed an excellent general tolerance in our patients.

As regards the various forms of tumours, we have obtained the best results in cases of cerebral tumours, in oesophageal and pulmonary tumours and in

                    B. Bellion and G. Lovera:

endothoracic tumours in general, as well as in tumours of the bladder and of the female genitals. Good palliative results were obtained in gastric tumours, though no appreciable regression of the roentgenologic picture has ever been observed.

In the course of our work we have often been struck by the high incidence of sterilization, also demonstrable histologically, of the irradiated neoplastic focus and also by a number of good results, even if only palliative, in patients in whom such results were unexpected on account of the advanced stage of the disease.

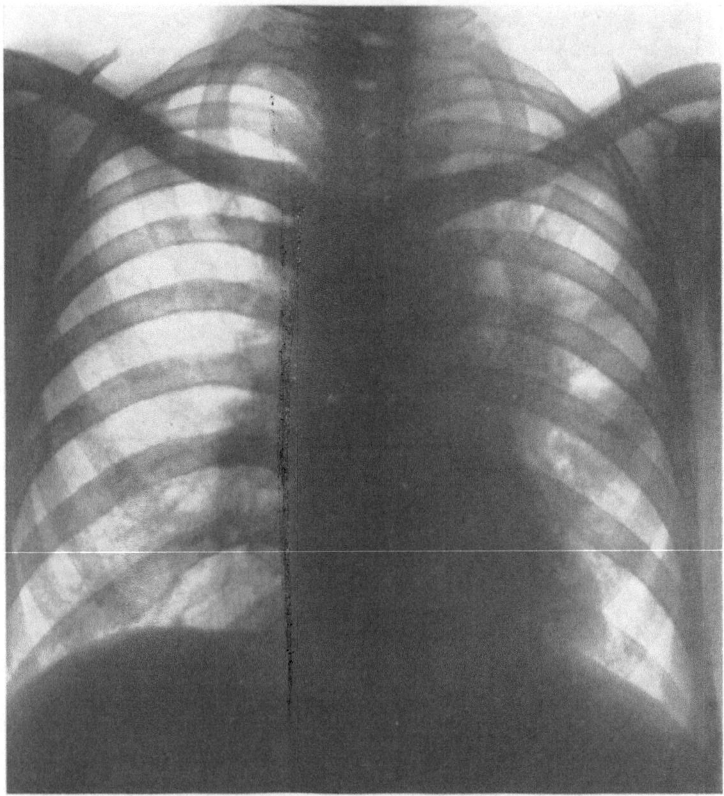

Fig. 11b

Similar observations have been made by many workers in the field of high energy radiations and all agree in admitting that, as well as the problems connected with the physical and geometrical aspects of irradiation, all the more so inasmuch as high energy radiations offer in this sense possibilities as were formerly unknown, importance must be attributed to the clinical concepts upon which the course of treatment must be based.

Favourable results such as those illustrated eloquently in the roentgenograms in Figs. 10a and 10b, and 11a and 11b, respectively, represent for us a fairly frequent occurrence; this fact induces us to attempt to improve in every way the standard of our work.

In conclusion, we feel we can affirm that the betatron represents a marked advancement, above all a technical improvement, in the field of radiotherapy that offers the possibility, as with our machine and with other similar ones, of using a beam of high energy gamma or beta rays, which is the best means for irradiating respectively deep or superficial neoplastic foci.

The greater possibilities in the treatment of these affections confers, in our opinion, as has already been stated, a decisive rôle to the clinical aspects of radiotherapy; this renders advisable the fullest collaboration of the various schools, on the basis of the uniformation of the dosimetric methods, of the classification of the pathological forms and, as far as possible, of the types of treatment, in order to obtain a series of sufficiently exact and extensive data which can allow the definition of the indications and possibilities of the treatment of neoplastic diseases with high energy radiations and to contribute to our knowledge of the pathology and clinical aspects of these affections.

Summary. The physical characteristics of the beta and gamma rays emitted by the betatron are discussed. The problems concerned with the dosimetry of high energy radiations are considered and the biological and therapeutical activity of these radiations is compared with that of conventional energy X-rays. A number of considerations are made on the clinical results obtained with high energy gamma rays.

## References


AUSTIN, V. T., T. G. KERLEY, E. F. LANZ, G. Y. McCLURE, E. A. THOMPSON and L. S. SKAGGS: Amer. J. Roentgenol. 61, 591—625 (1954).

BACQ, Q. M., and P. ALEXANDER: Principes de Radiobiologie. Paris: Masson et Cie. 1950.

BARTH, G., and F. WACHSMANN: Strahlenther. 93, 395—399 (1954).

BODE, H. G., W. PAUL and G. SCHUBERT: Strahlenther. 81, 251—266 (1950).

— D. HAMPEL and B. MARKUS: Strahlenther. 92, 563—575 (1953).

CARPENDER, J. W., S. T. CANTRIL, M. FRIEDMAN, R. J. GUTMANN and T. A. WATSON: Radiology 67, 481—516 (1956).

DITTRICH, W., and G. SCHUBERT: Strahlenther. 92, 532—554 (1953).

FRAIN, C., J. SURMONT, M. TUBIANA, B. PIERQUIN, R. MARLOIS, J. ABBATUCCI and A. DUTREIX: J. Radiol. et Electrol. 36, 792—794 (1955).

HAAS, L. L., and R. A. HARVEY: Amer. J. Roentgenol. 76, 905—919 (1956).

— — and J. S. LAUGHLIN: Amer. J. Roentgenol. 68, 544—653 (1952).

— and G. M. SANDBERG: Radiology 66, 102—104 (1956).

LAUGHLIN, J. S.: Physical Aspects of Betatron Therapy. Springfield, Ill., USA: C. C. Thomas 1954.

MARKUS, B., and W. PAUL: Strahlenther. 92, 599—611 (1953).

MOOS, W. S., J. B. FULLER, W. J. HENDERSON and R. A. HARVEY: Radiology 67, 697—703 (1956).

SCHINZ, H. R.: Fortschr. Röntgenstr. 80, 1—28 (1954).

— H. F. NIGGLER and K. SCHÄRER: Radiol. Clin. 24, 317—346 (1955).

— and R. WIDERÖE: Strahlenther. 95, 33—40 (1954).

SURMONT, J., M. TUBIANA, C. PIERQUIN, C. M. LALANNE and J. DUTREIX: J. Radiol. et Electrol. 36, 580 (1955).

TUBIANA, M., J. DUTREIX, B. PIERQUIN and P. JOCKEY: J. Radiol. et Electrol. 36, 507—522 (1955).

ZUPPINGER, A., W. MINDER, G. PORETTI, H. R. RENFER, C. TÜTSCH and A. SCHREIBER: Radiol. Clin. 24, 346—368 (1955).

## Diskussionsbemerkungen

### A. Zuppinger (Bern)*

### Klinische Erfahrungen mit energiereichen Strahlen

Wir Strahlentherapeuten haben, wenn wir von der wohlbegründeten Regel über 5-Jahres-Resultate zu berichten abweichen, immer Bedenken. Wenn aber, wie wir heute gehört haben, so viel Geist, Mühe und Zeit für die Entwicklung neuer Methoden verwendet werden, dann ist dies wohl nur möglich, weil wir alle überzeugt sind, daß das Neue gut und erfolgversprechend ist. Unsere Erfahrungen seien hier als Meinungsaustausch mitgeteilt, um möglichst rasch den Weg optimaler Anwendung energiereicher Strahlung finden zu können.

Im wesentlichen werde ich mich auf biologische Gesichtspunkte beschränken. Nur einleitend sei zur Frage Stellung genommen, ob die *Koppelung der Diagnostikapparatur mit dem Therapiegerät*, wie dies beim Asklepitron der Fall ist, zweckmäßig ist oder nicht. Die Koppelung ist praktisch sicher wünschenswert, weil sie uns immer gestattet, unmittelbar die Kontrolle der Feldeinstellung zu ermöglichen. Sie ist aber als solche nicht unbedingt notwendig. Wenn genügend Platz zur Verfügung steht, wird man mit Vorteil in einem eigenen Raum, in dem die Geometrie der Bestrahlung mit der Diagnostikapparatur kopiert ist, die Feldeinstellung vornehmen. Andernfalls wird man sich an die Koppelung halten müssen. Wesentlich ist auf jeden Fall nicht nur bei der Hochvolttherapie, sondern auch bei konventioneller Tiefentherapie, daß man überhaupt — wo es irgendwie möglich ist — die röntgenologischen Kontrollen der Feldeinstellung vornimmt.

Die auffallendste Erscheinung ist die Tatsache, daß mit der 31 MeV-Apparatur *die allgemeinen Reaktionen* im Sinne des Strahlenkaters und der Beeinflussung des Blutbildes *geringer ausfallen als bei der konventionellen Bestrahlung*. Diese Beobachtung wird auch bei anderen Bestrahlungsarten mit energiereichen Strahlungen gemacht, ohne daß wir bisher die Möglichkeit hatten, beim Patienten quantitative Vergleiche anzustellen.

Im Tierversuch läßt sich auf Grund der Taurinausscheidung ein Befund erheben, der diese Beobachtung dem Verständnis näher rückt. SH-haltige Substanzen werden bei Bestrahlung zu Taurin oxydiert und in dieser Form zum mindesten zum Teil im Urin ausgeschieden. Es zeigt sich in Versuchen, die wir zusammen mit Aebi, Lauber und Schmidli angestellt haben, daß selbst bei einer Verdoppelung der Röntgendosis mit dem Betatron die Taurinausscheidung noch nicht die gleiche Höhe erreicht. Wir können noch nicht sagen, weswegen diese hochenergetische Strahlung in dieser Beziehung einen geringeren Effekt aufweist.

Tabelle 1. *Taurinausscheidung nach Bestrahlung. Durchschnittswerte von je 6 Ratten*
(in mg Taurin/100 g Körpergewicht/24 Std.)

| Röntgen/Betraton r | vor Bestrahlung 72—0 Std I | nach Bestrahlung | | Signifikanz | |
|---|---|---|---|---|---|
| | | 0—72 Std II | 71—144 Std III | II—I | III—I |
| Röntgen 250 | 2,55 ± 0,15 | 4,35 | 2,9 | + | — |
| | | +1,8 ± 0,1 | +0,35 ± 0,3 | (p < 0,001) | |
| Röntgen 500 | 2,0 ± 0,35 | 5,05 | 2,6 | + | — |
| | | +3,05 ± 0,25 | +0,6 ± 0,3 | (p < 0,001) | |
| Betatron 250 | 2,3 ± 0,25 | 3,3 | 1,65 | + | + |
| | | +1,0 ± 0,45 | —0,65 ± 0,2 | (p < 0,05) | (p < 0,05) |
| Betatron 500 | 2,4 ± 0,4 | 4,0 | 0,7 | + | + |
| | | +1,6 ± 0,4 | —1,7 ± 0,4 | (p < 0,02) | (p < 0,001) |

Wenn wir eine neue Strahlung zur Anwendung bringen, interessiert uns in erster Linie die *relative biologische Wirksamkeit*. Dieser Faktor ist sicher bei verschiedenen Kriterien unterschiedlich. Für die Tumorrückbildung und die Toleranz der normalen Stützgewebe scheint er etwa um 0,75 herumzuliegen. Wir haben den Eindruck, daß die Schleimhautreaktion in ihrer Intensität und in ihrem zeitlichen Auftreten um etwas weniger differiert, d. h., daß dort der RBW möglicherweise etwas näher bei 1 liegt. Der von Bailey und Harvey angegebene Faktor von 0,6 ist für klinische Zwecke verwendet gefährlich und kann zur Überdosierung

---

* Zur Diskussion aufgefordert

führen. Bemerkenswert und für klinische Zwecke wichtig ist die Tatsache, daß die relative, biologische Wirksamkeit für Mitosen gleich ist bei konventioneller Bestrahlung von 250 kV und beim 31 MeV-Betatron.

Die Erholung der Mitosen erfolgt sogar bei konventioneller Strahlung rascher, doch ist die Differenz statistisch nicht gesichert. Da aber für die Tumorrückbildung der Faktor 0,75 beträgt, d. h. daß für gleiche Rückbildung mit der Betatronbestrahlung eine Dosiserhöhung von etwa 30% erfolgen muß, bedeutet dies, daß der Effekt der Tumorrückbildung offenbar nicht im wesentlichen über dem Mitoseeffekt abläuft, sondern durch eine Beeinflussung des Zellkernes außerhalb des direkten Teilungsstadiums. Andererseits aber ist uns die Beeinflussung der Mitose durch die Ausschaltung der vitalsten Zellen sehr erwünscht, besonders wenn wir die Vorbestrahlung von malignen Tumoren durchführen. Wenn wir diesen Effekt auf den Tumor exquisit ausnützen wollen, dann müßte man konsequenterweise den operativen Eingriff bald nach Bestrahlungsabschluß vornehmen, bevor im Tumor die Zellteilung wieder beginnt.

Die systematische Vornahme der Vorbestrahlung stößt auf erhebliche Schwierigkeiten, nicht zuletzt von Seiten vieler Chirurgen. Das Intervall zwischen Bestrahlung und Operation, das wir im allgemeinen fordern, bis die unmittelbaren Strahlenreaktionen abgeklungen sind, ist sicherlich einer der Hauptgründe, der der allgemeinen Einführung der Vorbestrahlung hindernd im Wege steht. Wenn wir die Vorbestrahlung bei einem an und für sich operablen Tumor durchführen, sollte aus theoretischen Gründen eine über wenige Tage sich erstreckende Vorbestrahlung mit sofort anschließender Operation empfehlenswert sein. Zum mindesten glauben wir, daß ein großangelegter Versuch in dieser Richtung anzuraten ist. Die Bestrahlung mit der Hochvolt-Therapie eignet sich für diese Zwecke ganz besonders, weil sie die Mitose spezifisch stärker schädigt und gleichzeitig zu geringeren Allgemeinreaktionen führt.

Aus physikalischen Gründen darf man annehmen, daß die *Hochvolt-Therapie bei Durchstrahlung des Knochens* besonders günstig ist. Die Indikation für die Hochvolt-Therapie bei Geschwülsten, die hinter Knochen liegen und bei denen besonders empfindliche Knochenpartien, wie der Unterkiefer, mit hohen Dosen belastet werden müssen, ist wohl kaum von irgend einer Seite ernstlich bestritten. Die bisherigen Erfahrungen bestätigen diese Auffassung. Die gleiche Überlegung sollte eigentlich auch bei *Knochensarkomen* gelten. Das osteogene Sarkom gilt als ein wenig strahlensensibler Tumor. Es war aber schon bei konventioneller Strahlung möglich, solche Geschwülste gelegentlich zu beseitigen. Es seien in dieser Beziehung die Erfahrungen von HOHLFELDER, BACLESSE und KOHLER erinnert, denen sich auch unsere eigenen Erfahrungen anschließen.

Es ist aber bei konventioneller Bestrahlung die Dosis so hoch zu wählen, daß eine spätere Knochennekrose nicht sicher vermieden werden kann. Wenn wir ein osteogenes Sarkom zum Verschwinden bringen wollen, dann ist sicher nicht die Dosis im Knochen selbst maßgebend. sondern diejenige im Tumorgewebe. Die Differenz ist bei konventioneller Strahlung ungünstig, Bei der Hochvolt-Therapie scheint nach unseren bisherigen Erfahrungen das Ziel der Beseitigung des Tumors leicht und mit geringerer Gefährdung des Knochens zu erzielen. Als recht drastisches Beispiel sei ein polymorphzelliges Sarkom erwähnt, das den Unterkiefer durchsetzt hat und bei dem die Probeexcision histologisch neben dem Sarkom die Symptome der Osteomyelitis in Form von kleinen Knochensequestern vor Behandlungsbeginn aufgewiesen hat. Der Tumor ging nach Verabreichung einer Dosis von 6000 r Betatron klinisch vollkommen zurück. Es stellte sich aber im Bereiche des Alveolarfortsatzes eine Knochennekrose ein, währenddem der Mandibularkörper sich gut erholte. Nach Entfernung der Knochensequester und nach anschließender Abstoßung von mehreren kleinen Sequestern hat sich das Ulcus epithelisiert. Die Patientin ist jetzt mehr als 2 Jahre symptomfrei. Mit konventioneller Bestrahlung hätte man dieses Resultat sicher nicht erreichen können. Wir verfügen noch über einen weiteren instruktiven Fall, bei dem ein 8jähriger Knabe mit einem polymorphzelligen Sarkom der Tibiametaphyse sich 1 Jahr nach durchgeführter Bestrahlung beim Fußballspiel eine pathologische Fraktur zuzog, dort wo zur Probeexcision die Osteotomie durchgeführt worden war. Die Fraktur heilte in normaler Zeit und es liegen jetzt 2 Jahre nach Behandlungsbeginn keine Symptome vom Tumor mehr vor. Wir glauben, daß gerade das osteogene Sarkom sich für die Behandlung mit dieser Strahlung besonders gut eignet.

Die *Dosierung* berücksichtigt zuerst den histologischen Befund, wobei die gleichen Regeln gelten wie bei der üblichen Tiefentherapie. In zweiter Linie ist die Tumorausdehnung zu berücksichtigen, indem größere Tumoren höhere Dosen benötigen. Die Grenze, bei der wir uns in den Gefahrenbereich von Spätschädigungen begeben, liegt bei 8000 r. Bei resistenteren Tumoren muß man gelegentlich höher gehen. Wenn die Reaktionen der gesunden Gewebe früh auftreten, muß die Dosis niedriger gehalten werden. Beim Auftreten der Reaktionen senken wir die Einzeldosis um 20—30%, weil entzündetes Gewebe empfindlicher ist. *Die Beobachtung der Strahlenreaktion der gesunden Gewebe und die Ansprechbarkeit des Tumors auf die Bestrahlung bestimmen in erster Linie die Dosishöhe.* In den letzten Jahren wurde die Dosierung auf die physikalischen Angaben aufgebaut, die wir durchaus respektieren, unter Hintansetzung der biologischen Reaktionen. Wir möchten mit besonderem Nachdruck auf die Bedeutung der Beobachtung der bestrahlten Gewebe hinweisen und deshalb bewußt von einer *biologisch-physikalischen Dosierung* sprechen.

Überblicken wir die einzelnen Lokalisationen, bei denen mit der Hochvolt-Therapie günstigere oder bessere Ergebnisse zu erwarten sind als mit der bisherigen Behandlung, so lassen sich nach unseren Erfahrungen die folgenden Aussagen machen:

*Epipharynx- und Mesopharynxtumoren* sprechen gut an. Die Herddosis kann auf relativ einfache Weise mit geringer Oberflächenbelastung erzielt werden, doch muß unbedingt die Dosis im regionären Drüsengebiet erhöht werden, weil man sonst mit dem Auftreten von regionären Metastasen rechnen muß. Es geschieht dies am einfachsten, indem wir eine Moulage vorbauen. Die Spätveränderungen im Sinne von Trockenheit und Schleimhaut-atrophie treten in ähnlichem Ausmaße auf wie bei der konventionellen Therapie.

Beim *Oesophagustumor* erreicht man die Durchgängigkeit im allgemeinen verhältnismäßig leicht. Da der Tumor aber ausgedehnt metastasiert, wird die Zahl der Symptomfreiheiten eine relativ begrenzte bleiben. Es treten aber auch lokale Komplikationen auf, die unter Umständen letal verlaufen können, die aber aller Wahrscheinlichkeit nach oft beherrscht werden können. Nach durchgeführter Bestrahlung bleibt bei ausgedehnten Tumoren, wenn die Wand durchsetzt und zerstört war, öfters ein unspezifisches Ulcus zurück, das den letalen Ausgang bedingen kann. Wenn man systematisch die Infektion bekämpft, sollte die Gefähr-lichkeit dieser Komplikation herabgesetzt werden können. Mehrfach haben wir Patienten an Kreislaufinsuffizienzen verloren, die besonders dann auftraten, wenn vorher schon Myokard-veränderungen vorgelegen haben. Durch systematische Beobachtungen und wenn nötig Behandlung des Kreislaufs scheint auch diese Komplikation beherrschbar zu sein. Die Lungeninduration läßt sich auf ein durchaus zu verantwortbares Maß reduzieren. Sie braucht nicht schicksalshaft abzulaufen, weil sich zeigen ließ, daß nicht nur die Infektion, sondern ein lokalisierter Bronchialverschluß pathogenetisch wichtig ist. Wir führen systematisch Atem-übungen durch und versuchen die Bronchien mit Aleudrin und Perphyllon zu öffnen.

Beim *Kardiacarcinom* zeigt sich, daß der Tumor ebenfalls gut anspricht. Da wegen der Histologie eine vollständige Rückbildung weniger wahrscheinlich ist als beim Plattenepithel-carcinom, beschränken wir uns in der Regel auf die Vorbestrahlung mit einer Dosis von etwa 4000 r B und lassen, wenn der Allgemeinzustand es zuläßt, 1 Monat nach Bestrahlungs-abschluß operieren.

Das *Bronchialcarcinom* spricht in der großen Mehrzahl der Fälle günstig an. Es mag wohl dann und wann gelingen, einen Tumor zum Verschwinden zu bringen, wobei der lokale Effekt beim kleinzelligen Carcinom besser ist als beim Plattenepithelcarcinom. Wir bestrahlen vor-läufig nur inoperable Fälle, bei denen man nicht erwarten kann, daß die Heilungsziffern günstig ausfallen werden; da die Ansprechbarkeit aber keineswegs eine schlechte ist, sollten auch hier bei systematischer Vorbestrahlung die bisher noch keineswegs befriedigenden Resultate der chirurgischen Behandlung sich verbessern lassen. Auch beim Bronchialcarcinom läßt sich die Lungenfibrose mit den erwähnten Maßnahmen anscheinend auf ein gut erträgliches Maß reduzieren.

Beim *Magencarcinom* ist die Strahlenbehandlung jetzt gut durchführbar. Der Tumor spricht häufig günstig an und kommt zu erheblicher Schrumpfung. Die alleinige Bestrahlung ist mit den heutigen Mitteln anscheinend wegen der relativ hohen Sensibilität der Schleimhaut nur ausnahmsweise als kurative Behandlung durchführbar. Wir beschränken uns entweder auf eine palliative Behandlung allein oder führen die Vorbestrahlung durch. Es gelingt in einem erheblichen Prozentsatz, inoperable Tumoren in ein operables Stadium überzuführen. Wenn man aber nur inoperable Tumoren vorbestrahlt, kann man keine günstigen Dauer-resultate erwarten, weil doch in einem gewissen Grade eine Parallelität besteht zwischen

Tumorausdehnung und Metastasierung. Die Beobachtung, daß der Tumor relativ gut anspricht, sollten aber ein Anreiz darstellen, um auch bei operablen Stadien die präoperative
Bestrahlung vorzunehmen.

Recht befriedigend sind die Resultate bei *Blasentumoren.* Sie sind besser als mit jeder
uns bisher bekannten Therapie. Die Frühergebnisse sind so ermutigend, daß man von der
Anwendung der Kobalttherapie bei allen infiltrativen Formen abraten muß, weil bei einmal
vorgenommener Kobaltbestrahlung eine zusätzliche Betatronbestrahlung nicht mehr durchführbar ist, weil sonst schwere Schleimhautreaktionen auftreten.

Bei den *Uterustumoren* waren wir bisher noch zurückhaltend, doch scheint auch hier die
ultraharte Strahlung zu günstigen Effekten zu führen.

Überblicken wir die bisherigen Erfahrungen, so glauben wir, zur Feststellung berechtigt
zu sein, daß die ultraharte Strahlung zu einem deutlichen Fortschritt in der Tumortherapie
führt, und daß jedes Tumorzentrum über eine Hochvoltmaschine verfügen sollte.

R. WIDERÖE (Baden/Schweiz):

Für die Anwesenden, die unser 31 MeV-Betatrongerät, das „Asklepitron", nicht kennen,
möchte ich einige kurze Angaben über dieses Gerät machen und einige Lichtbilder zeigen.
Eine ausführliche Beschreibung findet man im Sonderdruck aus Strahlenforschung und
Strahlenbehandlung, welcher anläßlich der 37. Tagung der Deutschen Röntgengesellschaft in
München 1955 herausgegeben wurde (Sonderband zu Strahlentherapie, Band 35).

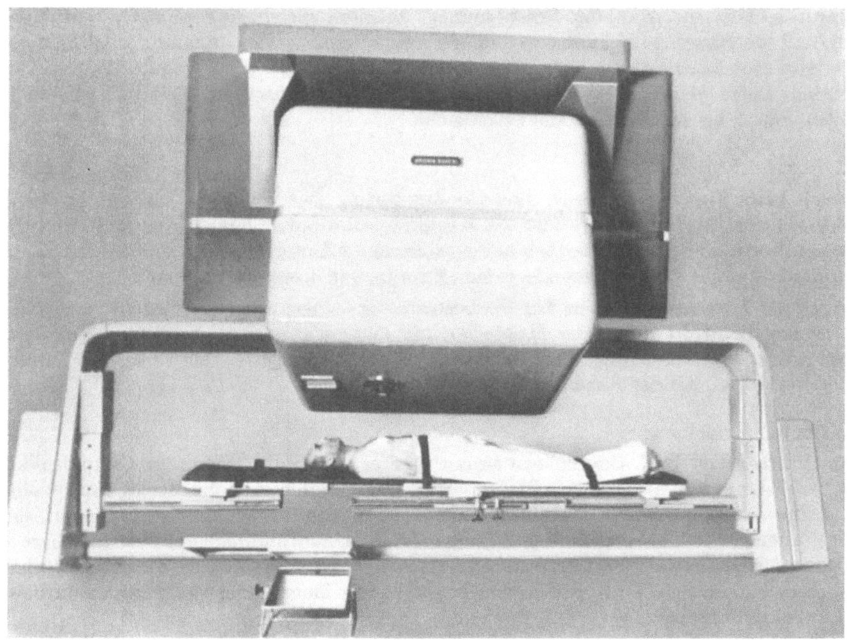

Fig. 1. Asklepitron von Brown-Boveri u. Co.

Das Asklepitron ist ein komplettes Bestrahlungsgerät für Tiefentherapie mit 31 MeV
Röntgenstrahlen oder 30 MeV Elektronenstrahlen. Es enthält ein Betatron mit motorisch
verstellbaren Blenden, ein Durchleuchtungsgerät mit einer 125 kV-Röntgenröhre, Bucky-
Blende, Fluorescenzschirm und Kassette sowie ein Bett für den zu bestrahlenden Patienten.
Das Bett kann mittels Servomotoren in drei Richtungen verstellt werden. Außerdem kann
das an der Decke montierte Betatron durch einen Motor gedreht werden, so daß die Strahlen
unter einem beliebigen Winkel auf den stets horizontal liegenden Körper des Patienten ein-
treffen. Das Asklepitron ist so eingerichtet, daß man nach Belieben mit festen oder mit
beweglichen Feldern (Pendeltharepie) bestrahlen kann. Bei Pendeltharepie beträgt der größte
Pendelwinkel ± 105°.

Fig. 1 (88216) zeigt einen schematischen Längsschnitt durch das Gerät. Der Focus der Durchleuchtungsröhre liegt 36 cm neben dem der Betatronröhre. Beide Fokusse haben denselben Abstand vom Patienten. Die Durchleuchtungsröhre ist *nicht* für diagnostische Untersuchungen vorgesehen, sondern dient vielmehr als Zielgerät für die genaue Einstellung des Patienten im Strahlenfeld des Betatrons.

Das durch die Blenden begrenzte Feld des Betatrons wird auch im Durchleuchtungsfeld markiert, und der Arzt kann deshalb im Durchleuchtungsfeld die Lage und die Größe des Betatronfeldes genau einstellen und kontrollieren. Dies ist besonders bei der Pendeltherapie von größter Wichtigkeit. Das Bett läßt sich motorisch in der Höhenlage, Längsrichtung und Querrichtung einstellen. Eine besondere Einrichtung verschiebt das Bett von der Durchleuchtungs- zu der Bestrahlungsstellung.

Das Gerät kann mit einem zweiten Strahl zur Behandlung von sitzenden Patienten geliefert werden.

Die ersten „Asklepitron"-Anlagen werden anfangs des nächsten Jahres nach Lausanne und Milano geliefert.

E. Scherer (Marburg):

Meine Damen und Herren, nur eine kurze Bemerkung zur Herdlokalisation und Herdbestimmung bei den Blasencarcinomen. Wir durchleuchten seit einiger Zeit die Blasencarcinome auf dem Müller-Gerät UGX, d. h. also im Liegen in der gleichen Lage wie sie bestrahlt werden im frontalen Strahlengang, machen gleichzeitig eine Cystographie und sehen, daß die Blasen ganz anders liegen, als man es sich vorher anhand der Ohlfelderschen, an der fixierten Leiche gewonnen, Situationsskizzen vorstellt. Beim Mann rutscht die Blase wesentlich tiefer in die Kreuzbeinhöhle herein. Die durchschnittlichen Herdtiefen liegen zwischen 10, 12 bis zu 16 cm an der Oberfläche.

H. R. Schinz (Zürich):

Auch beim Züricher Betatron sind für den Erfolg zwei Faktoren ausschlaggebend: die Strahlensensibilität des Tumors und des Krankheitsstadiums (Tumorgröße und Metastasen). Entgegen Herrn Schubert haben wir keine qualitativen Unterschiede gegenüber konventioneller Röntgentherapie bei unserem klinischen Krankengut feststellen können.

Auch die Züricher Schule ist für Vorbestrahlung — wenn der Chirurg einverstanden ist. Hier ist das Betatron wegen der Hautschonung eindeutig überlegen. Ferner wird der ausschließende operative, häufig verstümmelnde Eingriff kleiner. Vom Operateur muß man Geduld verlangen, bis der Strahleneffekt abgeklungen ist.

G. Weitzel (Heidelberg):

Im Vortrag von Prof. Cocchi war immer davon zu hören, Gesamtdosis soundsoviel r in 5 cm Tiefe. Da es sich hier um eine Tiefenbestrahlung handelt, und die Herdtiefen verschieden sind, nahm ich an, daß sich diese Dosisangaben auf das Dosismaximum beziehen, das bei 30 MeV ungefähr in dieser Tiefe liegt. Es würde mich nun interessieren, warum diese Form der Dosisangaben gewählt wurde, während allgemein die Angaben der Herddosis üblich ist, die nach meiner Ansicht auch anschaulicher und für die Beurteilung des Strahleneinflusses im Herd zweckmäßiger ist.

B. Rajewsky (Frankfurt a. M.):

Herr Zuppinger sagt, daß zwischen Hochvoltstrahlen und konventionellen Strahlen ein grundsätzlicher Unterschied der RBW besteht, und zwar mit Faktoren von 0,6—0,7. Ich möchte hierzu feststellen, daß nach dem heutigen Stand unserer Kenntnisse solche Feststellungen keine Bedeutung für die Klinik haben. Betrachtet man in der internationalen Literatur Mitteilungen für RBW, so finden sich für dieselben Objekte und dieselben Strahlenreaktionen Unterschiede im Verhältnis wie 1:22. Man kann daher bei einem Unterschied wie 1:0,7 nicht von einem wirklichen Unterschied sprechen, man sollte auch bei der Dosisbewertung vorsichtig sein, denn gerade im Hochvoltbereich kann die Dosismessung leicht große Fehler enthalten. Eine sichere physikalische Meßmöglichkeit für ultraharte Strahlen wird erst in 1—2 Jahren gegeben sein.

U. Cocchi (Zürich):

Soweit es möglich ist, müssen natürlich die Bestrahlungsfelder bei sämtlichen Patienten hinter dem Durchleuchtungsschirm eingezeichnet werden. Die Verwendung der relativ kleinen Felder bedingt natürlich beim Einstellen größte Genauigkeit, damit der Tumor auch genau im Strahl zu liegen kommt. Das Einzeichnen muß natürlich in der Körperlage vorgenommen werden, in der der Patient bestrahlt wird.

Auch wir haben im Vortrag auf den Wert der Vorbestrahlung mit dem Betatron hingewiesen.

G. Weitzel (Heidelberg):

Zum Teil sind die Dosisangaben für eine Tiefe von 4,5—5,5 cm angegeben; dies entspricht der maximalen Dosis. Zum Teil ist die genaue Tumordosis angegeben worden, und zwar hauptsächlich dann in den erwähnten Fällen, wenn infolge der Tiefenlage des Tumors ein mehr oder weniger großer Dosisabfall auftritt. Dieser beträgt beim 31 MeV-Betatron etwa in 10 cm Tiefe noch 92% der Maximaldosis in 4,5—5,5 cm Tiefe, in 12 cm noch 83%

A. Zuppinger (Bern):

Stände Rajewsky an meiner Stelle, dann sähe er sich tatsächlich dem Problem gegenüber, wie hoch darf ich bei diesen Patienten dosieren? Wir müssen eine Grundlage haben, sonst können wir überhaupt nicht arbeiten und wir können nicht warten, bis wir in 2 oder 3 Jahren eine bessere, gesicherte Grundlage haben, auch wenn die Werte, die wir nach unserer heutigen Erfahrung der Hochvolttheraphie zugrunde legen, keinen Anspruch auf größere Genauigkeit erheben können, so wirkt dies nur als Ansporn für andere, eigene Beobachtungen anzustellen und auf diese Weise kommen wir weiter, als wenn wir warten, bis die theoretischen Grundlagen besser erarbeitet sind.

# Klinische Erfahrungen mit schnellen Elektronen

## Die Betatrontherapie gynäkologischer Carcinome insbesondere des Vulvacarcinoms*

Von

H.-J. SCHMERMUND und F. OBERHEUSER, Hamburg

Wie die Erfahrung in der Strahlenbehandlung gynäkologischer Carcinome in den letzten Jahrzehnten gezeigt hat, ist eine Verbesserung der Heilungsergebnisse ganz allgemein auf 2 Wegen erreicht worden, und zwar 1. durch eine Individualisierung der Therapie, die sich gegenüber der früher üblichen Schematisierung der Strahlenbehandlung als überlegen erwiesen hat, und 2. durch eine Schonung des gesunden Gewebes, die durch eine Verbesserung des Quotienten zwischen der Dosis im Krankheitsherd und der Dosis im insgesamt durchstrahlten Gewebsvolumen erzielt werden kann. In beiden Richtungen bahnen sich heute durch die Anwendbarkeit sehr energiereicher Strahlen vermutlich weitere Fortschritte an.

Die Erfahrungen an der Univ.-Frauenklinik Hamburg beziehen sich auf die Strahlenbehandlung oberflächennaher gynäkologischer Prozesse mit schnellen Elektronen, wobei es sich vorwiegend um Vulvacarcinome handelt und fernerhin auf die Behandlung der im kleinen Becken gelegenen Collum- und Ovarialcarcinome mit ultraharten Röntgenstrahlen.

Die uns zur Verfügung stehende Elektronenschleuder der Siemens-Reiniger-Werke bietet die Voraussetzung für die Anwendung der genannten Strahlenarten bis zu einer Maximalenergie von 16 MeV. Wir verfügen damit über zwei verschiedene Strahlungen, die nach den jeweiligen Anforderungen benutzt werden können.

Die bisherigen Ergebnisse in der Verwendung sehr energiereicher Strahlungen lassen sich am besten an den Vulvacarcinomen objektivieren, die der Elektronentherapie zugeführt wurden. Wir hatten Gelegenheit, bereits im Zeitraum von 1947—1950 in dem Arbeitskreis von Professor SCHUBERT Erfahrungen an dem 6 MeV-Betatron in Göttingen zu sammeln. Da in der damaligen Zeit über die biologische Wirkung schneller Elektronen sehr wenig bekannt war, setzte ihre therapeutische Anwendung eine eingehende Überprüfung ihrer Wirkung voraus. Nach den umfangreichen Untersuchungen des genannten Arbeitskreises an der Göttinger Universitäts-Frauenklinik und gemeinsam mit Prof. PAUL kann heute als gesichert gelten, daß schnelle Elektronen im allgemeinen etwas weniger wirksam sind als die in der Therapie gebräuchlichen Röntgenstrahlen. Diese Ergebnisse wurden in der Folgezeit von anderen Untersuchern in den wesentlichen Punkten

---

* Aus der Univ.-Frauenklinik Hamburg-Eppendorf (Direktor: Prof. G. SCHUBERT).

vollauf bestätigt. Außerdem ergaben sich bereits damals Anhaltspunkte dafür, daß die sehr energiereichen Strahlen auch zu gewissen qualitativ anderen und vielleicht sogar günstigeren Wirkungen auf Tumorgewebe führten. Nach diesen summarisch erwähnten Ergebnissen der Grundlagenforschung an der Göttinger Univ.-Frauenklinik konnte 1949 mit der Anwendung schneller Elektronen zu therapeutischen Zwecken begonnen werden. Es lag nahe, gerade bei den Vulvacarcinomen frühzeitig einen Versuch mit der Elektronentherapie zu wagen. Maßgebend dafür war die Beobachtung, daß die bis dato geübte kombinierte chirurgisch-strahlentherapeutische Behandlung in Form der elektrochirurgischen Entfernung des Primärtumors mit nachfolgender Röntgenbestrahlung des Operationsgebietes und der Leistenregionen nur eine Kompromißlösung darstellt. Wie die Erfahrung zeigt, handelt es sich beim Vulvacarcinom im allgemeinen um einen schnell metastasierenden, ziemlich strahlenresistenten Tumor. Eine Röntgenbestrahlung des gesamten Tumorgebietes und seiner Lymphabflußwege einschließlich der Leistenlymphknoten führt zu einer beträchtlichen Beanspruchung des gesunden Gewebes mit einer hohen Volumendosis. Um hochgradige Schäden zu vermeiden, mußte man sich daher hinsichtlich der Größe des Bestrahlungsfeldes und der Höhe der Strahlendosis Beschränkungen auferlegen.

Bei der Verwendung schneller Elektronen liegen demgegenüber die Verhältnisse günstiger. Da die Elektronendosis von der Oberfläche aus zunächst nach der Tiefe zu ansteigt oder zumindest der Oberflächendosis entspricht und dann in Abhängigkeit von der Energie rasch den Nullwert erreicht, ist die räumliche Dosisverteilung außerordentlich günstig. Hieraus ergibt sich eine genau lokalisierbare massierte Strahlenwirkung im Tumor und eine sehr geringe Strahlenbelastung des gesunden Gewebes. Allein diese Gesichtspunkte legten nahe, auf die Exstirpation des Primärtumors zugunsten der Strahlenbehandlung zu verzichten.

Infolge der begrenzten Elektronenreichweite des damals zur Verfügung stehenden Göttinger Betatrons von 2—3 cm mußte sich die Behandlung auf die Bestrahlung oberflächlicher Geschwülste beschränken.

Die nächsten Figuren zeigen einige Vulvacarcinome aus der Gruppe der Patienten, die im Jahre 1949 von SCHUBERT, KEPP und Mitarbeitern an der Göttinger Universitäts-Frauenklinik behandelt wurden. Die Fig. 2 zeigt ein carcinomatöses Ulcus an der rechten Labi vor der Bestrahlung. Nach einer Bestrahlung mit 1500 r, die aus technischen Gründen einzeitig mit einer Dosisleistung von 1000 r/min und mit einer Elektronenenergie von 4 MeV durchgeführt wurde, heilte das Carcinom nach Abklingen einer starken erosiven Reaktion glatt ab. Hierbei sei erwähnt, daß die Leistenregionen beiderseits mit Röntgenstrahlen unter den Bedingungen der Halbtiefentherapie bestrahlt wurden. Fig. 2 zeigt den Lokalbefund 3 Monate nach der Bestrahlung und läßt die glatte Narbenbildung erkennen.

Trotz dieser ermutigenden Ergebnisse, die auch bei anderen Patienten gewonnen wurden, sofern die Tiefenausdehnung der Carcinome etwa 1 cm nicht überschritt, bestand Klarheit darüber, daß die eigentlichen Schwierigkeiten in der Behandlung der Vulvacarcinome auch weiterhin durch die Carcinomausbreitung in den Lymphwegen bestimmt wurden. Die begrenzte Elektronenreichweite

der 6 MeV-Elektronenschleuder erwies sich z. B. für die Behandlung von Leisten-
lymphknotenmetastasen nicht mehr als ausreichend.

Mit Hilfe des 16 MeV-Betatrons sind wir seit 2 Jahren in der Lage, auch die
krebsbefallenen regionären Lymphknoten einer intensiven Strahlenbehandlung
zu unterziehen, da die maximale Reichweite von 16 MeV-Elektronen etwa 8 cm
beträgt. Unsere Erfahrungen in dieser Zeit beziehen sich auf die Behandlung
von insgesamt 27 Vulvacarcinomen. Entsprechend den modernen Grundsätzen

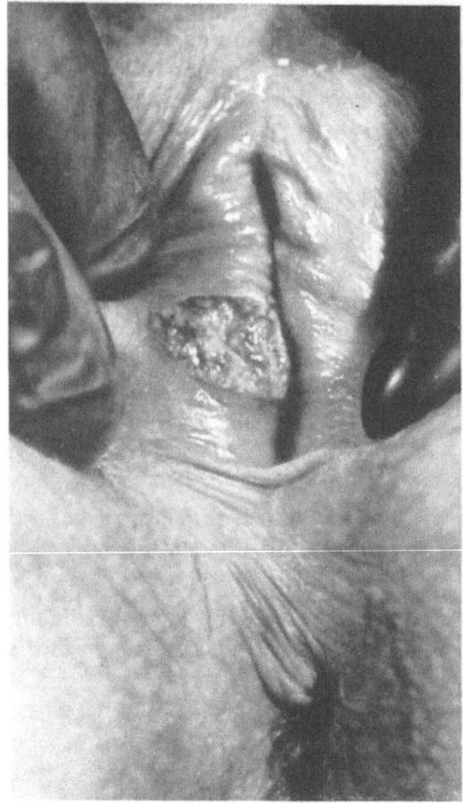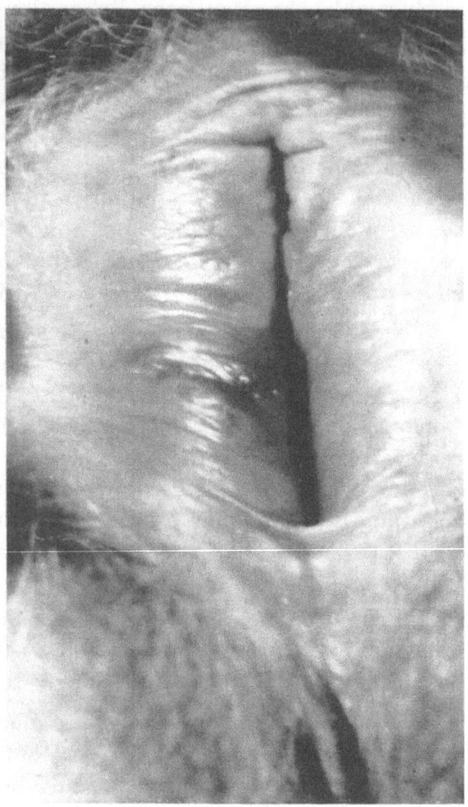

Fig. 1. Carcinomatöses Ulcus der rechten Labie
vor der Behandlung

Fig. 2. Ulcus der rechten Labie (Fig. 1) 3 Monate
nach Bestrahlung mit 1500 r 4 MeV-Elektronen

der Strahlentherapie wurde die Bestrahlung fraktioniert, und zwar mit Einzeldosen von 300 r, vorgenommen. Die Gesamtdosen im Primärtumor lagen zunächst
knapp unter 4000 r im Dosismaximum. Nach den beobachteten Gewebsreaktionen
erwies es sich jedoch als zweckmäßig, den Primärtumor und das Lymphabfluß-
gebiet, speziell aber die Leistenregionen von relativ großen Feldern aus mit
höheren Dosen, und zwar mit 5500—6000 r zu bestrahlen. Außerdem erschien es
zweckmäßig, auch die verhältnismäßig oberflächlichen Tumoren mit hohen
Elektronenenergien zu behandeln. Dieses Vorgehen bezweckt, auch die in die
Tiefe führenden Lymphbahnen mit in das Bestrahlungsgebiet einzubeziehen.
Trotz der dosis- und volumenmäßig stärkeren Strahlenbelastung des Gewebes ist

eine weitgehende Schonung des gesunden Gewebes gewährleistet. Nach unseren bisherigen Beobachtungen heilen selbst starke erosive Reaktionen der Haut in kurzer Zeit glatt ab, ohne wesentliche Veränderungen zu hinterlassen. Ein Beispiel kommt in Fig. 3 zur Darstellung. Nach der Bestrahlung mit schnellen Elektronen war der Prozeß 3 Wochen später.

Nach unseren bisherigen Erfahrungen glauben wir den Schluß ziehen zu können, daß die Elektronentherapie der Vulvacarcinome, wie auch ganz allgemein

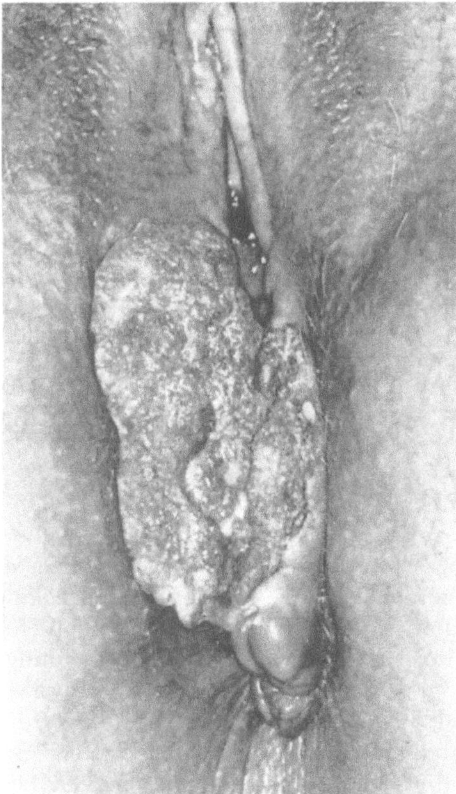
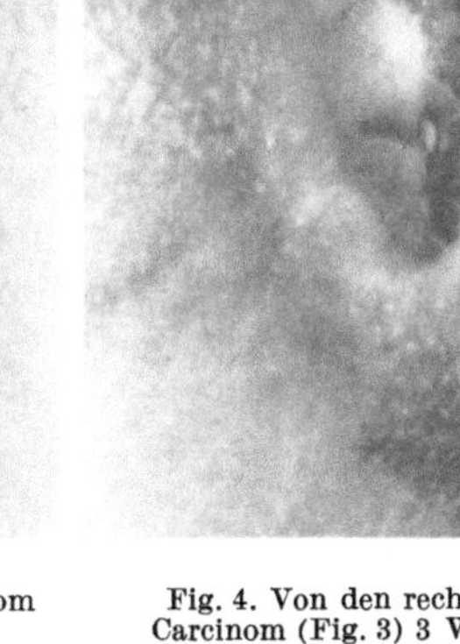

Fig. 3. Von den rechten Labien ausgehendes Carcinom
mit Übergang auf Damm und Vagina
vor der Behandlung

Fig. 4. Von den rechten Labien ausgehendes
Carcinom (Fig. 3) 3 Wochen nach Bestrahlung
mit 5400 r 12 MeV-Elektronen

der Oberfläche nahegelegenen Tumoren, heute als das schonendste und erfolgversprechendste Behandlungsverfahren angesehen werden kann.

Gegenüber diesen Erfolgen der Elektronentherapie sind die Ergebnisse der Behandlung tief im kleinen Becken gelegener carcinomatöser Prozesse mit 16 MeV-Röntgenstrahlen noch nicht so sicher zu beurteilen, da erst die Verlaufsbeobachtung über längere Zeit weitere Schlußfolgerungen zuläßt. Wie wir aber nach den Erfahrungen bei der Behandlung von insgesamt 44 primären und rezidivierenden bzw. metastatischen Collum- und Ovarialcarcinomen feststellen können, bedeutet allein die außerordentlich günstige Tiefendosisverteilung einen wesentlichen Gewinn. Nach dem Tiefendosisverlauf von 16 MeV-Röntgenstrahlen steigt die Dosis von der Oberfläche aus zunächst an und erreicht in 3 cm Gewebstiefe ein Maximum. Nach einem nahezu exponentiellen Abfall beträgt die Tiefendosis in 12 cm Gewebstiefe noch 100% der Oberflächendosis. Zieht man die Kreuzfeuerbestrahlung eines etwa in Körpermitte gelegenen Herdes in Betracht, so ergibt sich bei einem Körperdurchmesser von 30 cm eine Tiefendosis von 120% gegenüber etwa 30% bei 200 kV (Fig. 5). Es gelingt demnach, auch bei sehr adipösen Patienten einen günstigen Dosisquotienten zu erzielen, was mittels der „klassischen Röntgentherapie" nicht möglich ist. Die damit erzielbare Steigerung der Herddosis bei der Kreuzfeuerbestrahlung z. B. mit runden Feldern von 12 cm Durchmesser ist daher zweifellos von erheblichem Nutzen und hat den Vorteil,

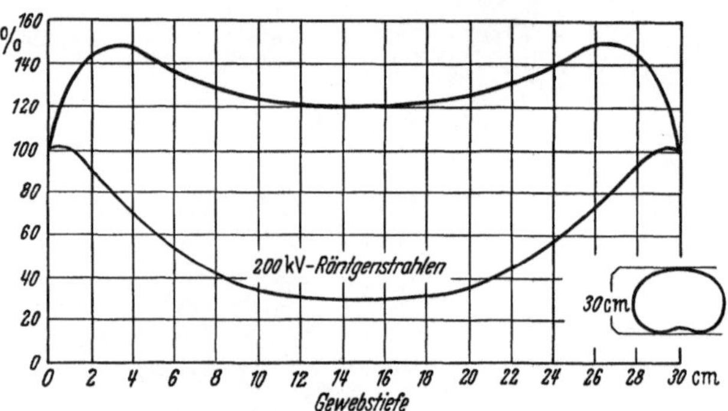


Fig. 5. Tiefendosisverteilung von 16 MeV-Röntgenstrahlen bei Kreuzfeuerbestrahlung. Durchmesser 30 cm

daß bei der Bestrahlung parametraner Infiltrate auch die übrigen Absiedlungen entlang der großen Gefäße einer intensiven Bestrahlung unterzogen werden können. Zieht man in Betracht, daß schon beim Portiocarcinom der Gruppe I in 10—20% der Fälle Lymphknotenmetastasen im Becken bestehen, so ist die Bedeutung dieses Gesichtspunktes nicht zu unterschätzen.

Bei der Strahlenbehandlung des Portiocarcinoms bildete bisher die kombinierte Radium-Röntgenbestrahlung die Grundlage der Therapie. Die Radiumapplikation ist für die Behandlung des Primärtumors an der Cervix uteri unentbehrlich. Der zusätzlichen Röntgenstrahlenbehandlung muß auch bei der Verwendung ultraharter Strahlen nach wie vor die Aufgabe zufallen, das Geschwulstgewebe im Parametrium und an der Beckenwand ausreichend zu schädigen. Hierfür kommt die Stehfeldbestrahlung von rechten und linken Bauch- und Rückenfeldern aus in Betracht, die auch bei Dosen in einer Größenordnung von 5000 bis 6000 r gut vertragen wird.

Bei besonderen Tumorlokalisationen, z. B. bei isolierten Beckenwandrezidiven, läßt sich eine weitere Verbesserung des Dosisquotienten zwischen dem oft erheblich strahlenvorbelasteten gesunden Gewebe und dem Infiltrat durch die Bewegungsbestrahlung mit 16 MeV-Röntgenstrahlen erzielen. In Abhängigkeit von der Feldgröße und dem Pendelwinkel kann nach Messungen in einem Plexiglasphantom ein kreisrundes, nahezu scharf begrenztes und sehr hohes Dosismaximum erzielt werden. Fig. 6 zeigt z. B. ein Dosismaximum an der seitlichen Beckenwand, das in einem Plexiglasphantom bei einem Pendelwinkel von 180° über eine Seite und einer Feldbreite von 6 cm auf einem Röntgenfilm dargestellt wird. Wir haben zwar heute bereits die Möglichkeit, die hier gelegenen Prozesse mittels der Pendelbestrahlung von 200 kV-Röntgenstrahlen bei einer guten räumlichen Dosisverteilung anzugehen. Immerhin liegen hier die Verhältnisse wegen der unübersichtlichen Strahlenabsorption im knöchernen Beckenring nicht so ganz einfach. Infolge der fast gleich hohen Strahlenabsorption in den verschiedenen Gewebsarten ergibt sich gegenüber der Bewegungsbestrahlung mit 200 kV für die Anwendung ultraharter Röntgenstrahlen in vielen Fällen, z. B. bei isolierten Befunden von Ovarialcarcinomen, ein günstiges Indikationsgebiet. Die bisherigen Erfahrungen haben gezeigt, daß es ohne Schwierigkeiten gelingt, in jede Region der Beckenhöhle hohe Strahlendosen zu applizieren. Selbst wenn man die gegenüber der 200 kV-Röntgenstrahlung nachgewiesene allgemein etwas geringere biologische Wirksamkeit der Millionenvoltbestrahlung in Rechnung stellt, ist die Verträglichkeit sehr hoher Dosen besonders augenfällig.

Nicht immer ist jedoch die Erzielung eines kreisrunden Dosismaximums bei der Pendelbestrahlung von Vorteil. In der letzten Zeit haben wir uns deshalb mit der Methodik der Pendelbestrahlung mit aus dem Drehpunkt ausgelenktem Strahlenkegel besonders befaßt, und zwar in der Überzeugung, daß sich hierdurch neue Gesichtspunkte für die Bestrahlung der Portiocarcinome ergeben. Wie die Erfahrung zeigt, kommt es bei der Strahlenbehandlung primärer Cervixcarcinome darauf an, 1. die Radiumisodosen mit den Röntgenstrahlenisodosen im kleinen Becken so abzustimmen, daß Primärtumor und Parametrien bis zur Beckenwand optimal belastet werden, und 2., daß mit dem parametranen Gewebe an der Beckenwand gleichzeitig auch die regionären Lymphknoten an den großen Gefäßen mit hohen Dosen bestrahlt werden können. Wird der Zentralstrahl bei der Pendelbestrahlung um wenige Grad seitlich ausgelenkt,

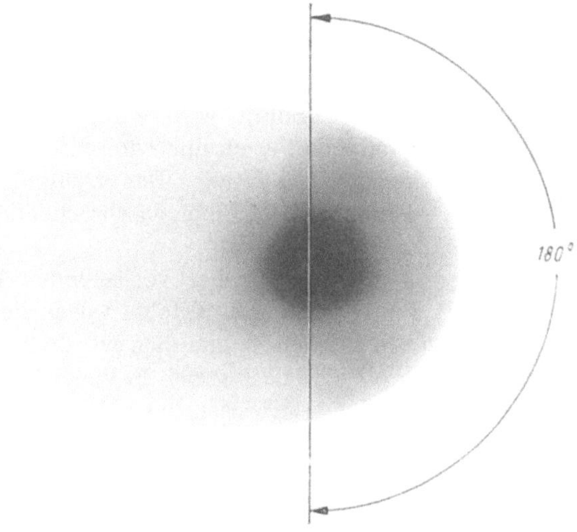

Fig. 6. Dosisverteilung in einem Plexiglasphantom bei Pendelbestrahlung mit 16 MeV-Röntgenstrahlen, Pendelwinkel 180°, Feldbreite 6 cm

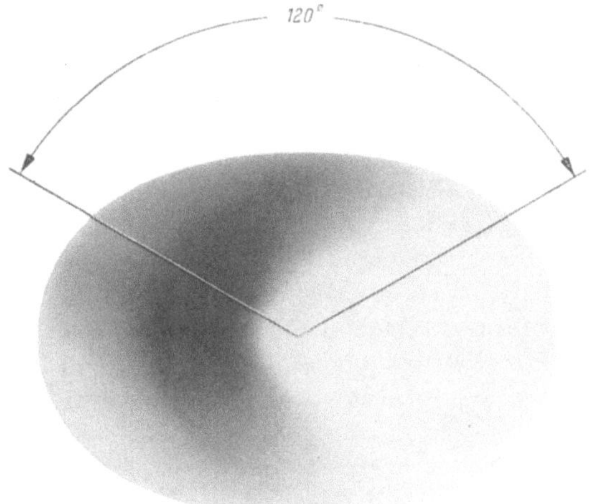

Fig. 7. Dosisverteilung in einem Plexiglasphantom bei Pendelbestrahlung mit 16 MeV-Röntgenstrahlen, Zentralstrahl um einen Winkel von 5° aus dem Drehpunkt ausgelenkt, Pendelwinkel 120°, Feldbreite 6 cm

so können diese Forderungen sehr viel besser als mit der einfachen Pendelbestrahlung erfüllt werden.

Die Fig. 7 zeigt die Form des Dosismaximums an der seitlichen Beckenwand in einem Plexiglasphantom bei einem Pendelwinkel von 120°, einer Feldbreite von 6 cm und einer seitlichen Auslenkung des Zentralstrahls um 5°. Man erkennt, daß auf diese Weise das Dosismaximum nach vorn und hinten ausgezogen und der Krümmung der Beckenwand angepaßt wird. Median wird ein Bereich ausgespart, der der Strahleneinwirkung des intracervical applizierten Radiums vorbehalten bleibt. Somit wird ein guter Anschluß an die Radiumisodosen gewährleistet, ohne daß es zu unerwünschten Überlappungen hoher Radium- und Röntgenstrahlendosen kommt. Das Lymphausbreitungsgebiet an den großen Gefäßen wird durch Wahl einer entsprechenden Feldhöhe in den Bestrahlungsbereich einbezogen.

Wenn wir nach den bisher vorliegenden Ergebnissen der Anwendung von ultraharten Röntgenstrahlen Erfolge wegen der Kürze der Zeit noch nicht endgültig übersehen können, so glauben wir doch, den vorläufigen Schluß ziehen zu dürfen, daß ihre Verwendung sich in vielen Fällen der konventionellen Röntgentherapie als überlegen erwiesen hat. Bei der Behandlung gynäkologischer Carcinome bedeutet die individualisierende Anwendung schneller Elektronen und ultraharter Röntgenstrahlen nach unserer Auffassung ohne Zweifel einen Fortschritt.

# Erfahrungen mit der Elektronentherapie oberflächlicher Tumoren*

Von

G. WEITZEL, Heidelberg

Unsere erste Tumorbestrahlung mit schnellen Elektronen erfolgte im Herbst 1953. Die einzigen Erfahrungen, die damals auf diesem Gebiet vorlagen, stammten aus Göttingen und waren von SCHUBERT, BODE und KEPP mit Elektronen von 2,5—5,5 MeV an Vulva-, Haut- und Lippencarcinomen gewonnen worden. Das von uns in Betrieb genommene 15 MeV-Siemens-Betatron bietet bekanntlich die Möglichkeit der Anwendung ultraharter Röntgenstrahlen und schneller Elektronen, wobei uns damals die Röntgenstrahlen für eine radiologische Klinik von größerem Interesse zu sein schienen. Die Elektronentherapie hielten wir für eine begrüßenswerte Zugabe, die sicherlich manche Möglichkeiten in sich barg, deren Anwendungsbereich aber vermutlich beschränkt war. Daß wir zunächst dennoch mit Elektronenbestrahlungen begannen, lag einzig daran, daß zu diesem Zeitpunkt für die Bestrahlungen mit ultraharten Röntgenstrahlen noch einige technische Vorbedingungen nicht erfüllt waren; insbesondere waren die Dosismessungen noch nicht abgeschlossen und das Problem der exakten Tumorlokalisationen noch nicht zufriedenstellend gelöst.

Daß der erste Behandlungsfall ein enttäuschendes Ergebnis hatte, traf uns infolgedessen auch nicht besonders schwer. Wir hatten hierfür eine Leistendrüse bei einem Peniscarcinom ausgewählt, wobei zu Vergleichszwecken die kontralaterale Drüse in üblicher Art mit 200 kV-Röntgenstrahlen behandelt wurde. Die elektronenbestrahlte Seite wurde ein glatter Mißerfolg: die Drüse wuchs unter der Bestrahlung, während das röntgenbestrahlte Vergleichsobjekt bald nicht mehr zu tasten war. Rückblickend dürfen wir als Erklärung für diesen Versager annehmen, daß die verabfolgte Gesamtdosis zu gering war.

Inzwischen hat sich unsere Einstellung zur Elektronentherapie grundlegend geändert. Von den bis zum 1. Juni 1957 mit dem Betatron bestrahlten 1000 Patienten wurden 861, das sind 86,1%, mit schnellen Elektronen behandelt.

Von den in dieser Zeit bestrahlten 1656 Tumorherden erhielten 1495, das sind 90,5%, Elektronenbestrahlungen, von den hierbei durchgeführten 28646 Einzelbestrahlungen entfielen 23632, das sind 80,8%, auf schnelle Elektronen. Diese eindeutige Bevorzugung der Elektronentherapie hat ihre Ursache nicht nur darin, daß wir uns inzwischen von der guten therapeutischen Wirksamkeit dieser Behandlungsform überzeugt haben, sondern auch darin, daß sich ihr Indikationsbereich sehr wesentlich erweitert hat. Wir konnten inzwischen feststellen, daß sich unter den täglich zur Behandlung in unsere Klinik kommenden Patienten

---

* Aus dem Czerny-Krankenhaus für Strahlenbehandlung der Universität Heidelberg (Direktor: Prof. Dr. J. BECKER)

viel mehr für eine Elektronentherapie geeignete Fälle finden, als wir ursprünglich angenommen hatten. Dazu kommt noch der nicht zu übersehende wirtschaftliche Faktor der mit 200—300 r/min etwa 10mal höheren Dosisleistung gegenüber der Gammabestrahlung. Die entsprechend erheblich kürzeren Bestrahlungszeiten ermöglichen mehr Bestrahlungen pro Arbeitstag.

Fig. 1. Bestrahlungstubusse für schnelle Elektronen für runde und rechteckige Felder von 5—12 cm Durchmesser bzw. Feldgrößen von 4,5 × 7 cm bis 8 × 12 cm

Ich erwähnte vorhin, daß sich der Indikationsbereich für die Elektronentherapie immer mehr erweitert hat. Das beruht naturgemäß einmal auf der zunehmenden Erfahrung. Andererseits verdanken wir diese erweiterten Möglich-

Fig. 2. Kleine Rundtubusse und Schrägtubusse zur Elektronentherapie bei Intrakavitärbestrahlung und kleinen, schwer zugänglichen oder in empfindlicher Umgebung gelegenen Herden

keiten auch einer immer mehr ausgebauten und den jeweiligen Bedürfnissen angepaßten *Bestrahlungstechnik,* die es uns heute ermöglicht, mit geringem Zeit- und Personalaufwand den meisten Anforderungen routinemäßig gerecht zu werden. Hatten wir in der Anfangszeit unsere Felder mit Bleischablonen ausgeblendet, so erwies sich dieses Verfahren bald für die fraktionierte Behandlungsform als zu zeitraubend. Wir haben daher für die wichtigsten Feldgrößen Blendentubusse

entwickelt, die heute von der Herstellerfirma des Betatrons serienmäßig mit-
geliefert werden. Ein Satz dieser Tubusse für runde und rechteckige Felder von
5 cm bis maximal 12 cm Felddurchmesser (Fig. 1) erfüllt die meisten Anforderungen bei der Oberflächentherapie. Kleinere Feldgrößen als 5 cm sind nach unseren
Erfahrungen nicht zweckmäßig, da sie die Rezidivgefahr beträchtlich erhöhen.
Außerdem verringert sich bei kleinen Feldern bekanntlich die Tiefenreichweite
des Strahlenkegels. Nur bei besonderen Lokalisationen, wie z. B. in Augennähe,
an der Ohrmuschel, an den Lippen und anderen ähnlichen Stellen verwenden wir
auch kleinere — runde oder ovale — Tubusse, wobei jedoch auf die Fixierung am
Herd besonderes Augenmerk gerichtet werden muß. Diese Tubusse, die an der

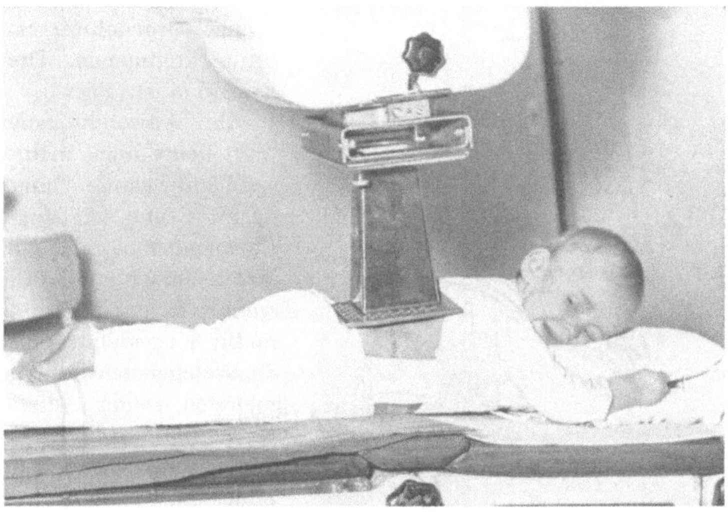

Fig. 3. Gitterbestrahlung mit Elektronen von 9—15 MeV mit einem 5 mm dicken Bleigitter. Lochdurchmesser
7 mm (Postoperative Nachbestrahlung eines Nierensarkoms)

Austrittsstelle gerade oder abgeschrägt sein können (Fig. 2), sind in der Hauptsache für intrakavitäre Bestrahlungen vorgesehen, d. h. im wesentlichen für
intraorale und intravaginale Bestrahlungen, unter Umständen, wenn ein operativ
geschaffener Zugang besteht, auch in der Orbita oder Nasen- und Kieferhöhle.
Die intrakavitären Bestrahlungen bedeuten eine wesentliche Erweiterung des
Indikationsbereiches auf dem gynäkologischen und Hals-Nasen-Ohren-Gebiet;
bei intravaginaler Applikation bieten sie in manchen Fällen sogar Gelegenheit,
an den Blasenboden heranzukommen. Wegen ihrer leichteren Einführbarkeit
und besseren Anpassung an die anatomischen Gegebenheiten haben sich bei der
Intrakavitärbestrahlung gerade unsere Schrägtubusse besonders bewährt. Sie
sind außerdem auch in den Fällen angezeigt, wo es auf eine besondere Schonung
des unter dem Herd liegenden Gewebes ankommt, wie z. B. über den Wachstums-
zonen des kindlichen Knochens. Der schräge Strahlenaustritt bringt hierbei eine
um den Cosinus des Abschrägungswinkels geringere Reichweite des Strahlenkegels senkrecht zur Oberfläche mit sich, ohne daß die Energie der Strahlung
herabgesetzt werden muß. Der Durchmesser der Intrakavitärtubusse sollte 4 cm
nicht übersteigen, da die Einführung stärkerer Tubusse meist nicht mehr möglich
ist. Sehr zweckmäßig ist die Anbringung kleiner Öffnungen von etwa 7 mm

Durchmesser in der Tubuswand, wodurch sich nach Einführung einer geeigneten Optik — wir verwenden hierzu ein ausgedientes Cystoskop — die Einstellung auf den Herd gut kontrollieren läßt.

Ein ansehnlicher Teil unserer Patienten ist bereits vorher mehr oder weniger hoch bestrahlt worden. In manchen Fällen finden sich dabei Feldbelastungen, die eine erneute Bestrahlung nach herkömmlichen Gesichtspunkten nicht mehr ratsam erscheinen lassen. Solche Fälle veranlaßten uns zur Entwicklung der *Elektronen-Gitterbestrahlung*. Die Erfahrungen, die wir in der Zwischenzeit mit dieser Methode machten, waren so gut, daß aus der ursprünglichen Verlegenheitslösung jetzt eine immer ausgedehnter zur Anwendung kommende Therapieform geworden ist (Fig. 3).

Ihr Anwendungsbereich betrifft heute nicht mehr nur stark strahlenbelastete Tumoren, sondern wir sind seit längerem dazu übergegangen, auch völlig unbelastete Felder damit zu bestrahlen, wenn es darauf ankommt, flächenmäßig ausgedehntere Herdbereiche anzugehen, wie das bei Drüsenregionen oder disseminierten Knotenbildungen der Fall ist. Die letzte Stufe dieser Entwicklung bildet die sog. Oberflächengitterbestrahlung mit einem dünnen feingelochten Gitter (Fig. 5), das bei ausgedehnten Intracutanherden, also bei der Lymphangiosis carcinomatosa, erfolgreich angewendet werden kann. Wir haben hiermit Fälle behandeln können, deren Ausdehnung vorher ein

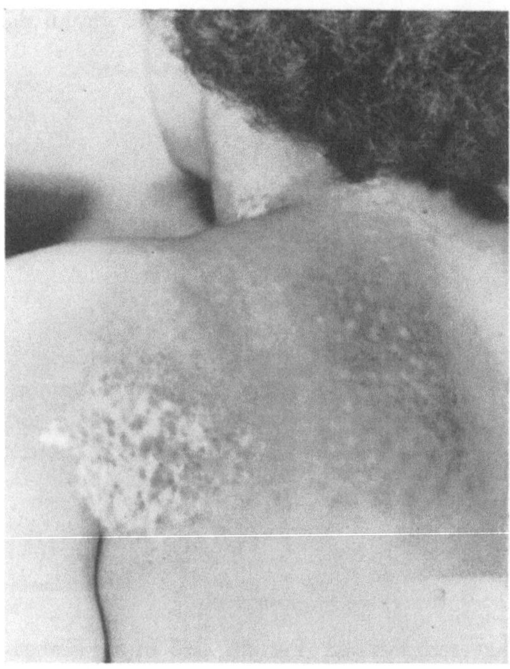

Fig. 4. Pigmentreaktionen der Haut nach Elektronenbestrahlung mit und ohne Gitter. *Links:* 12 cm-Feld, 15 MeV, 5200 r in 21 Tagen, Zustand 16 Mon. nach Behandlung. *Rechts:* 8 × 12 cm-Feld, 15 MeV, 5 mm-Bleigitter mit 7 mm Lochdurchmesser, Aperturenverhältnis Loch: bedeckt 40:60, 9000 r in 20 Tagen. Zustand 13 Wochen nach Behandlung

radiologisch nicht mehr lösbares Problem dargestellt hatte. Welche Bedeutung die Gitterbestrahlung für uns gewonnen hat, geht daraus hervor, daß ihr Anteil bei den letzten 100 Fällen 47,5% der Bestrahlungsfelder betrug. Der wesentliche Vorzug dieser Methode liegt in der ausgezeichneten örtlichen Verträglichkeit, bei gutem Effekt auf den Tumor. Die ablaufende Hautreaktion ist auch bei hohen Dosen sehr mild, und die spätere Narbenbildung beeindruckt durch ihre Zartheit und ihre geringen funktionellen Störungen (Fig. 4). Die Wirkung auf die Tumorrückbildung ist auch bei sehr oberflächlichen Tumoren, also auch bei völlig inhomogener Durchstrahlung, vollbefriedigend. Als Nachteil sehen wir lediglich die Verringerung der therapeutisch wirksamen Tiefenreichweite auf etwa 4 cm an, die ihre Verwendung auf oberflächennahe und flache Tumorherde einschränkt.

Durch eine relativ einfache Modifikation der Kupplungseinrichtung am Betatronstativ wurde es möglich, die Mittelachse des Elektronenstrahlenbündels auf den Drehpunkt zu zentrieren und damit die bei unserem Gerät vorhandene Pendeleinrichtung auch für die Elektronentherapie zugänglich zu machen. Die

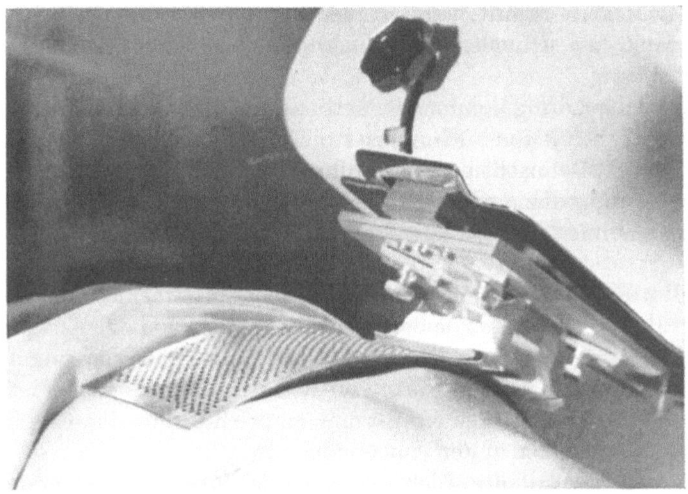

Fig. 5. Elektronen-Pendelbestrahlung mit 2 mm-Bleigitter, Lochdurchmesser 3 mm, Aperturenverhältnis 40:60 für Elektronenenergien bis 6 MeV, bei intracutanen Hautmetastasen bei Mammacarcinom post op.

Elektronen-Pendeltherapie gestattet die Bestrahlung gürtelförmiger Streifen in der Körperoberfläche und ist für die Behandlung multipler, über Rumpfabschnitte verteilter Oberflächenmetastasen vorgesehen. Bei Verwendung des bereits erwähnten Oberflächengitters lassen sich sehr ausgedehnte Hautflächen bestrahlen

Fig. 6. Dosisverteilung bei Gitter-Pendelbestrahlung mit schnellen Elektronen von 5 MeV. Pendelwinkel 90°. Länge des Pendelfelds etwa 30 cm. Filmschwärzungsaufnahme im Plexiglasphantom

(Fig. 5 und 6). Bei kleinem Pendelradius, also bei Körperabschnitten, deren Durchmesser kleiner ist als die doppelte maximale Reichweite des Strahlenkegels, erhält man den von der Röntgenpendelbestrahlung her bekannten Effekt einer Dosiskonzentration im Bewegungszentrum, bei gleichzeitiger Verringerung der Oberflächenbelastung.

Reicht die Tumorausdehnung nach der Tiefe zu weiter als 6 cm der therapeutischen Reichweite des 15 MeV-Elektronenstrahlenkegels, so läßt sich durch

Kombination mit einem entsprechend eingestellten Gamma-Pendelfeld ein zusätzliches Dosisvolumen anbauen. Diese Methode stellt eine Notlösung dar, die bei Verwendung energiereicherer Elektronen, als sie uns zur Verfügung stehen, natürlich keine Bedeutung besitzt.

Auf Grund unserer bisherigen Erfahrungen, die naturgemäß auch mit zahlreichen Fehlschlägen erkauft wurden, möchten wir für die *Bestrahlungsplanung* bei der Therapie mit schnellen Elektronen die Beachtung folgender allgemeiner Regeln empfehlen:

1. Die zur Anwendung kommende Methode sei immer die einfachste unter den vorhandenen Möglichkeiten. Komplizierte Einstellungen, die unter Umständen den Aufbau von Hilfseinrichtungen erfordern, sind häufig sehr elegant, aber meist auch schwer zu reproduzieren und zeitraubend, was sich in einem ausgedehnten Routinebetrieb immer negativ auswirkt. Das gleiche gilt für die Lagerung des Patienten.

2. Das Bestrahlungsfeld sollte immer möglichst groß bemessen werden, soweit die örtlichen Bedingungen es zulassen, und dabei sollte von der Gitterbestrahlung weitmöglichst Gebrauch gemacht werden. Zu exakte Feldanpassung an den tastbaren Herd rächt sich häufig in Form von Randrezidiven oder Metastasen in der Nachbarschaft. Auch bei kleinen Herden empfiehlt sich die Einplanung prophylaktischer Riegelfelder in der Umgebung.

3. Auch in der Wahl der Elektronenenergie, also der Tiefenreichweite, ist Kleinlichkeit verfehlt. Soll wirklich nur ganz oberflächlich bestrahlt werden, so ist die Herabsetzung der Elektronenenergie auf den Minimalwert wegen der Verringerung der Oberflächendosis und der erhöhten Belastung der Röhrenwandung die schlechteste Methode. Geringe Tiefenreichweiten lassen sich besser durch Verwendung von Schrägtubussen oder die Zwischenschaltung von Gittern oder entsprechend dicken Absorberscheiben aus Plexiglas oder Aluminium erreichen.

4. Von der Direktmessung der Herddosis mit einer Kleinkammer sollte nach Möglichkeit Gebrauch gemacht werden. Besonders bei Bestrahlungen im Kopfbereich läßt sich die wirkliche Herddosis meist nur unvollkommen auf Grund der am Phantom gemessenen Isodosen abschätzen. Als Meßgerät für diese Zwecke hat sich uns das „Simplex"-Dosimeter (P. T. W. Freiburg) mit Schlauchkammer gut bewährt.

Hinsichtlich der *Dosierung* können wir auch jetzt noch kein abschließendes Urteil abgeben. Wir treten für eine fraktionierte Bestrahlung ein, da hierbei die Reaktion des Tumors und seine Strahlenempfindlichkeit laufend beobachtet und die Gesamtdosis entsprechend bemessen werden kann. Als Regeldosis für die Einzelsitzung verabfolgen wir 400 r, bezogen auf das dicht unter der Oberfläche liegende Dosismaximum. Bei Gitterbestrahlungen gehen wir auf 500 r, zuzüglich eines vom verwendeten Gitter abhängigen, zwischen 60 und 100 r betragenden Oberflächenstreuzusatzes, bezogen auf die Hautoberfläche im Lochbereich. Abweichungen von dieser Dosierung nach oben und unten können unter Berücksichtigung der Feldgröße, Vorbelastung und der Strahlenempfindlichkeit von Herd und Umgebung aus Gründen angebracht sein, die auf die psychische oder wirtschaftliche Situation des Patienten Bezug haben. Behandlungsziel der ersten Bestrahlungsserie ist die Vernichtung des Tumorherdes, falls der Tumorprozeß nicht von vornherein eine Beschränkung auf palliative Maßnahmen notwendig macht. Wir verabfolgen im allgemeinen hierzu in 10—15 Sitzungen zwischen 4000 und 6000 r, bei Gitterbestrahlung zwischen 6000 und 9000 r. Als Palliativdosis sehen wir 2000—3000 r an.

Die Hautreaktionen können aus noch nicht völlig geklärten Gründen in ihrer Stärke und Dauer recht unterschiedlich sein und sind im Einzelfall kaum vorauszusagen. Als stärkste Reaktion beobachtet man bei voller Dosierung — ohne Gitter — an der Haut eine fast das gesamte Feldgebiet umfassende Epidermolyse, die meist nach 4—5 Wochen völlig reepithelisiert ist und unter glatter Vernarbung mit Pigmentreaktionen abheilt. Als Endzustand findet sich im stärksten Fall eine das Feldzentrum betreffende völlige Depigmentierung mit später einsetzender leichterer Teleangiektasiebildung und einem peripheren hyperpigmentierten Randsaum. Geringere Grade äußern sich in kleinfleckigen Pigmentverschiebungen, die dem Feld ein marmoriertes Aussehen verleihen (Fig. 4). Vielfach ist aber auch nur eine leichte gleichmäßige Bräunung der Haut anzutreffen. Starke Reaktionen sind meist mit einer stärkeren subcutanen, als derbe Platte tastbaren Verschwielung verbunden, die gelegentlich für den Patienten unangenehme Nebenerscheinungen hervorruft, vor allem, wenn es zur Einmauerung von Nerven und Gefäßen kommt. Diese Spätzustände sind mit ein Grund, weshalb wir jetzt auch primär die wesentlich zarter abheilende Gitterbestrahlung bevorzugen (Fig. 4).

An der Schleimhaut läuft die Strahlenreaktion meist in Form fibrinöser Beläge ab.

Ich möchte nunmehr noch in gedrängter Form auf die wichtigsten Indikationen für die Therapie oberflächennaher Tumoren mit schnellen Elektronen eingehen. Dabei stütze ich mich ausschließlich auf die an unserer Klinik bisher gewonnenen Erfahrungen, die selbstverständlich keinen Anspruch auf Vollständigkeit erheben, sondern lediglich ein Zwischenresümee darstellen.

Mit 40% aller bestrahlten Tumorherde stellen die oberflächennahen *Lymphknotenmetastasen* unser ausgedehntestes Anwendungsgebiet dar. Diese Indikation hat so überzeugende Behandlungserfolge aufzuweisen, daß wir seit längerem dazu übergegangen sind, Drüsenmetastasen der prä- und retroauriculären, der submandibulären, cervicalen, nuchealen, supra- und infraclaviculären, axillären und inguinalen Region sowie im Bereich der Extremitäten systematisch der Betatrontherapie zuzuweisen. Bevorzugte Bestrahlungsmethode ist dabei aus bereits erwähnten Gründen die Gitterbestrahlung.

Als zweitwichtigste Gruppe müssen die verschiedenen Tumorformen im *Hals-Nasen-Ohren-Bereich* erwähnt werden, die etwa 16% der Gesamtfälle umfassen. Hier ist eine enge Zusammenarbeit mit der Fachklinik unumgänglich. Topographisch günstige Bestrahlungsbedingungen für Elektronen bieten vor allem die Tumoren der Nebenhöhlen, der Mundhöhle — einschließlich Mundboden und Zunge —, des Hypopharynx und des Larynx. Wir sind auch deshalb sehr für die Elektronentherapie in diesem Bereich eingenommen, weil hierbei die Gefahr der mit Recht so gefürchteten Knorpel- und Knochennekrosen, eine Crux radiologica an Ohrmuschel, Larynx, Unterkiefer und Schädel, nach unseren Erfahrungen auf ein Minimum reduziert ist. Wir konnten bisher kein solches Ereignis beobachten. Mehr in der Tiefe gelegene Herde, wie im Bereich des Zungengrundes, der Tonsillen und der Rachenwand, sollten nach Möglichkeit von außen und von intraoral her angegangen werden, damit eine ausreichende Herddosis gewährleistet wird. Messungen mit der Schlauchkammer in Kieferhöhle und Mundhöhle ergaben, daß bei äußerer Bestrahlung durch die Wange oder von der Submentalregion aus etwa 60—80% der Maximaldosis im Herdgebiet erreicht werden können. Für den Behandlungserfolg bei Zungen- und Mundbodentumoren ist eine sorgfältige antiphlogistische örtliche Nachbehandlung ebenso wichtig, wie die Ausschaltung der Gefahr einer lebensbedrohlichen Arrosionsblutung durch eine vorherige Gefäßligatur. Bestrahlungen im Mundhöhlen- und Larynxbereich bringen unvermeidlich stärkere reaktive Beschwerden mit sich, über die dem Patienten wegzuhelfen eine ebenfalls wichtige ärztliche Aufgabe darstellt. Auch sonst bieten sich im *Kopfbereich* zahlreiche dankbare Therapiemöglichkeiten für schnelle Elektronen. Besonders eindrucksvolle Ergebnisse, auf die ich kurz hinweisen möchte, haben wir erzielt bei Bestrahlungen in Augennähe, bei Herden des Schädeldachs und der Kopfschwarte, bei rindennah gelegenen Hirntumoren und bei Parotistumoren. All diese diffizilen Lokalisationen betrachten wir als eine Domäne der Elektronentherapie.

109 Fälle betrafen *Mammacarcinommetastasen* im Bereich der Operationsnarbe oder deren Umgebung. Sie bieten technisch keine Schwierigkeiten und reagieren fast ausnahmslos gut. Wegen der großen Rezidivneigung sollten die Bestrahlungsfelder sehr großzügig bemessen werden, auch wenn der augenblickliche Lokalbefund ausdehnungsmäßig gering ist. Ebenso sollte die Tiefenreichweite so gewählt werden, daß die in der Tiefe gelegenen retrosternalen und pleuralen Lymphgebiete gleich miterfaßt werden. Umschriebene *Pleuratumoren* bringen gleichfalls gute Ergebnisse, vor allem die Solitärmetastasen. Eine ausgesprochene Elektronen-Indikation ist für uns die *Struma maligna* geworden, deren Ergebnisse bisher sehr befriedigt haben. Große inoperable Tumoren erfordern unbedingt eine stationäre Behandlung, da die Gefahr reaktiver Komplikationen nie ausgeschlossen werden kann. Wenn wir bei der Elektronentherapie bisher solche Zwischenfälle auch noch nie beobachtet haben, sondern im Gegenteil durch die gute Verträglichkeit der Bestrahlungen beeindruckt waren, sind wir dennoch bei großen Tumoren für die Anlegung eines Tracheostomas vor Bestrahlungsbeginn. Postoperative Nachbestrahlungen und kleinere Herde eignen sich gut für Gitterfelder.

Knochentumoren sind in vielen Fällen ebenfalls gut für Elektronen erreichbar. Zwar haben wir röntgenologisch faßbare Besserungen bisher nur in vereinzelten Fällen gesehen, doch sollte der Versuch einer Elektronentherapie im Hinblick auf die gute Gewebsverträglichkeit und die reelle Chance, einen Wachstumsstillstand des Tumors zu erreichen, immer wieder gemacht werden. Wirbelherde stellen bei Kindern und Schwangeren eine ausgesprochene Indikation für schnelle Elektronen dar. Von großer Bedeutung, schon in Anbetracht der Schonung des Keimgewebes, sind Tumoren im Bereich des *männlichen Genitale*. Besonders beim Peniscarcinom besteht eine deutliche Überlegenheit der Elektronentherapie gegenüber anderen Therapieformen, wenigstens nach unseren Erfahrungen, wobei allerdings die Zahl unserer Beobachtungen wegen der relativen Seltenheit dieser Fälle noch gering ist.

Schließlich verdienen noch die *tiefsitzenden inoperablen Rectumcarcinome* und die Tumoren im *Analbereich* eine Erwähnung. Sie sind nicht nur günstig lokalisiert, sondern bieten bei Anwendung schneller Elektronen auch gute Aussichten auf Erhaltung der für das Wohlbefinden der Patienten wichtigen Sphincterfunktion des Anus.

Die bisher genannten Indikationen machten in unserem Material 57% der Fälle aus. Die verbleibenden 43% verteilen sich auf

183 Hauttumoren,

73 Tumoren des weiblichen Genitale und

318 sonstige Weichteiltumoren verschiedenster Art und Lokalisation.

Von einer *statistischen Aufschlüsselung* unserer Behandlungsergebnisse haben wir abgesehen, da bei unserem Krankengut dafür die Zahl der vergleichbaren Fälle zu gering ist. Der überwiegende Teil unserer Patienten stellt weit fortgeschrittene und unterschiedlich vorbehandelte Fälle dar, deren quo ad vitam-Prognose von vornherein schlecht war. Unter diesen Voraussetzungen ist eine Beurteilung der Wirksamkeit der Elektronentherapie auf der Grundlage einer Erfolgsstatistik nicht möglich. Andererseits konnten wir uns aus ärztlichen Gründen nicht entschließen, die Auswahl der für die Betatrontherapie vorgesehenen Fälle nach statistischen Gesichtspunkten vorzunehmen.

Worauf der Wirkungsunterschied zwischen schnellen Elektronen und Röntgenstrahlen üblicher Energie beruht, ist noch umstritten. Wir können anhand unserer klinischen Beobachtungen zu diesem Problem keine Stellung nehmen, sondern nur die Tatsache eines solchen Unterschiedes als solche bestätigen.

Wir sind auf Grund unserer, bei der Behandlung von rund 800 Patienten im Lauf von $3^1/_2$ Jahren gemachten Beobachtungen zu der Ansicht gelangt, daß die Therapie mit schnellen Elektronen infolge ihrer eigentümlichen Dosisverteilung und ihrer spezifischen klinisch-biologischen Wirkung auf das bestrahlte Gewebe einen echten Fortschritt darstellt. Ihre praktische Durchführung ist sowohl von der technischen wie von der klinischen Seite her als Routinemethode möglich. Sie nimmt innerhalb der Strahlentherapie einen festen Platz ein und besitzt einen eigenen Indikationsbereich, in dessen Rahmen sie anderen radiologischen Methoden überlegen ist.

## Diskussionsbemerkungen

A. ZUPPINGER (Bern):

Herr SCHMERMUND hat bei der Elektronentherapie des Vulvacarcinoms in einem Fall von einer Dosis von 5400 r gesprochen. Wie ist diese Dosis gemessen und in welcher Zeit wurde sie gegeben?

Wir haben bei der konventionellen Bestrahlung uns immer gescheut, auf die Vulva höhere Dosen zu applizieren, weil unangenehme Früh- und Spätreaktionen auftraten. Diese Situation hat sich offenbar bei der Elektronentherapie grundlegend geändert. Welche Erklärung läßt sich geben?

H. J. MAURER (Bern):

Da der Verlauf der ableitenden Lymphwege (s. bei MATRINS, G. S. STÜTZ und H. J. MAURER, TRAUTMANN und A. J. MAURER) angedeutet S-förmig nach medial verläuft und außerdem noch das Niveau wechselt, erscheint eine Verlängerung der Parametrienfelder, wie sie von H. L. KOTTMEIER seinerzeit vorgeschlagen wurde, unzweckmäßig. Statt dessen muß ein zweites nach medial und dorsal versetztes Feld auch bei ultraharten Röntgenstrahlen angewandt werden.

R. KEPP (Gießen):

Ich glaube auch, daß die Therapie des Vulvacarcinoms mit schnellen Elektronen die zweckmäßigste Behandlungsmethode darstellt, da es sich ja hierbei um eine räumlich genau begrenzte intensive Strahlenwirkung handelt, also der gleiche Effekt erzielt wird, wie bei der elektrochirurgischen Abtragung in der betreffenden Tiefe. Den Prüfstein für die therapeutische

Überlegenheit der schnellen Elektronen werden die sehr ausgedehnten Vulvacarcinome bilden
müssen, die für eine operative Behandlung nicht mehr in Frage kommen. Bei solchen Vulvacarcinomen lassen sich, wie mein Mitarbeiter D. Hofmann kürzlich auf dem Österreichischen
Gynäkologenkongreß in Wien mitgeteilt hat, durch eine entsprechend hoch dosierte Bestrahlung unter den Bedingungen der Nahbestrahlung sehr befriedigende Ergebnisse erreichen.
Voraussetzung ist allerdings, daß das gesamte Tumorgebiet mit entsprechend großen und
geformten Feldern belegt werden kann. Die guten Ergebnisse der Nahbestrahlung müßten
sich durch die Elektronenbestrahlung noch übertreffen lassen.

Was die gynäkologische Tiefentherapie mit ultraharten Röntgenstrahlen betrifft, so bin
ich der Meinung, daß sich mit der klassischen Röntgenbestrahlung unter Verwendung der
Bewegungsbestrahlung sowohl dosismäßig als auch dosisverteilungsmäßig alle gewünschten
Bedingungen herstellen lassen. Ob sich die ultraharten Röntgenstrahlen hier als günstiger
erweisen, kann nur davon abhängen, ob sie eine gesteigerte elektive Wirkung auf bösartiges
Gewebe besitzen, was allerdings anzunehmen ist.

H. R. Schinz (Zürich):

Beim Vulvacarcinom ist wahrscheinlich die Supervolt-Elektronentherapie-Methode die
der Wahl. Dies gilt sicher nicht für das Cervix- und das Corpuscarcinom. Hier bringen wir
auch ohne Radium, Kobalt oder Caesiumröhren die cancericide Dosis leicht an den Wirkort
(eigene Versuche mit dem 31 MeV-Betatron). Vor Komplikationen, z. B. Perforationen,
schützt zum Unterschied von Speiseröhrenkrebs der dicke Muskelmantel.

P. Krahl (Heidelberg):

Schnelle Elektronen (schn. E.) dürften wegen ihrer bekannten physikalischen und biologischen Eigenschaften auch mit Vorteil für die Strahlentherapie von Larynx- und Hypopharynx-
Tumoren zu verwenden sein. Da in diesem Bereich aber erst begrenzte Erfahrungen vorliegen,
haben wir als Laryngologen in Zusammenarbeit mit der Heidelberger Universitäts-Strahlen-
klinik experimentelle Untersuchungen am gesunden Kehlkopf von Meerschweinchen und
Kaninchen angestellt, um die Besonderheiten der Elektronenstrahlenwirkung im Vergleich mit
Röntgenbestrahlungsfolgen im histologischen Bild zu studieren.

In einer Serie wurden Einzeitbestrahlungen an über 100 Meerschweinchen mit jeweils
4000 r schn. E. (15 MeV) oder Röntgenstrahlen (200 kV) durchgeführt, die Tiere jeweils nach
Tagen, Wochen oder Monaten getötet und ihre Kehlköpfe histologisch untersucht.

Nach beiden Strahlenarten entwickelten sich im Laufe von mehreren Wochen etwa gleich-
artige, schwere regressive und nekrobiotische Veränderungen an allen Geweben. Leichter
waren die Alterationen nur am Binde- und Knorpelgewebe. Gleichzeitig kam es zu schweren
akut-entzündlichen Reaktionen, vor allem in der Submucosa und im Perichondrium schleim-
hautnahe gelegener Knorpel. Als Beispiele sollen einige Abbildungen dienen, welche die
Epithelveränderungen und die entzündliche Infiltration der Knorpelhaut am II. Tag nach
Elektronenbestrahlung zeigen. Nach einigen Monaten klangen die degenerativen und ent-
zündlichen Prozesse wieder ab und liefen in ein Narbenstadium aus.

In einer anderen Untersuchungsreihe wurden bei 20 Kaninchen in mehreren Sitzungen,
ähnlich der fraktionierten Bestrahlung zusammen 8000—10000 r schn. E. oder Röntgenstrahlen appliziert. Die erste Bestrahlung lag einige Monate, die letzte einige Wochen zurück.
Hierbei waren im Gegensatz zu den Beobachtungen nach Einzeitbestrahlung die Reaktionen
und Veränderungen nach Bestrahlung mit schn. E. deutlich geringer als nach Röntgenbestrahlung. Auch an Bindegewebs- und Knorpelzellen fanden sich bei dieser Dosierung
auffallende Unterschiede der biologischen Veränderungen. Auf die möglichen Ursachen dieses
unterschiedlichen Verhaltens kann hier nicht näher eingegangen werden.

Für die Belange der Klinik ergeben sich aus den Resultaten unserer Untersuchung bei
fraktionierter Bestrahlung einige Hinweise. Abgesehen von der offenbar größeren Schonung
von Schleimhaut, Schleimdrüsen und Gefäßbindegewebe besitzt die geringere Strahlenbelastung des Knorpels im Hinblick auf die Entwicklung der Perichondritis eine besondere
Bedeutung. Die Strahlenschädigung am Knorpel kann sich offenbar auf zwei Wegen entwickeln, nämlich an der Knorpelzelle selbst und an der Knorpelhaut, deren Ernährungs-
funktion durch allmähliche Umwandlung in Narbengewebe gestört wird.

Nach den heutigen Auffassungen von der Entstehung der nekrotisierenden Perichondritis bei Bestrahlung kommt es dabei zu einem Zusammenwirken von Strahlenschädigung und Infektion. Bei der Frühperichondritis nach wenigen Einzeldosen dürfte die Infektion im Vordergrund stehen, denn die Strahlenschädigung des Knorpels ist noch gering.

Das umgekehrte Verhalten liegt wohl bei der Spätperichondritis vor, die erst nach ein oder mehreren Bestrahlungsserien entsteht. Wird nun durch Verwendung schn. E. die Strahlenbelastung des Knorpelgewebes geringer, so dürfte die Spätperichondritis wahrscheinlich seltener auftreten, die Entwicklung der Frühperichondritis hingegen dadurch kaum beeinflußt werden.

Durch Schonung der Haut, des Knorpels und der tiefer liegenden gesunden Gewebe kann am Kehlkopf mit schn. E. möglicherweise ein ähnlicher Effekt erzielt werden, wie er durch Schildknorpelfensterung bzw. -abtragung und Weichteilaufklappung zur Belastung angestrebt wird.

H.-J. SCHMERMUND (Hamburg):

Zur Frage von Herrn Prof. ZUPPINGER möchte ich antworten, daß die Elektronendosis mit einer Fingerhutkammer in einem Plexiglas-Phantom gemessen wurde, wobei sich die Dosisangabe auf den Wert im Dosismaximum bezieht. Die Gesamtdosis wurde in Einzelfraktionen von 300 r verabfolgt. Die rasche Abheilung der Frühreaktionen nach Elektronentherapie der Vulvacarcinome ist in erster Linie auf die günstige räumliche Dosisverteilung zurückzuführen. Spätreaktionen sind nach unseren bisherigen Erfahrungen weit weniger ausgeprägt als nach der konventionellen Röntgentherapie. Diese Beobachtung wird von uns dahingehend gedeutet, daß auch gewisse qualitative Wirkungsunterschiede zwischen ultraharten Strahlen und 200 kV Röntgenstrahlen eine Rolle spielen.

# Betatron- und Telekobalttherapie

## Expérience clinique avec le Bétatron*

Par

M. Tubiana, Paris

Un Bétatron émettant des rayons X de 22 MeV d'énergie maximum est utilisé pour la radiothérapie des cancers profonds depuis novembre 1953 à l'Institut Gustave Roussy.

Si trois ans et demi d'expérience clinique n'autorisent pas de conclusion formelle, ils permettent déjà de recueillir quelques enseignements.

I. La plupart des avantages pratiques que l'on pouvait espérer des caractères physiques des rayonnements de grande énergie ont été observés.

1. A la diminution de la dose intégrale correspondent de moindres réactions générales pendant le traitement.

Avec le rayonnement X du Bétatron 22 MeV la décroissance de la dose est après le maximum, assez lente. La distance nécessaire pour que la dose décroisse de moitié dans les tissus est d'environ 22 cm alors qu'elle est d'environ 5 cm pour des rayons X de 200 kV, de 10 cm pour les rayons gamma du cobalt et de 26 cm pour les rayons de 70 MeV d'énergie maximum (pour une même DFP de 80 cm).

Pour une même dose à la tumeur, les tissus sains situés entre la tumeur et la peau absorbent donc une dose moindre, lorsque l'on utilise le Bétatron, que pour un rayonnement d'une énergie plus faible. En revanche, les tissus situés au-delà de la tumeur sont irradiés dadvantage, mais l'examen des courbes montre que cet inconvénient a relativement moins d'importance. L'augmentation du rendement en profondeur permet donc pour une même dose à la tumeur, de diminuer l'irradiation globale de l'organisme.

La dose intégrale absorbée dans tout l'organisme est égale à la somme des doses délivrées au tissu sain (dose nuisible) et au tissu tumoral (dose utile). Dans le cas de rayons X de faible énergie, une fraction très importante de la dose intégrale est délivrée à des tissus situés en dehors des limites géométriques du faisceau du fait de la diffusion très importante à ces énergies.

En prenant l'exemple particulier d'un cancer de l'oesophage dans lequel le volume à irradier est approximativement un cube de 6 cm de côté et que l'on suppose traité par 4 champs opposés 2 à 2 et croisés à angle droit, nous avons calculé les doses intégrales (distance foyer-peau: 100 cm, surface d'entrée: 35 cm²), grâce aux méthodes de calcul de Mayneord et Johns. La Fig. 1 montre comment la dose nuisible diminue, quand croît l'énergie des rayons X.

---

* Travail du service des isotopes et du Bétatron de l'Institut Gustave Roussy (Villejuif). Les résultats dont il est fait état dans ce travail ont été obtenus en collaboration avec les Drs. B. Pierquin, J. M. Dutreix et Mme. A. Dutreix, radiophysicienne.

Pour une même dose tumorale et des champs de dimensions identiques, la dose intégrale varie en fonction de 3 facteurs:

la position de la tumeur,

les dimensions de l'organisme irradié,

le rendement en profondeur du faisceau utilisé.

La Fig. 2 montre que l'épaisseur du segment irradié augmente, plus grand est l'avantage lié à l'emploi des radiations les plus pénétrantes.

La parfaite tolérance qui est le fait le plus remarquable observé au cours du traitement, l'absence de réactions générales, de nausées, de vomissements, de troubles graves de la formule sanguine sont sans doute la conséquence de cette diminution de la dose intégrale. Pour des doses tumorales importantes les réactions générales qui sont

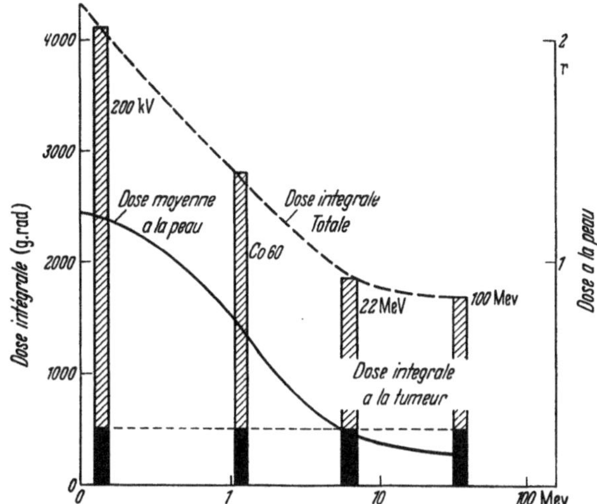


Fig. 1. Variation de la dose intégrale, pour 1 rad au centre de la tumeur. Dose calculée dans le cas d'un cancer de l'oesophage irradié par 4 champs 6 × 6 opposés 2 à 2, avec une DSP de 100 cm (l'épaisseur traversée par les faisceaux est de 26 cm). A mesure que l'énergie du faisceau augmente la dose intégrale diminue et le rapport dose intégrale à la tumeur (zone ombrée) à la dose intégrale aux tissus sains (en clair) augmente cependant que diminue la dose moyenne aux tissus sains et à la peau

en radiothérapie classique d'une intensité telle qu'elles nécessitent souvent l'interruption du traitement, sont en haute énergie généralement négligeables.

Il est malheureusement difficile de traduire en données numériques l'intensité des réactions générales d'un malade soumis à une irradiation thérapeutique, ce

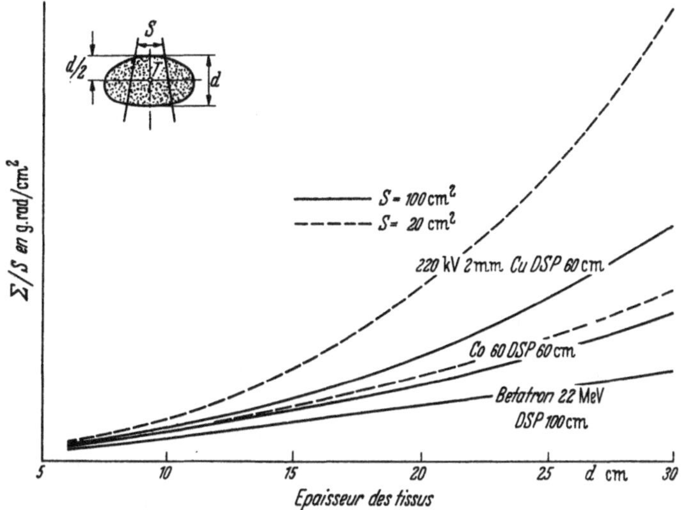


Fig. 2. Variation de la dose intégrale en fonction de l'épaisseur du sujet. Les doses intégrales sont ici rapportées à 1 cm 2 de surface irradiée, pour 1 rad au centre du volume irradié. On constate que la dose intégrale augmente beaucoup plus rapidement en fonction de l'épaisseur du sujet, pour des rayonnements de faible énergie

qui serait nécessaire pour permettre la comparaison systématique des réactions notées au cours des divers types de traitement. L'étude de l'évolution de la formule sanguine fournit néanmoins des données comparatives intéressantes. Nous avons comparé les diminutions du nombre de globules blancs observées au cours du traitement des cancers de l'oesophage ou des cancers du poumon selon qu'ils étaient irradiés par rayons, X de 220 kV ou par le Bétatron, (Fig. 3 et 4). Dans les 2 cas la diminution est, pour une même dose tumorale, nettement plus importante avec le 220 kV qu'avec le 22 MeV.

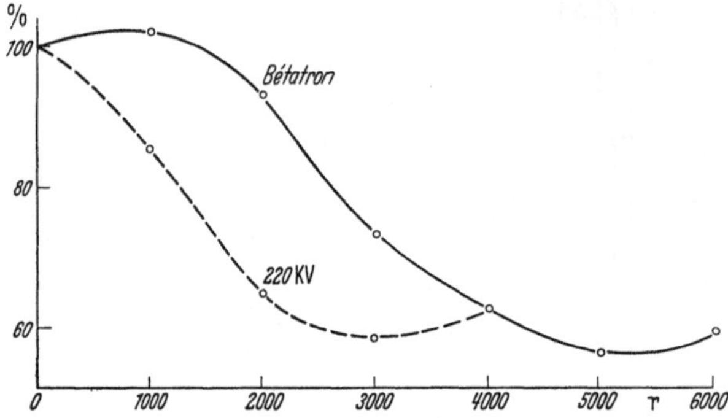

Fig. 3. Variation du nombre des globules blancs (en % de la valeur initiale) en fonction de la dose tumorale, chez des malades atteints de cancer de l'oesophage. Chez 22 malades traités par le Bétatron, la baisse du nombre des globules blancs est en moyenne, pour une même dose tumorale nettement plus faible que chez 16 malades traités par RX de 220 kV

Pour rechercher si cet effet est bien lié à une diminution de la dose intégrale, nous avons comparé les variations du nombre des globules blancs non plus en fonction de la dose tumorale, mais en fonction de la dose intégrale. Les courbes (Fig. 5 et 6) sont alors pratiquement superposables, ce qui montre bien le role clinique joué par la diminution de la dose intégrale. En comparant enfin les diminutions du nombre des globules blancs au cours des irradiations des cancers du poumon ou de l'oesophage, on constate qu'elles sont analogues bien qu'il s'agisse de cancers dont les retentissements biologiques et les réactions radio-biologiques sont très différentes (Fig. 7). Ce fait constitue un dernier argument soulignant l'intérêt pratique de la notion de dose intégrale lorsqu'il s'agit de comparer les irradiations d'un même segment de l'organisme.

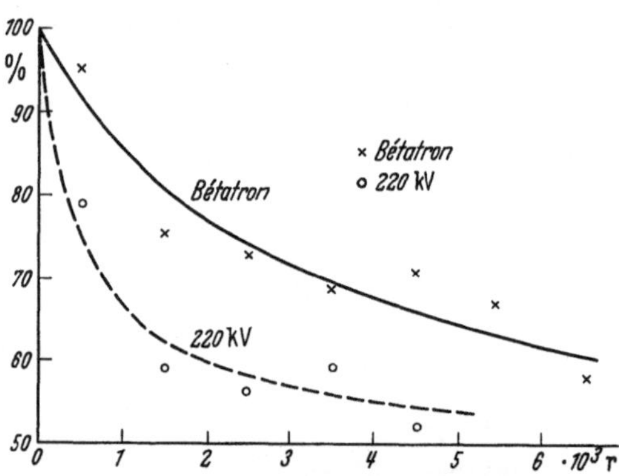

Fig. 4. La baisse du nombre des globules blancs est également plus faible, pour une même dose tumorale chez 45 malades atteints de cancer du poumon traités par Bétatron que chez 25 malades traités à 220 kV

2. L'absence de diffusion facilite l'établissement du plan de traitement.

Dans toutes les interactions entre les photons et le milieu matériel apparaissent de nouveaux photons qui diffèrent des photons incidents par leur énergie et leur direction et que l'on groupe sous le nom de photons diffusés. C'est à cause de cette diffusion que les limites d'un faisceau ne sont jamais précises et s'étalent hors de ses limites géométriques, que le rendement en profondeur varie avec les dimensions des champs et que les isodoses ne sont jamais planes en profondeur.

Dans l'effet Compton, qui joue un rôle prédominant pour les rayons X du Bétatron, une partie de l'énergie du photon incident est transmise à un électron de recul et est absorbée localement. Le reste de l'énergie est emportée par le photon diffusé. Lorsque l'énergie des photons croît, la proportion d'énergie absorbée croît par rapport à celle qui est diffusée. C'est ainsi que pour des photons de 100 keV, l'énergie absorbée ne représente que le 1/10 de celle perdue par les photons primaires; pour des photons de 1 MeV il y a égalité entre énergie absorbée et diffusée, et pour des photons de 10 MeV il y a 2 fois plus d'énergie absorbée que diffusée. Tout ceci explique que l'accroissement de la dose en profondeur qui est liée à la diffusion, soit plus faible à haute énergie (Fig. 8). Cependant, si la diffusion est faible, elle est loin d'être nulle (Fig. 9) mais comme, contraitement à ce qui se passe avec des énergies plus faibles, la direction des photons diffusés s'écarte peu de celle du rayonnement primaire, le rôle joué par la diffusion est pratiquement très minimisé et la diffusion ne provoque pas comme aux basses énergies l'irradiation inutile, sinon nuisible, des tissus péritumoraux.

Au point de vue dosimétrique les limites de la zone irradiée sont nettes, même à une profondeur relativement considérable par rapport aux dimensions de l'organisme, et se confondent pratiquement avec les bords géométriques du faisceau. Ce fait qui traduit le peu d'importance de la diffusion latérale oppose les

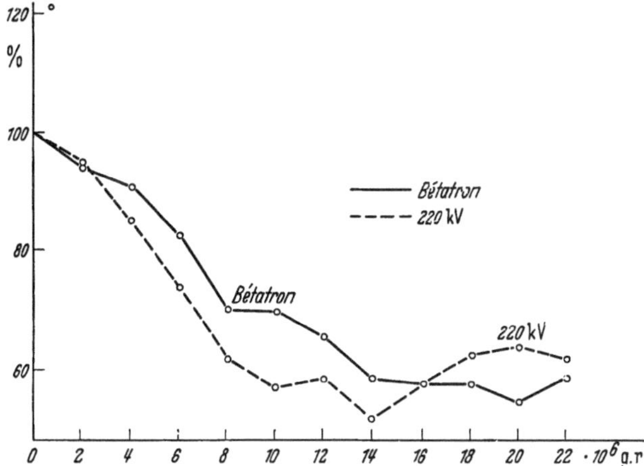


Fig.5. Si l'on suit la variation du nombre des globules blancs en fonction de la dose intégrale, la baisse est, pour les malades atteints de cancer de l'oesophage, comparable, qu'ils soient traités par des rayons X de 22 MeV ou de 220 kV

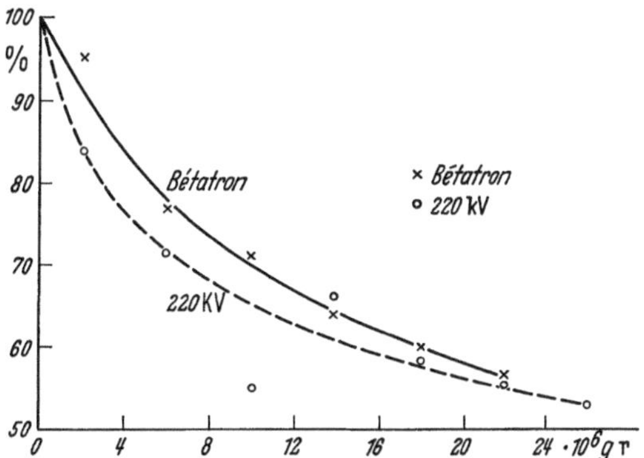


Fig. 6. Les résultats précédents se retrouvent chez les malades atteints de cancer du poumon

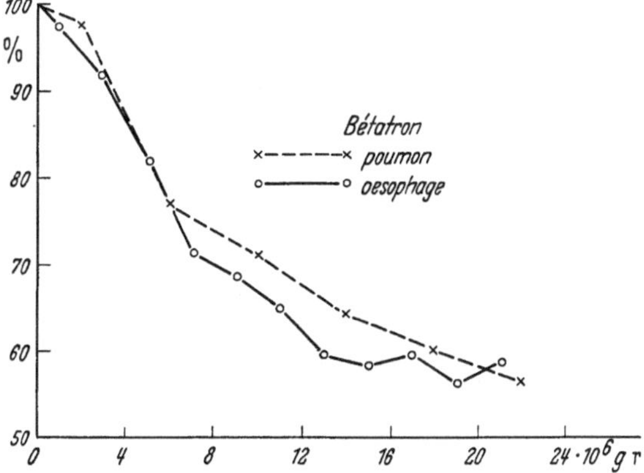


Fig. 7. A dose intégrale égale, la baisse du nombre des globules blancs est comparable chez les cancers de l'oesophage et les cancers du poumon traités par le Bétatron

courbes isodoses obtenues avec le Bétatron, à celles obtenues avec de plus basses énergies.

Lorsque l'on compare les isodoses obtenues avec des faisceaux de section très différente, si l'on exprime, comme on le fait usuellement, les doses en % de la

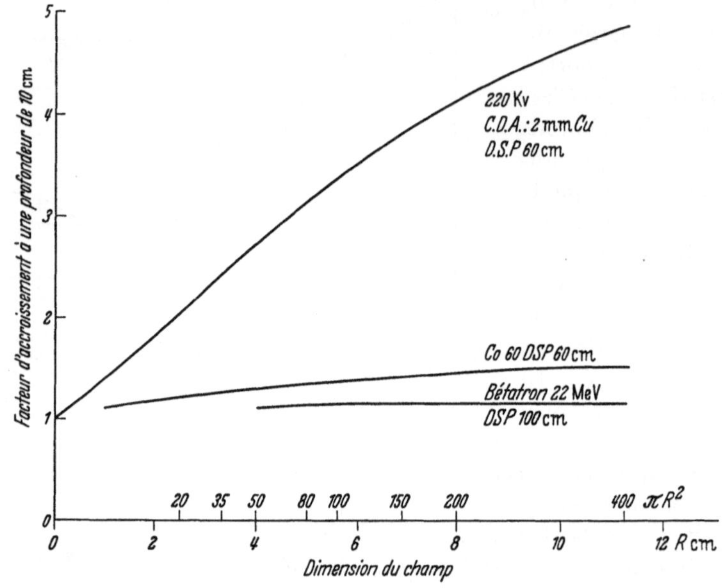


Fig. 8. Variation en fonction de la surface des champs, du facteur d'accroissement qui exprime le rapport de la dose totale (primaire + diffusé) à la dose due aux photons primaires. Il est beaucoup plus élevé pour des rayonnements de faible énergie

dose maximum, la répartition des doses en profondeur ne varie que très peu avec la dimension des champs (Fig. 10).

Les isodoses restent planes, même profondément, ce qui est encore une conséquence du peu d'importance de la diffusion.

L'intérêt clinique de ces faits réside en ce que l'on peut obtenir des doses assez homogènes dans toute la section d'une tumeur profonde, en utilisant un champ dont les dimensions sont égales à celles de la tumeur alors qu'avec des rayons X de 200 kV il serait nécessaire d'utiliser pour obtenir ce même résultat, un champ nettement plus large (cf. tableau 1).

Des avantages radiothérapiques en découlent immédiatement.

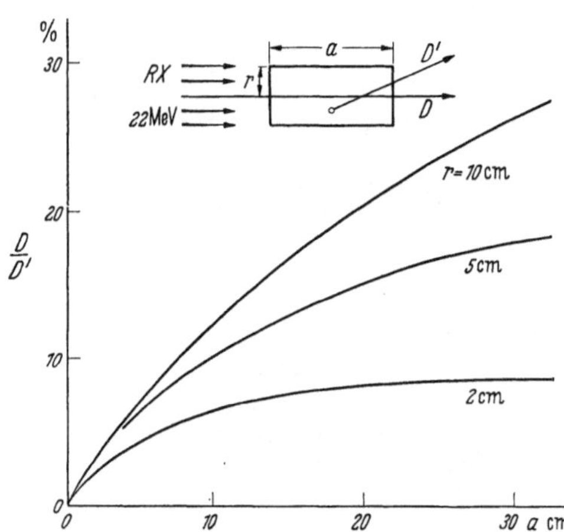


Fig. 9. Dose due a la diffusion dans le faisceau de 22 MeV. Le rapport dû à la diffusion à la dose primaire croît en fonction de la profondeur a et du rayon r du faisceau supposé ici cylindrique. Il est néanmoins loin d'atteindre les valeurs importantes observées avec des énergies plus faibles

a) Il est relativement facile d'irradier une tumeur, même située dans le voisinage immédiat d'un tissu noble que l'on doit respecter. Ceci est dû à la netteté des limites du faisceau et à la possibilité d'obtenir une dose homogène dans tout le volume cible avec un petit nombre de faisceaux.

*Tableau 1*

Pour obtenir une dose homogène (à $\pm$ 5% près) en tous les points d'une section droite de la tumeur située à 10 cm en profondeur et supposée avoir un diamètre de 5 ou 10 cm, il faut utiliser des champs plus larges lorsque la diffusion (200 kV) ou la pénombre (60 Co) sont importantes (la DFP est 60 cm)

| Diamètre de la tumeur | 5 centimètres | | 10 centimètres | |
|---|---|---|---|---|
| Diamètre du champ | Entrée | à la tumeur | Entrée | à la tumeur |
| 220 kV (CDA 1,5 Cu) . . . . . . . . . . . | 7 | 8,2 | 14 | 16 |
| Co$^{60}$ | | | | |
| Distance source-collimateur 26 cm . . . . | 7 | 8,2 | 12 | 14 |
| Distance source-collimateur 46 cm . . . . | 6 | 7 | 10 | 12 |
| Bétatron . . . . . . . . . . . . . . . . | 4,3 | 5 | 8,5 | 10 |

b) Pour délivrer une dose homogène à une tumeur on est obligé d'utiliser un champ dont la section est nettement plus grande si les surfaces isodoses sont convexes, comme dans la radiothérapie classique; un même résultat peut être obtenu avec des faisceaux dont la section est égale à celle de la tumeur, si les

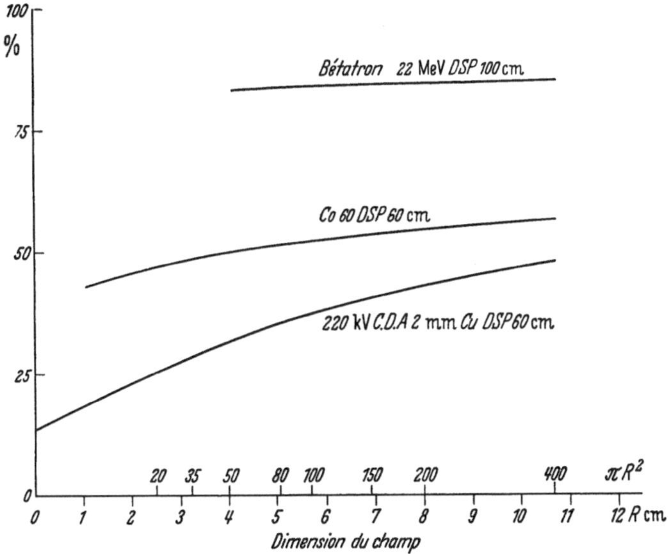


Fig. 10. Rendement en profondeur en fonction de la surface du champ a l'autre. Il est pratiquement constant pour le faisceau du Bétatron tandis que l'accroissement de la diffusion le fait augmenter sensiblement pour les énergies plus faibles

surfaces isodoses sont planes. Le tableau 1 indique des largeurs de champs nécessaires pour une tumeur de 5 cm de diamètre, située à 10 cm de profondeur, en fonction de l'énergie du rayonnement utilisé et en supposant que l'on tolère une variation de $\pm$ 5% de la dose d'un bord à l'autre de la tumeur. On remarquera l'influence de la distance source-collimateur, c'est-à-dire de la pénombre sur ces dimensions lorsque la dimension de la source n'est pas négligeable (cas du cobalt par exemple).

*3.* Le long parcours des électrons secondaires explique l'accroissement de la dose dans les premiers centimètres.

L'intérêt thérapeutique est lié à la possibilité de délivrer des doses importantes aux tissus profonds tout en irradiant de façon minime le tissu superficiel. Effectivement on ne constate sur les malades arrivés en fin de traitement que des réactions cutanées à peine visibles.

Ce facteur devrait théoriquement permettre d'aumenter la dose en profondeur. En fait, la technique des feux croisés et de la cyclothérapie permettent déjà dans le cadre de la radiothérapie classique d'atteindre des doses en profondeur telles que la peau cesse d'être un obstacle et que seule la tolérance des tissus profonds intervient pour fixer le choix de la dose tumeur. Il serait cependant, même avec ces techniques, difficile de délivrer des doses supérieures à 5000 rad à de larges volumes.

L'absence de réaction cutanée à d'autres avantages.

a) Elle permet de limiter le nombre de champs nécessaires pour traiter une tumoricides. Ceci peut être intéressant pour des tumeurs situées au voisinage de tumeur profonde, voir même d'atteindre avec deux ou trois champs des doses tissus radiosensibles qu'il sera possible d'épargner si l'on n'utilise qu'un petit nombre de champs.

b) Elle permet d'effectuer dans de meilleures conditions une intervention chirurgicale post-radiothérapique, l'incision se faisant à travers des tissus peu lésés.

c) Enfin, l'existence d'altérations cutanées même sévères, liées à des irradiations antérieures de la région, ne constitue plus un obstacle majeur à l'acte radiothérapique.

Cependant, l'absence de réactions superficielles ne doit pas faire oublier le risque de lésion dans la profondeur du tissu conjonctif sous-cutané. Ceci est surtout important avec le cobalt où l'on a déjà rapporté des réactions sévères en cuirasse, situées à quelques millimètres de profondeur apparaissant quelques mois après la fin d'une irradiation qui avait peu lésé la peau. Au point de vue pronostic ces lésions intéressant en bloc une épaisseur importante de tissus, sont plus grave, que des lésions superficielles.

*4.* Influence de la composition atomique.

Des trois types d'interaction entre photons et milieu matériel, seul l'effet Compton est pratiquement indépendant de la composition atomique du milieu traversé et ne dépend que de la densité.

Ce phénomène a une importance radiothérapique due à la différence de composition atomique entre les tissus mous et l'os, qui, du fait du phosphore et du calcium qu'il contient a un numéro atomique moyen nettement plus élevé. Quand le rôle joué par l'effet photo-électrique ou la formation de paire n'est pas relativement négligeable, il reçoit donc des doses supérieures à celles administrées aux tissus mous.

L'os et le cartilage posent à la radiothérapie classique un double problème. En absorbant plus de radiations que les tissus mous voisins, soumis à un flux de même intensité, ils sont exposés à des risques plus grands de radiolésions. De plus, lorsque l'os est interposé entre la peau et la tumeur (par exemple dans les cas d'une tumeur située derrière le maxillaire inférieur) une proportion importante du faisceau est absorbée dans l'os qui joue le rôle d'un écran pour les tissus situés derrière lui. A ces deux points de vue, les rayons X du Bétatron représentent un progrès certain.

a) Rôle d'écran. — Ce rôle existe dans tous les cas, puisque l'os ayant une densité supérieure à celle des tissus, atténue de toutes façons le faisceau de radiations plus que ne le ferait une épaisseur équivalente de muscles ou de tissus. Une façon simple d'apprécier le phénomène est de comparer la distance sur laquelle la dose diminue de moitié dans le cas de l'os et des tissus mous avec des faisceaux circulaires de 10 cm de diamètre.

Avec le 200 kV (CDA 1,5 mm de Cu), elle est dans les tissus mous d'environ 5 cm, de 2,5 cm dans l'os compact (densité 1,85) et de 4 cm pour l'os spongieux (densité 1,15). Avec les rayons gamma du cobalt, ces valeurs seraient de 11 cm, 6,5 cm et 10 cm. Enfin, pour les rayons X de 22 MeV du Bétatron, elles seraient de 22 cm, 12,5 cm et 20 cm.

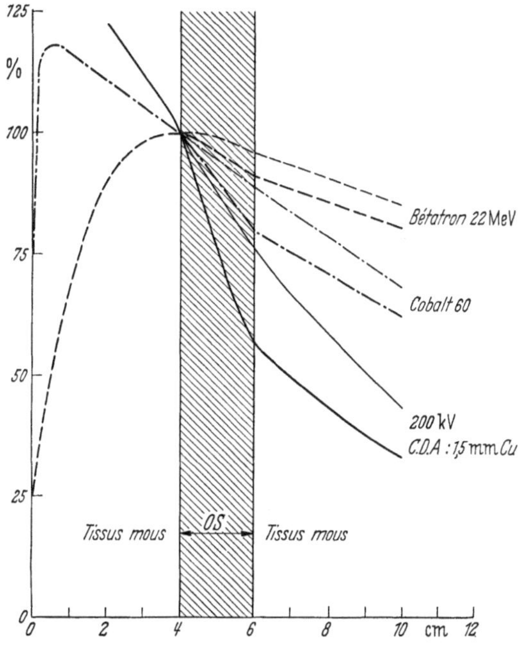


Fig. 11. Influence d'un écran «osseux». L'atténuation du faisceau à la traversée d'une épaisseur d'os entraîne une réduction sensible de la dose reçue par la tumeur située derriere lui lorsque l'énergie des photons est peu élevée (200 kV) tandis qu'elle est moins marquée aux énergies élevées

Un os de 2 cm d'épaisseur, si on le suppose entièrement composé d'os compact de densité 1,85, réduirait donc l'intensité du faisceau d'environ 45% dans le cas du 200 kV (cette réduction n'aurait été que de 23% s'il s'était agi de tissus mous), de 20% dans le cas du cobalt (11% dans le cas de tissus mous), et de 9% avec le Bétatron (4% dans le cas de tissus mous) (Fig. 11).

Ce phénomène joue un rôle thérapeutique important dans le cas des tumeurs de l'amygdale et de la base de la langue où la traversée du maxillaire inférieur est difficilement évitable. Il joue un rôle moindre dans le cas de tumeurs du petit bassin.

b) Risques d'ostéo-radionécroses.

Le danger de lésion osseuse dépend de la quantité de radiations reçues non par la substance osseuse proprement dite, mais par les tissus cellulaires inclus dans les os, en particulier par les vaisseaux nourriciers.

Même dans un os spongieux, ceux-ci traversent la corticale de l'os et s'y trouvent au contact d'un os compact, c'est donc le cas des tissus mous situés dans une cavité d'os compact qu'il convient de considérer, cavité dont le diamètre peut varier entre 10 et 50 $\mu$.

Pour des radiations d'énergie très faible, les électrons secondaires ont des trajectoires très courtes et le flux d'électrons ainsi que la dose varient très rapidement au sein d'une cavité de cette taille. Dès que l'énergie des photons atteint ou dépasse 200 kV on peut considérer que la dose y est pratiquement homogène, car le flux électronique ne varie pratiquement pas sur de telles distances.

Dans ce cas on peut admettre que les énergies respectivement absorbées par unité de volume d'os et de tissu mou situé dans une cavité creusée dans l'os dépendent du pouvoir relatif d'arrêt de l'os et des tissus vis-à-vis des électrons, donc pratiquement du rapport du nombre d'électrons.

Le rapport des doses absorbées par cm³ est donc:

$$\frac{\text{Nombre d'électrons par cm}^3 \text{ d'os}}{\text{Nombre d'électrons par cm}^3 \text{ d'eau}} = \frac{5,83 \cdot 10^{23}}{3,34 \cdot 10^{23}} = 1,77.$$

La dose absorbée par cm³ d'os est égale au produit de la dose absorbée par gramme d'os, par la densité (1,85).

La dose absorbée par cm³ de tissu mou inclu dans l'os est donc:

$$\frac{\text{Dose absorbée par cm}^3 \text{ d'os}}{1,77} = \text{dose absorbée par g. d'os} \times \frac{1,85}{1,77}$$
$$= 1,05 \times \text{dose absorbée par g. d'os.}$$

Les tissus mous, au contact de la cloison osseuse ne reçoivent dans un faisceau de photons de basse énergie les électrons secondaires provenant de cette cloison osseuse car ces électrons ont un faible parcours. Dans un faisceau de photons de haute énergie, la cavité est traversée par des électrons secondaires qui ont un long parcours et ont pris naissance dans un volume épais situé devant la cavité.

Dans le 1er cas le flux d'électrons correspond à la cloison osseuse donc à l'os compact et il est élevé; dans le 2ème cas le flux correspond à la composition moyenne du milieu antérieurement traversé, mélange d'os et de tissus mous, et il est plus faible.

A 200 kV les doses absorbées par du tissu mou situé dans l'os sont de 80% supérieures à celles reçues par des tissus mous hors de l'os. Cette différence s'estompe pour des énergies supérieures. Elle devrait s'élever à nouveau sous l'influence de la formation de paires mais ce phénomène reste relativement peu important aux énergies du Bétatron. De plus, le flux d'électrons est moins élevé puisqu'il faut dans ce cas considérer non la couche d'os compact au contact de laquelle est situé le tissu, mais l'ensemble des tissus voisins dans un rayon de plusieurs centimètres dont la densité moyenne ne dépasse pas 1,15 à 1,20 puisqu'il s'agit alors d'os spongieux.

Remarquons enfin, au point de vue clinique, que l'augmentation des risques liés à l'emploi du 200 kV, lors de l'irradiation de l'os, n'est pas compensée par une efficacité thérapeutique plus grande lors de l'irradiation des tumeurs primitives ou secondaires de l'os. En effet, le surdosage des tissus mous voisins de l'os est surtout important pour les couches de quelques dixièmes de millimètre d'épaisseur situées au contact des travées osseuses. Ce surdosage ne joue pas pour l'ensemble de la tumeur dont les dimensions sont beaucoup plus considérables et dont la composition chimique est voisine de celle des tissus mous. Ainsi, la dose absorbée par la tumeur osseuse est finalement plus faible aux basses énergies qu'aux énergies élevées, pour une même dose au tissu osseux normal.

L'importance clinique de ces faits est surtout notable pour les cancers O.R.L., ils ne jouent qu'un rôle beaucoup moins important pour les cancers situés dans les autres régions de l'organisme. Certes, l'os et le cartilage ont avec des rayons X de haute énergie, une tolérance supérieure à celle qui est la leur au 200 kV; mais si celle-ci facilite le traitement elle ne saurait faire négliger l'existence de l'os. Les risques d'ostéonécroses et de nécrose du cartilage sont atténués, mais subsistent et des accidents ont déjà été observés (WATSON).

**II.** Si les caractéristiques physiques des rayons X de haute énergie ont des avantages cliniques indiscutables, aucun de ceux-ci n'est de nature telle qu'il puisse modifier les données fondamentales de la radiothérapie. L'introduction des hautes énergies en thérapeutique ne saurait en aucune façon justifier des espoirs déraisonnables. Il ne s'agit pas et il ne saurait s'agir d'une révolution thérapeutique. On peut simplement espérer que les avantages pratiques qu'elles présentent permettent:

1) De traiter certains cancers considérés jusque là comme radiorésistants parce qu'il était impossible de leur délivrer des doses suffisantes.

2) D'améliorer les résultats statistiques obtenus dans les cancers radiocurables grâce, soit à une augmentation de la dose locale, soit à une extension du volume cible, soit enfin à une altération moindre des tissus voisins et de l'état général du malade.

3) De développer largement les indications de traitements palliatifs, les malades avec un état général médiocre pouvant bénéficier d'une irradiation sans mal des rayons.

L'expérience thérapeutique destinée à mettre ces résultats en évidence peut donc choisir:

1) Des cancers exceptionnellement radiocurables.

L'obtention d'un résultat, même très limité, suffirait dans ces cas à établir immédiatement de façon éclatante la valeur de la méthode.

2) Des cancers radiocurables et rechercher une amélioration des courbes de survie. Cette seconde méthode impose le traitement de séries importantes suivies pendant de longues années.

**III. Resultats cliniques.** Nous avons, depuis 1953, choisi l'étude de 4 types de cancers: cancer de l'oesophage, cancer du poumon dont le Dr. PIERQUIN expose par ailleurs les résultats, cancer de la vessie, cancer de la base de langue.

Le choix de ces localisations a été dicté par:

*1. Des raisons d'ordre physique.* — Le meilleur rendement de la dose en profondeur invitait à choisir une tumeur profonde.

L'homogénéité de la dose dans la section droite des faisceaux et la netteté des limites du faisceau qui permettent de délivrer la dose désirée à une tumeur profonde avec des faisceaux de faible section et au prix d'une irradiation minimum des organes voisins prennent tout leur intérêt dans une région telle que le médiastin.

*2. Des raisons d'ordre médical.* Le repérage géométrique de ces lésions est relativement précis, leur fréquence est assez élevée pour acquérir rapidement une expérience statistiquement valable, le mauvais résultat de la radiothérapie conventionnelle pour ces lésions donne permettre de conclure rapidement si les avantages physiques des hautes énergies permettent ou non d'améliorer des résultats cliniques immédiats et lointains.

6*

Technique du traitement. — Le principe du traitement est de délivrer une dose homogène à la totalité du volume tumoral en respectant au maximum les tissus sains environnants.

Ceci implique:

1) Un repérage exact du volume tumoral (volume cible).
2) Une détermination précise de la direction et de la dimension du faisceau.
3) Une parfaite contention du malade pendant l'irradiation.
4) Une vérification des volumes irradiés.

Pour satisfaire ces conditions, un corset en matière plastique (verplex) a été moulé sur chaque malade, qui est ainsi maintenu dans une position reproductible depuis les premières radiographies de centrage jusqu'à la dernière séance de son traitement. Le corset constitue d'autre part, un système de référence fidèle et commode pour le repérage de la lésion, et un support plus fixe que la peau pour le tracé des champs.

Nous avons également adopté la position assise qui donne plus de liberté dans l'orientation des faisceaux.

Pendant l'irradiation, le malade est maintenu dans un cadre rigide; l'alignement des faisceaux est réglé par centreurs lumineux sur les repères portés par le corset; l'immobilité du malade est vérifiée par une cellule photo-électrique à quadrants éclairés par le rétro-centreur.

Les opérations de centrage ont pour but de déterminer:
1) La position et les dimensions du volume cible.
2) La direction et les dimensions des faisceaux.
3) Les points d'entrée et de sortie de l'axe de chaque faisceau qui seront portés sur le corset.

**Oesophage.** Le contour interne de la lésion est visualisé par contraste baryté qui donne aussi les limites supérieures et inférieures de la tumeur. Le contour externe est habituellement inaccessible à l'examen radiologique et doit être construit arbitrairement sur les considérations anatomo-pathologiques.

La position de la tumeur est repérée par rapport à un système de référence de forme géométrique simple qui est solidarisé au corset; ce système est constitué de deux plaques de plexiglas portant deux grillages de plomb; deux clichés pris sous deux incidences différentes permettent de reconstituer la position dans l'espace du filet de baryte moulé par l'oesophage.

Nous avons utilisé 4 faisceaux à angle droit dont la direction est choisie de manière à éviter le rachis et à réduire au minimum le volume cardiaque traversé par les faisceaux.

La largeur du champ est de 6 à 7 cm environ au centre de la tumeur, et le volume cible est un prisme à base carrée de 6 à 7 cm de côté; sa hauteur a été, dans une première étape, de l'ordre de 17 cm. L'observation de récidives sus- ou sous-lésionnelles nous a conduits dans une deuxième étape, à augmenter la hauteur à 22 cm, englobant pratiquement tout l'oesophage thoracique.

La distribution des doses est satisfaisante, car elle remplit les conditions d'homogénéité et de protection des organes voisins, sauf pour le coeur où le volume cible englobe la zone postérieure.

Les malades reçoivent dans la règle, 1000 r à la tumeur par semaine, répartis en 5 séances de 200 r. Le débit de l'appareil est environ 50 r/mn à 100 cm. La distance de traitement varie de 90 à 110 cm selon les dimensions choisies pour le volume irradié.

Nous avons traité 56 malades de décembre 1953 à décembre 1955. Il n'y a eu aucune sélection particulière et cette série comprend presque tous les malades examinés à la consultation de l'Institut G. ROUSSY pendant cette période; les seules contre-indications absolues que nous ayions retenues étaient, à la bronchoscopie, une extension à la trachée et aux bronches, avec bourgeonnement ou voussure rigide de la paroi, ainsi que les métastases à distance, ganglionnaires ou viscérales évidentes.

L'état général n'a pas été pris en considération. Le traitement est d'ailleurs aidé par une diététique qui permet d'administrer sous forme fluide un régime équilibré au point de vue calorique et comportant une quantité suffisante de protides.

Sur ces 56 malades, 47 ont été jusqu'au terme de leur irradiation et ont reçu une dose tumorale de 5500 r à 6500 r. Les 9 autres cas sont évidemment considérés comme des échecs du traitement, mais ne sont pas retenus dans cette étude qui a pour objet d'analyser les résultats de l'irradiation à la dose de 5500 r et 6000 r.

Sur ces 47 malades, se trouvaient 44 hommes (âge moyen 59 ans) et 3 femmes (âge moyen 45 ans). Le temps perdu depuis le premier symptôme présente une valeur médiane[1] de 4,5 mois qui doit être ajoutée à la survie moyenne après traitement pour obtenir la survie totale depuis le début clinique de la maladie.

Le siège des lésions se distribue de façon à peu près égale entre la moitié supérieure de l'oesophage thoracique, la moitié supérieure, et à cheval sur les deux parties.

La longueur apparente se distribue de façon symétrique entre 2 à 14 cm avec une valeur moyenne de 5 à 6 cm qui représente 1/4 de la longueur de l'oesophage.

**Résultats.** *1. Immédiats.* Le traitement a été bien toléré du point de vue général et nous n'avons pas observé de «mal des rayons».

La formule sanguine ne montre pas de modifications importantes et 6 cas seulement ont présenté une leucopénie modérée et transitoire. La comparaison statistique des modifications sanguines avec celles observées à 200 kV montre qu'elles sont nettement diminuées pour une même dose tumorale (Fig. 3).

Les courbes de poids effectuées chez 40 malades ont montré au cours du traitement:

un état stationnaire dans 15 cas,

une augmentation de 2 à 4 kg dans 8 cas et de 10 kg dans 1 cas,

une diminution de 2 à 4 kg dans 7 cas, de 4 à 9 kg dans 8 cas et de 18 kg dans 1 cas.

L'amaigrissement a été en particulier observé dans 4 cas où des métastases se sont manifestées, et dans 10 cas où la dysphagie tenace n'a cédé qu'à la fin du traitement.

---

[1] La valeur médiane est plus significative que la moyenne en raison de la prédominance des petites valeurs de 1 à 5 mois

Au point de vue biologique nous avons, sur un groupe de 11 cancers de l'oesophage, trouvé les moyennes suivantes:

|  | Avant traitement par le Bétatron | Après traitement par le Bétatron |
|---|---|---|
| Protéines totales . . . . . | 80,4 | 74,3 |
| Sérum albumine . . . . . | 49,4 | 39,9 |
| Globuline . . . . . . . . | 32,2 | 33 |
| Rapport S/G . . . . . . | 1,5 | 1,3 |
| Cholestérol total . . . . . | 2,05 | 2,17 |
| Cholestérol estérifié . . . . | 1,3 | 1,3 |
| Rapport E/T . . . . . . | 65 | 64 |
| Lipides totaux . . . . . . | 5,7 | 6,06 |

Les différences entre les taux de protéines totales ne sont pas significatives statistiquement; par contre la différence des taux de sérum albumine est très significative, on peut donc affirmer l'existence d'une chute légère de leur taux pendant le traitement. Nous n'avons malheureusement pas pu étudier de façon comparative ce qui s'est produit pendant le traitement conventionnel.

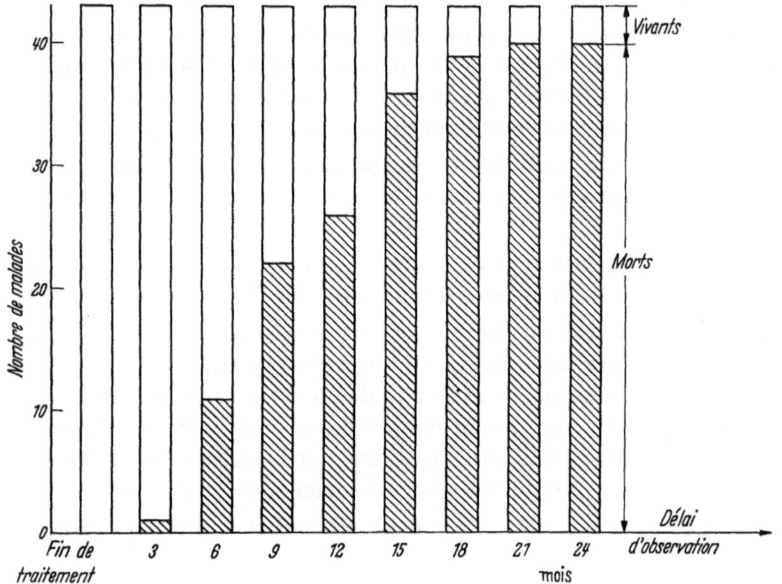


Fig. 12. Cancer de l'oesophage. 43 malades traités. Nombre de malades décédés après des délais d'observation de 3, 6, 9, 12, 15, 18, 21 et 24 mois

L'amélioration de l'état général et l'amélioration fonctionnelle ont été cependant la règle: la dysphagie régresse de façon très nette à partir de la dose de 2500 r, c'est-à-dire au cours de la 3ème semaine de l'irradiation, et parallèlement, la sténose oesophagienne apparaît moins serrée sur les clichés. La reprise de l'alimentation solide s'accompagne fréquemment de douleurs qui cèdent aux antibiotiques et aux pansements oesophagiens.

Les résultats immédiats sont donc assez favorables dans l'ensemble, malgré des accidents précoces de perforation médiastinale qui ont causé le décès de 3 malades dans les 8 semaines qui ont suivi l'irradiation.

*2. à distance.* 4 malades sont encore en vie (31, 29, 28 et 19 mois après la fin du traitement) dont 3 en guérison apparente.

La valeur médiane de la survie est de 10 mois, et pour 19 malades, elle dépasse 12 mois.

L'histogramme de la Fig. 12 représente la distribution des survies chez les 39 malades décédés (4 cas sont actuellement perdus de vue). Parmi ceux-ci, la persistance du néoplasme était certaine dans 10 cas et probable dans 7 cas: chez 16 malades décédés loin de l'Institut, l'état des lésions lors du décès n'a pu être affirmé.

Tableau 2. *Oesophage*

| Nom | Siège | Stade | Survie (mois) | Oesophage | | Etage sous-diaphrag. métastase | | Radiolésion au niveau | |
| --- | --- | --- | --- | --- | --- | --- | --- | --- | --- |
| | | | | Etat local | Récidive autre point | Ganglion. | Hé-patique | Myocarde | Poumons ou bronches |
| Cheen. | 1/3 inf. (7) | 3 | 3 | 0 | 0 | + | 0 | + | + |
| Bett. | 1/3 inf. (7) | 3 | 2 | + | | + | 0 | + | + |
| Pati. | 1/3 inf. (7) | 3 | 2 | 0 | + | + | + | + | + |
| Schne. | 1/3 moy. (6) | 3 | 5 | + | | 0 | 0 | 0 | ± |
| Bla. | 1/3 moy. (6) | 3 | 5 | + | | 0 | 0 | ± | ± |
| Papo. | 1/3 moy. (6) | 2 | 14 | + | | | | | |
| Koi. | 1/3 moy. (4) | 3 | 8 | + | | 0 | 0 | 0 | 0 |
| Four. | 1/3 inf. (6) | 3 | 4 | + | | + | 0 | 0 | 0 |
| Emel. | 1/3 inf. (7) | 13 | 12 | 0 | récidive lymph. juxta-oeso-phagienne | + | + | 0 | 0 |

*3. Accidents post-radiothérapiques.* Le décès est imputable à un accident post-radiothérapique dans 2 cas de fistules bronchiques, et 1 cas de myocardite survenue moins de 12 mois après l'irradiation.

Dans l'ensemble, nous avons observé parmi les manifestations précoces, 8 cas de perforation dont 3 dans les 2 mois qui suivirent l'irradiation.

L'examen autopsique a pu être effectué dans 9 cas: le tableau 2 montre que dans 3 cas seulement le tissu néoplasique avait disparu dans le volume-cible; on notera également la fréquence des lésions du myocarde (4 cas). Il faut noter à ce sujet que la surveillance du coeur par électrocardiographies effectuées systématiquement n'a apporté aucun élément de pronostic: parmi ces malades, le tracé s'est modifié dans un sens défavorable dans quelques cas, mais inversement une amélioration du tracé a été observée dans d'autres cas, sans qu'il y ait de corrélation avec les constatations nécropsiques.

**Cancers de la vessie.** Nous avons traité, depuis le 1er décembre 1955, 20 cancers de la vessie. Leur nombre est donc faible et le recul trop court pour pouvoir tirer de cette série autre chose qu'une impression clinique.

Techniquement, il est relativement facile, avec le Bétatron, d'irradier de façon homogène tout le petit bassin, soit dans un même temps la vessie, les lymphatiques et relais ganglionnaires. Nous utilisons pour cela 3 champs en T se recoupant au niveau du petit bassin. La dimension et la position des champs est déterminée en fonction des repères osseux.

Une dose de 7500 r délivrée en 8 semaines semble parfaitement supportée et nous n'avons observé aucun accident immédiat dû à l'irradiation de la région.

Des rectites peuvent s'observer, elles sont d'intensité modérée, surtout si l'on prend soin de limiter en arrière le champ et de ne pas irradier la face postérieure du rectum.

Les réactions de cystite sont généralement modérées et ne nécessitent pas d'interruption du traitement.

L'absence de mal des rayons et d'anémie, la conservation de l'état général sont les faits notables pour cette localisation comme pour les autres.

Au point de vue résultats lointains, nous ne disposons encore que de 2 résultats anatomiques obtenus par intervention chirurgicale. Dans 1 cas, opéré 4 mois après la fin de l'irradiation, il a été impossible de retrouver, malgré des coupes en séries, du tissu cancéreux au niveau de la vessie; dans l'autre cas, opéré 6 semaines après la fin du traitement, il subsistait du tissu cancéreux en évolution.

Ces irradiations larges du petit bassin soulèvent le problème difficile de la conservation de la fonction vésicale. Les vessies largement irradiées sont souvent sclérosées et intolérantes et des séquelles fonctionnelles sérieuses sont à redouter. De plus, le risque existe d'infection ascendante des urétères et des bassinets. C'est pourquoi on est en droit de soulever le problème de la mise au repos systématique de la vessie, avant irradiation, par abouchement des urétères à la peau. L'avenir montrera si cette précaution améliore les résultats immédiats et lointains du traitement.

Plusieurs autres auteurs ont eu l'occasion d'irradier des vessies à doses élevées avec des générateurs de haute énergie; les résultats dans l'ensemble, confirment notre première impression: bons résultats immédiats, amélioration des résultats lointains, sans modification profonde du pronostic.

**Cancers de la base de langue.** La situation profonde des cancers de la base de la langue, leur extension locorégionale souvent très étendue dès le premier examen, la nécessité où l'on se trouve de traverser les maxillaires inférieurs pour les irradier par voie transcutanée expliquent que ces cancers constituent une indication intéressante pour les radiations de haute énergie.

La plupart des 25 cas traités depuis 1954 avaient une extension locale et ganglionnaire très considérable. Ceci explique peut-être que malgré les doses locales qui ont atteint 9000 à 7500 rads, étalées sur 8 semaines, les résultats ont été médiocres et les récidives locales ou ganglionnaires nombreuses. Nos quelques résultats actuellement favorables (plus d'un an de survie sans récidive apparente) sont des cas où la tumeur locale ne dépassait pas les limites de la base de la langue et où le curage ganglionnaire chirurgical a pu être effectué dans les suites immédiates de l'irradiation.

Néanmoins, même dans les cas les plus avancés, les résultats palliatifs immédiats ont été remarquables: régression ou disparition totale des lésions locales, suppression des troubles fonctionnels.

La surveillance des cas traités a par ailleurs montré que l'intensité et la date d'apparition de la radiomucite étaient comparables à ce que l'on observe en cours de traitement avec les énergies dites classiques.

Pour cette localisation encore, l'extension du volume des territoires traités et la possibilité de leur délivrer des doses importantes semble donc intéressante. Mais si les résultats immédiats sont excellents, les résultats lointains sont moins satisfaisants. Seule la surveillance de séries importantes de malades permettra de savoir si une amélioration notable du pronostic résulte de l'emploi de ces nouveaux agents d'irradiation.

**Conclusions.** Pour toutes les localisations étudiées les résultats immédiats sont excellents et l'utilisation des radiations de haute énergie représente de ce point de vue, un progrès considérable.

La tolérance locale et générale est bonne même pour de fortes doses dans de grands volumes. L'absence de mal des rayons, d'anémie, la conservation de l'état général du malade facilitent la tâche du radiothérapeute et la réalisation du protocole de traitement qu'il s'était fixé. Un effet palliatif est toujours obtenu et le malade se trouve à la fin du traitement toujours amélioré.

Il n'est pas encore possible d'évaluer de façon définitive la conséquence de ces faits sur les résultats lointains. Ceux-ci en effet demeurent modestes et quelque soit la localisation, la fréquence de récidive locale ou à distance reste considérable. Il est néanmoins possible qu'ils soient supérieurs à ceux obtenus avec le 200 kV. Il faut attendre les résultats statistiques pour en juger.

De plus, si les réactions immédiates des tissus sains avoisinants la tumeur sont moins violentes, rien n'autorise à affirmer qu'à dose égale la fréquence des radiolésions séquelles est moins grande.

Ainsi avec les hautes énergies comme avec la radiothérapie classique, il n'existe pas de méthode idéale capable de stériliser toutes les tumeurs sans provoquer d'accidents au niveau des tissus sains. A cette notion trop optimiste il faut substituer celle de dose optimum susceptible de stériliser le plus grand nombre de tissus en provoquant le moins d'accidents. De la même façon il n'existe pas de volume-cible idéal et il faut rechercher le volume optimum.

Seules des études patientes et la comparaison de séries étendues de malades comparables permettront de déterminer les meilleures méthodes de traitement.

L'avènement des méthodes utilisant les radiations de haute énergie facilitent la tâche du radiothérapeute et le libèrent de nombreux problèmes techniques. Les problèmes fondamentaux de la radiothérapie conservent cependant leurs caractères et il ne faut pas attendre de ces méthodes des miracles

## Bibliographie

Pierquin, B., A. Dutreix, J. Dutreix et M. Tubiana: Conditions techniques d'irradiation des cancers bronchiques par le Bétatron. J. de Radiol. **38**, 504 (1957).

Surmont, J., M. Tubiana, B. Pierquin, C. Lalanne et A. Dutreix: Considérations sur quelques procédés de centrage et de dosimétrie en radiothérapie transcutanée. J. de Radiol. **36**, 580 (1955).

Tubiana, M., J. Dutreix, B. Pierquin et P. Jockey: Conditions pratiques d'utilisation d'un Bétatron de 22 MeV. J. de Radiol. **36**, 507 (1955).

— B. Pierquin, B. Dutreix et P. Jockey: Discussion Symposium Betatron. Radiol. Clin. **24**, 371 (1955).

# Das Kilocurie-Kobaltgerät des Radiumhemmet, genannt das Gammatron*

## Konstruktionsprinzipien, Behandlungsabteilung, Einstellungstechnik und Strahlendosismessungen

Von

S. HULTBERG, O. DAHL, R. THORAEUS und K. J. VIKTERLÖF, Stockholm

Apparate für Behandlung mit Strahlen von hoher Energie hat es in der einen oder anderen Form seit mehr als 30 Jahren gegeben. Es mag deshalb überraschen, daß nicht schon früher Millionenvoltgeräte für Behandlungsabstände von der Größenordnung 50—100 cm am Radiumhemmet installiert worden sind. Die Entwicklung der ausländischen Apparatekonstruktionen ist natürlich genau verfolgt worden. Es wurden dabei Geräte mit solchen Eigenschaften gesucht, mit denen sowohl die Behandlung als auch die damit verbundenen Kontrollmaßnahmen ebenso sicher und geschmeidig durchgeführt werden konnten wie mit den bisher üblichen Behandlungseinheiten. Andernfalls bestand die Gefahr, daß die Vorteile, die in strahlenphysikalischer und klinischer Hinsicht durch Übergang zur Millionenvolt-Therapie gewonnen werden konnten, in hohem Grade zufolge der verschlechterten praktischen Arbeitsbedingungen verlorengingen.

Unsere damals aufgestellten Forderungen können folgendermaßen zusammengefaßt werden:

1. Die Strahlung soll eine Tiefendosiswirkung haben, die derjenigen von einigen Millionen Volt Röntgenstrahlen entspricht.

2. In klinischer Praxis soll das Gerät genauso leicht zu handhaben sein wie ein gewöhnlicher 200 kV-Apparat.

3. Bewegungsbestrahlung soll durchgeführt werden können.

4. Um kostspielige Verstärkungen des Fußbodens sowie Neubauten zu vermeiden, darf das Gewicht des Gerätes nicht ein paar Tonnen überschreiten und der Raumbedarf soll möglichst klein sein.

5. Felder verschiedener Größe und Form sollen einfach erzielt werden können.

Da wir keinen Millionenvolt-Apparat fanden, der allen diesen Anforderungen entsprach, beschlossen wir, eine auf eigene Erfahrungen basierte Behandlungseinheit auszuarbeiten. In dem Energiebereich, einigen wenigen Millionen Volt Röntgenstrahlung entsprechend, wo die wesentlichen Vorteile der MeV-Therapie erreicht werden können, kann man durch Verwendung energiereicher $\gamma$-Strahlung von Isotopen mit hoher spezifischer Aktivität praktische Therapie mit hoher Betriebssicherheit, geringem Bedarf an Bedienungspersonal und verhältnismäßig guter Betriebsökonomie durchführen. Es wurde deshalb beschlossen, ein Kobaltgerät in der Kilocurie-Klasse zu konstruieren.

---

* Aus dem Radiumhemmet (Vorstand: Prof. Dr. S. HULTBERG) und dem Radiophysikalischen Institut (Vorstand: Prof. Dr. R. M. SIEVERT), Karolinska Sjukhuset, Stockholm

Seit dem Sommer 1954 haben wir zusammen mit Vertretern der *Siemens-Reiniger-Werke* daran gearbeitet, eine geeignete Konstruktion zu finden. Das Gerät ist auf der Grundlage der von uns definierten klinischen und physikalischen Bedingungen ausgeformt worden. Betreffend der Ausführung von Einzelheiten hat während der ganzen Konstruktionszeit eine enge Zusammenarbeit stattgefunden. Unserer Meinung nach haben die *SRW* unsere Intentionen gut verwirklicht.

**Konstruktionsprinzipien.** Als Schutzmaterial im Strahlerkopf des Gamma-trons haben wir eine 95%ige Wolframlegierung mit einem Volumgewicht von

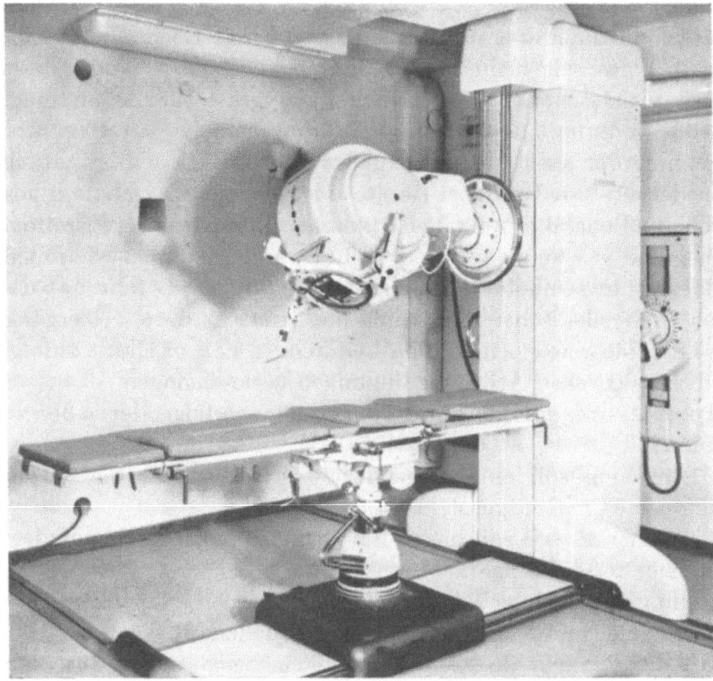

Fig. 1. Das Gammatron mit dem Behandlungstisch

etwa 17 gewählt. In mehreren früheren Kobaltgeräten hat man wenigstens teil-weise Blei mit einem spezifischen Gewicht von nur 11 als Strahlenschutzmaterial verwendet. Die Wolframlegierung ermöglichte bei ausreichendem Strahlenschutz eine erhebliche Reduktion sowohl der Größe als auch des Gewichtes des Kopfes. Das Gewicht des Strahlerkopfes beträgt nur etwa 500 kg. Das *SRW*-Betatron-stativ konnte dadurch verwendet werden. Der ganze Apparat wurde damit verhältnismäßig klein (Fig. 1).

Mit dem genannten Stativ kann man auf einfache Weise Pendelbestrahlung sowie die praktisch erforderlichen Einstellungen bei festen Feldern vornehmen. Der Strahlerkopf ist sehr gut beweglich und es ist leicht, nahe heranzukommen, um die Einstellungen vorzunehmen und die Handräder und übrigen Anordnungen zu bedienen.

Das Gewicht des ganzen Gammatrons beträgt nur etwa 1,5 t. Es ist darum nicht erforderlich, umfassende Verstärkungen des Fußbodens für die Apparatur selbst vorzunehmen. Der notwendige Strahlenschutz kann natürlich gewisse Verstärkungen erfordern. Um den Strahlerkopf mit dem $Co^{60}$-Präparat laden zu können, sind dagegen keine besonderen Verstärkungen erforderlich. Der Transportbehälter, in dem die Quelle geliefert wird, wiegt zwar etwas mehr als 2 t. Wenn die Konstruktion des Fußbodens es nicht erlaubt, bei der Ladung den Transportbehälter neben dem Gammatron aufzustellen, kann man aber statt

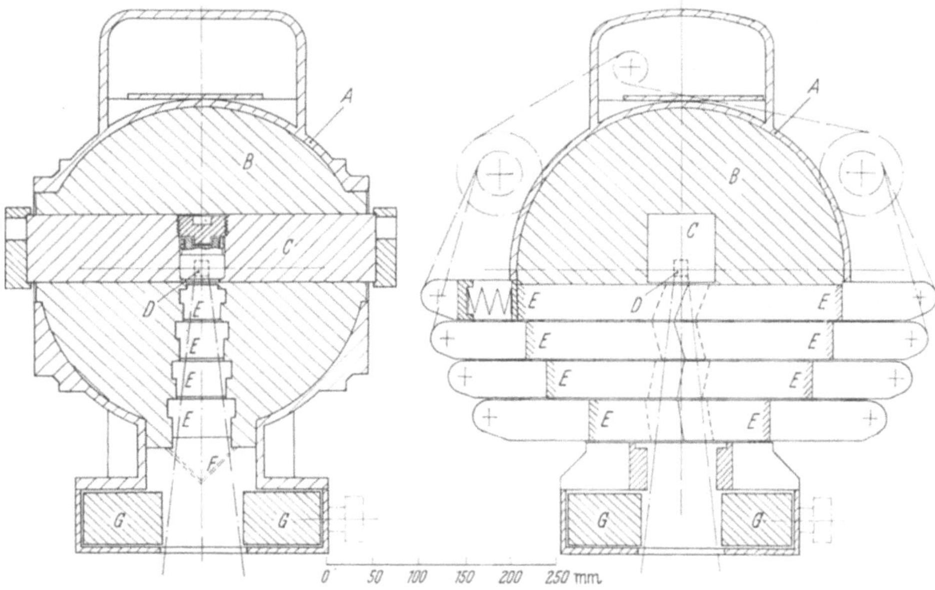


Fig. 2. Zwei gegeneinander winkelrechte schematische Schnitte durch den Strahlerkopf. *A* Das Stahlgehäuse. *B* Der Strahlenschutzkörper. *C* Die Einsteckkassette. *D* Die $Co^{60}$-Strahlenquelle. *E* Der Verschlußmechanismus. *F* Das Spiegelsystem des Lichtvisiers. *G* Die Lamellenblende

dessen den Strahlerkopf, dessen Gewicht ja verhältnismäßig niedrig ist und der leicht vom Stativ abgenommen werden kann, in einen für die Durchführung der Ladung geeigneten Raum bringen. Es hat sich auch gezeigt, daß dieses Verfahren die Ladung selbst wesentlich erleichtert.

Im Strahlerkopf (Fig. 2) bestehen sowohl das fest eingebaute Strahlenschutzmaterial als auch die Einsteckkassette, der Verschlußmechanismus und die Lamellen der Blendenanordnung aus der erwähnten Wolframlegierung. Die Strahlungsquelle, deren Ausdehnung winkelrecht gegen die Achse des Strahlenbündels 2 cm ist, besteht aus millimetergroßen $Co^{60}$-Perlen, die gemäß internationalem Standard in einer Kapsel gasdicht eingeschlossen sind. Die Quelle ist von der *Atomic Energy of Canada Ltd.* geliefert worden.

Der Verschlußmechanismus (Fig. 2) besteht aus vier unabhängig voneinander arbeitenden Paaren von Vierkantstäben. Der Mechanismus ist so konstruiert, daß sich die Stäbe sowohl am Ende einer Behandlung als auch bei Störungen irgendwelcher Art, z. B. Stromunterbrechung oder Behinderung der Pendelbewegung, durch Federeinwirkung automatisch schließen. Mit Hilfe von zwei Signallampen auf dem Bedienungstisch kann man kontrollieren, ob sich sämtliche Stäbe in der offenen bzw. geschlossenen Lage befinden.

Bei der Konstruktion der Blendenanordnung (Fig. 2 u. 3) haben wir ein System vermeiden wollen, bei dem man gezwungen ist, zwecks Änderungen der

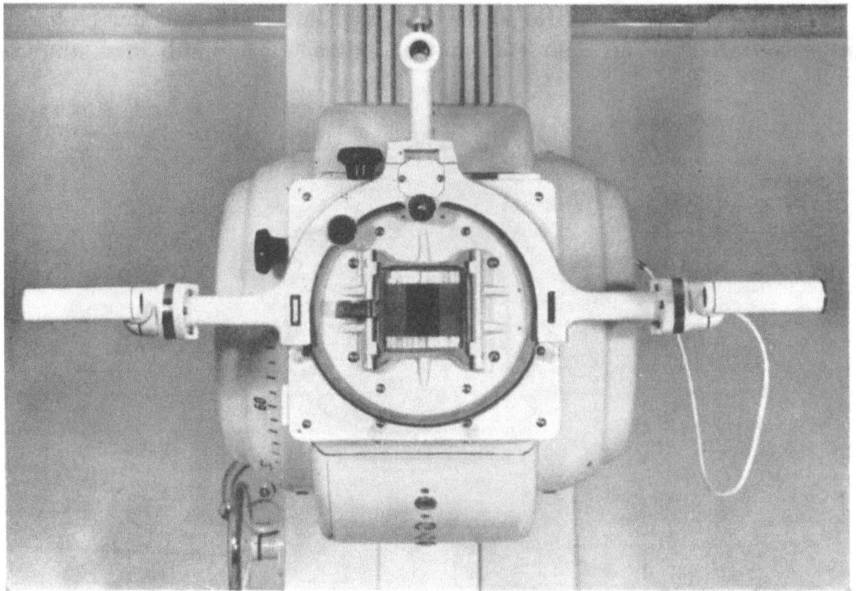

Fig. 3. Ein Frontalbild des Strahlerkopfes mit der Lamellenblende

Feldgröße Tubusse auszutauschen. Wir haben weiterhin angestrebt, die Möglichkeit zu erhalten, rechteckige Felder wahlfreier Größen und bei Bedarf auch mehr unregelmäßige Felder einstellen zu können. Die gewählte Konstruktion der

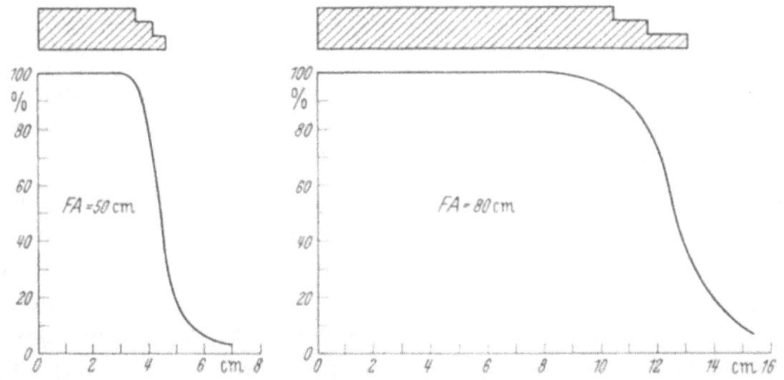


Fig. 4. Die Korrelation zwischen den durch das Lichtvisier angezeigten Feldgrenzen und dem frei in Luft gemessenen Dosisabfall bei zwei verschiedenen geometrischen Verhältnissen

Blende ermöglicht die einfache Einstellung gewünschter Bestrahlungsfelder von maximal 14 × 14 cm Größe auf 50 cm Focusabstand. Die Lamellen der Blende lassen weniger als 1% der Primärstrahlung hindurch.

Im Strahlerkopf ist ein Lichtvisier (Fig. 2) eingebaut, das auf der Oberfläche des Bestrahlungsobjektes Form und Größe des Bestrahlungsfeldes sowie den Randschatteneffekt mit guter Lichtstärke anzeigt. Wie aus Fig. 4 hervorgeht, ist der Dosisabfall, frei in Luft gemessen, zwischen etwa 100 und 50% optisch sichtbar.

Noch zwei Lichtvisiere sind, wie aus Fig. 1 u. 3 hervorgeht, seitlich am Strahlerkopf befestigt. Sie sind vor allem für eine genauere Reproduzierung der Einstellung von Behandlung zu Behandlung bei dem gleichen Patient vorgesehen.

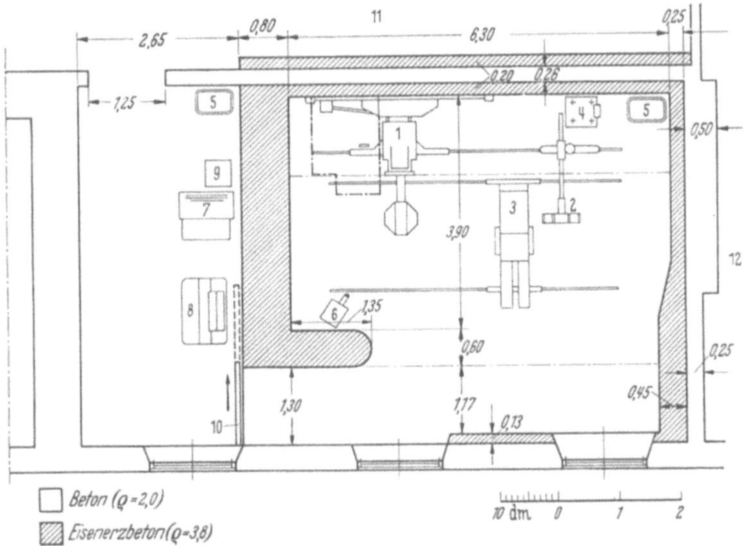


Fig. 5. Grundriß der Behandlungsabteilung. *1.* Das Gammatron. *2.* Das Lokalisations- und Einstellungsgerät. *3.* Der Lagerungstisch. *4.* Der Spannungserzeuger für das Lokalisationsgerät. *5.* Waschbecken. *6.* Die Fernsehkamera. *7.* Der Schalttisch des Gammatrons. *8.* Der Schalttisch des Lokalisationsgerätes. *9.* Der Fernsehempfänger. *10.* Schiebetür. *11.* Flur. *12.* Flur

Mit Hinsicht auf den Strahlenschutz ist das Gammatron des Radiumhemmet, das mit 1035 effektiven Curie Co⁶⁰ geladen worden ist, umfassenden Messungen unterworfen worden. Dabei konnte festgestellt werden, daß die Dosisleistungen auf 1 m Focusabstand überall unter dem Wert liegen, welcher im *National Bureau of Standards Handbook 1954* als Grenze angegeben wird.

**Behandlungsabteilung.** Die Planlösung der Behandlungsabteilung ist aus Fig. 5 ersichtlich. Es handelt sich um die Einrichtung eines schon vorhandenen Raumes. Sämtliche Wand- und Deckenverstärkungen sind aus vibriertem Eisenerzbeton ausgeführt worden. Es war dadurch möglich, ein Strahlenschutzmaterial mit einem Volumengewicht von 3,8 herzustellen. Da die Wand zwischen dem Behandlungsraum und dem Bedienungsraum von Primärstrahlung getroffen werden kann, ist diese Wand entsprechend dick gebaut worden. Die Einsicht vom Bedienungsraum in den Behandlungsraum geschieht deshalb mit Hilfe einer Fernsehanlage. Das Gammatron und das Lokalisations- und Einstellungsgerät sind, wie ersichtlich, an derselben Längswand des Behandlungsraumes plaziert worden.

Im Bedienungsraum sind die Schalttische für das Gammatron und das Lokalisationsgerät sowie der Fernsehempfänger untergebracht. Während der Behandlung steht man mit Hilfe eines Lautsprechers mit dem Patienten in Verbindung.

Eine Strahlenschutzuntersuchung der ganzen Behandlungsabteilung hat ein befriedigendes Resultat gegeben.

**Einstellungstechnik.** Bei der Ausarbeitung unserer Apparatur haben wir den am Radiumhemmet üblichen Grundsatz befolgt, bei der Strahlenbehandlung von

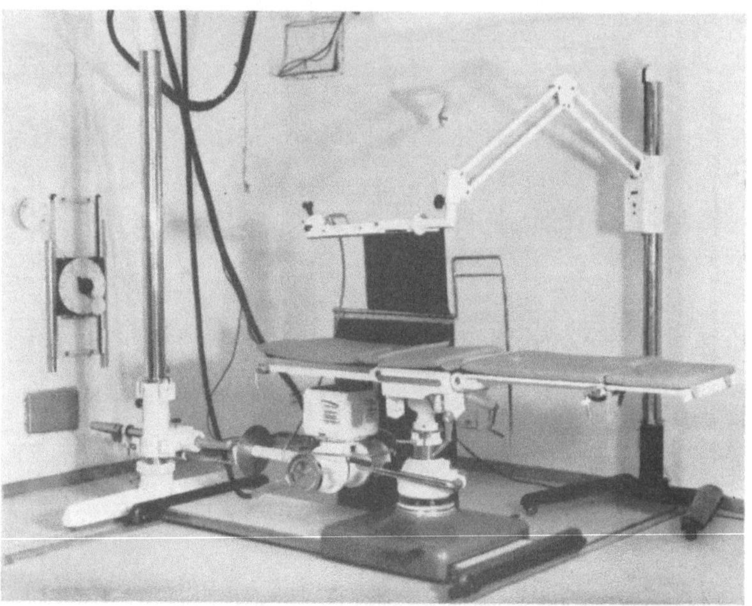

Fig. 6. Das Lokalisations- und Einstellungsgerät

tiefliegenden, abgegrenzten Tumoren möglichst die Einstellungskontrolle mit Röntgendurchleuchtung durchzuführen, nachdem der Patient für die Behandlung in adäquater Lage plaziert worden ist. Ein Patient, der auf einem Tisch für Pendelbestrahlung gut plaziert, gestützt und wenn notwendig fixiert worden ist, behält im allgemeinen seine Lage ausreichend bei. Wir finden deshalb, daß eine genaue Durchleuchtungskontrolle zu Beginn jeder einzelnen Pendelbestrahlung ausreichend ist.

Für diese Durchleuchtungskontrolle haben wir angestrebt, gute röntgen-diagnostische Arbeitsverhältnisse zu erhalten. Wir wünschen dabei in erster Linie einen kleinen Brennfleck, einen kurzen Abstand zwischen Patient und Durch-leuchtungsschirm, eine veränderliche Feldgröße und einen ausreichenden Strahlenschutz.

In Zusammenarbeit mit Civ. ing. Alvin von *A. B. Elema* haben wir unser Lokalisations- und Einstellungsgerät (Fig. 6) konstruiert. Im Prinzip ist dieses so ausgeführt worden, daß es in Hinsicht auf die Ebene, in der sich der Brennfleck bewegen kann, und auf die Strahlenrichtungen, die zustande gebracht werden können, mit dem Gammatron vollständig kongruent ist. Um die Überführung einer an dem Lokalisationsapparat vorgenommenen Einstellung auf das Gammatron zu vereinfachen, sind beide Apparate mit vollständig gleichzeigenden Abstands- und Winkelskalen sowie Lichtvisieren versehen. Auf einem auf dem Fußboden angeordneten Schienensystem kann der Behandlungstisch mit dem darauf gelagerten Patienten von dem einen Apparat zu dem anderen parallel verschoben sowie in kongruenten Punkten vor jedem Gerät mit Hilfe von Zentimeterskalen eingestellt werden. Bei Pendelbestrahlung machen wir im allgemeinen die Einstellungsdurchleuchtung mit der Strahlenrichtung von unten, wie aus Fig. 6 hervorgeht.

An dem Behandlungstisch der *SRW* haben wir weiter einige Änderungen machen lassen. Durch diese Änderungen und durch einen neukonstruierten Halter für den Durchleuchtungsschirm (Fig. 6) ist es möglich geworden, innerhalb weiter Grenzen verschiedene Körperteile mit verschiedenen Strahlenrichtungen zu durchleuchten. Ohne sich unnötigerweise Strahlungen auszusetzen, kann man die Feldgröße mit Hilfe einer verstellbaren Blende, die vom Durchleuchtungsschirm aus bedient wird, so ändern, daß man optimale Durchleuchtungsbedingungen erhält. Der Strahlenschutz wird durch Bleiglas am Leuchtschirm, durch am Behandlungstisch herabhängende Bleigummigardinen, und schließlich durch am Leuchtschirm befestigte Bleigummilappen gewährleistet (Fig. 6).

Durch die genannte Art der Einstellungskontrolle wird die Zeit, die sich das Personal am Behandlungsgerät aufhalten muß, wesentlich verkürzt, was im Hinblick auf den Strahlenschutz von großer Bedeutung sein dürfte.

**Strahlendosismessungen.** In Zusammenhang mit der Planung der Gammatronanlage des Radiumhemmet vor etwa drei Jahren wurde die Frage der Messung der Gammastrahlung des $Co^{60}$ für uns äußerst aktuell. Diese Frage ist von fundamentaler Bedeutung für die ganze Arbeit und forderte darum eine schnelle Lösung, welche folgendermaßen ermöglicht wurde.

In einer besonderen Arbeit wurde 1953—1956 ein Vergleich der in Schweden seit 1931 im Betrieb gewesenen Standardkammer mit den englischen, westdeutschen und amerikanischen Standardkammern durchgeführt. Eine Übereinstimmung binnen $\pm 0,5\%$ wurde dabei gefunden. Der Vergleich wurde sowohl mit Hilfe einer transportablen Standardkammer-Einheit als auch einem Substandard, mit einer Fingerhutkammer versehen, ausgeführt.

Dieser Substandard ist schon seit 1946 im Gebrauch und hat dabei gute Stabilität und Genauigkeit gezeigt. Er hat auch beim Vergleich mit den vier nationalen Standardkammern gute Übereinstimmung gezeigt. Die letzte dieser Eichungen wurde bei einem Besuch am *National Bureau of Standards* in Washington, USA, im Februar 1956, mit Strahlenqualitäten zwischen HWS 0,06 und 2,0 mm Cu ausgeführt. Bei dieser Gelegenheit konnten auch einige Eichwerte mit $Co^{60}$-Strahlung erhalten werden. Um direkten Anschluß an die übrigen Eichreihen zu erhalten, wurde die $Co^{60}$-Eichung in gleichartiger Weise wie die Eichung mit Röntgenstrahlen ausgeführt. Um Elektronengleichgewicht in der Ionisationskammer zu erhalten, war aber eine 3,5 mm dicke, fingerhutähnliche Hülle aus Graphit über die Fingerhutkammer geschoben.

Alle Strahlendosismessungen am Gammatron des Radiumhemmet sind entweder mit dem genannten Substandard ausgeführt worden oder mit anderen

Instrumenten, die gegen diesen Substandard geeicht worden sind. Sämtliche Instrumente zeigten für den in Frage kommenden Zweck zufriedenstellende Meßeigenschaften.

Während der Durchführung der Bauarbeiten für die Gammatronanlage wurde der Gammatronkopf nebst dem Schalttisch an einem anderen Ort provisorisch aufgestellt. Dadurch konnten die erforderlichen, umfassenden Strahlungsmessungen wesentlich früher begonnen werden, als es andernfalls möglich gewesen wäre. Fig. 7 zeigt eine Photographie der entsprechenden Meßaufstellung, bei der das horizontal eingestellte Strahlenbündel gegen ein Wasserphantom gerichtet ist, dessen Wände aus 2 mm Cellon bestehen.

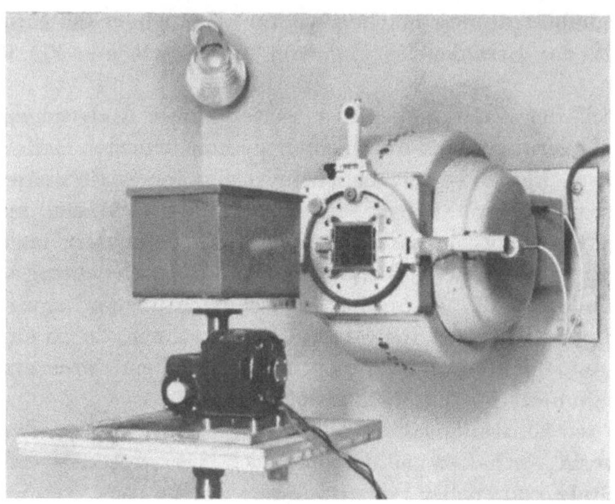

Fig. 7. Die Meßaufstellung für die Tiefendosenbestimmung

Eine Zusammenstellung einiger der wichtigsten Resultate der Standardmessungen wird zuerst gegeben.

Tab. 1 zeigt einige Werte der Strahlungsintensität, frei in Luft und an der Oberfläche des Wasserphantoms gemessen. Sämtliche Werte beziehen sich auf die maximale Feldgröße. Die Oberflächenwerte wurden erhalten, indem sich das Kammerzentrum 5,5 mm oberhalb

Tabelle 1

| Focusabstand cm | Strahlungsintensität bei größter Blendenöffnung in r/min | | Beziehung zwischen Oberflächenwert und freier Luft-Wert |
|---|---|---|---|
| | Frei in Luft | An der Oberfläche des Wasserphantoms. Kammer nicht eingetaucht | |
| 50 | 79,0 | 85,0 | 1,08 |
| 60 | 53,7 | 59,1 | 1,10 |
| 70 | 39,7 | 43,2 | 1,09 |
| 80 | 30,0 | 32,7 | 1,09 |
| 100 | 19,0 | 20,9 | 1,10 |

der Oberfläche befand, wobei aber die Kammer nicht in das Wasser eingetaucht war, sondern nur die Wasserfläche berührte.

Als Resultat einer HWS-Bestimmung wurden die Werte 14,5 mm Cu ($\varrho = 8,93$) und 10,0 mm Blei ($\varrho = 11,39$) erhalten.

Wie unsere Messungen in Fig. 8 zeigen, fängt bei Tiefen von etwa 4 cm die Überlegenheit der $Co^{60}$-Gammastrahlung bezüglich der Tiefenwirkung an deutlich zu werden und tritt bei größeren Tiefen immer mehr in Erscheinung. Nach früheren Literaturangaben ist die $Co^{60}$-Strahlung in dieser Hinsicht einer 3 MeV-Röntgenstrahlung gleichwertig.

In Tab. 2 wird das Resultat unserer Tiefendosenmessungen am Gammatron mit dem entsprechenden kanadischen, an der *Saskatoon-Universität* erhaltenen, verglichen. Um einen direkten Vergleich zu erhalten, haben wir hier in Übereinstimmung mit dem kanadischen System die Tiefendosen 5 mm unter der Oberfläche als 100% bezeichnet. Wie ersichtlich, ist die Übereinstimmung sehr gut.

Fig. 9 zeigt, wie man im Körper beim Gammatron infolge der bei Co[60]-Strahlung überwiegend nach vorn gerichteten Streustrahlung trotz der etwas ungünstigeren Strahlengeometrie flachere und am Feldrand steiler abfallende Dosiskurven erhält.

Tabelle 2. *Vergleich der prozentualen Tiefendosen in Wasser bei einem Focus-Oberflächenabstand von 50 cm und einer Oberflächen-Feldgröße von 10 × 10 cm*

| Tiefe | Messungen am Radiumhemmet | Kanadische Messungen |
|---|---|---|
| cm | % | % |
| 0,5 | 100 | 100 |
| 1 | 97,5 | 97,5 |
| 2 | 91,5 | 91,4 |
| 3 | 85,3 | 85,4 |
| 4 | 79,7 | 79,6 |
| 6 | 68,7 | 68,6 |
| 10 | 49,3 | 49,7 |
| 13 | 38,3 | 39,0 |
| 16 | 30,5 | 30,6 |

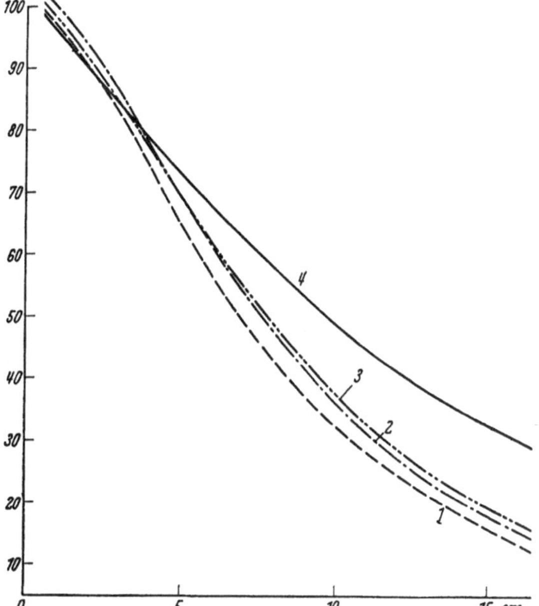

Fig. 8. Tiefendosenkurven, welche sich auf einen Focus-Oberflächenabstand von 50 cm und eine Feldgröße von 10 × 10 cm beziehen. Gemessene Oberflächendosen sind mit 100% bezeichnet. Kurve 1: 175 kV, 1,0 mm Cu HWS. Kurve 2: 200 kV, 2,0 mm Cu HWS. Kurve 3: 400 kV, 4,0 mm Cu HWS. Kurve 4: Co[60], 14,5 mm Cu HWS

Um in klinischer Arbeit sich die erforderliche Auffassung von der wahrscheinlichen Dosisverteilung bilden zu können, muß man über eine genügend große

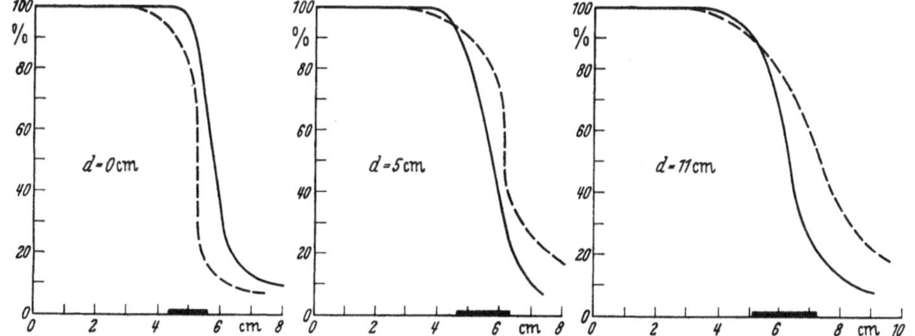

Fig. 9. Die Dosisverteilung senkrecht zur Strahlenrichtung, in verschiedenen Tiefen an einem „Mix D"-Phantom gemessen. Focus-Oberflächenabstand = 50 cm. Gestrichelte Linie — — — — 200 kV. (Ausblendung an der Oberfläche. Focusgröße = 1,2 cm.) Ausgezogene Linie ——— Co[60]. (Ausblendung in 30 cm Abstand vom Focus, Focusgröße = 2,0 cm.) Das Halbschattengebiet bei Co[60] ist an der Abszisse angedeutet

Anzahl von Isodosen verfügen. Um die erforderliche Anzahl von Dosisverteilungen hinreichend schnell ausmessen zu können, muß man aber über eine Ausrüstung verfügen, die schnelle und zuverlässige Dosismessungen ermöglicht. Bei den Messungen der folgenden Dosisverteilungen haben wir die Sievert-Kondensatorkammern verwendet. Dank ihrer kleinen Dimensionen ermöglichen sie eine

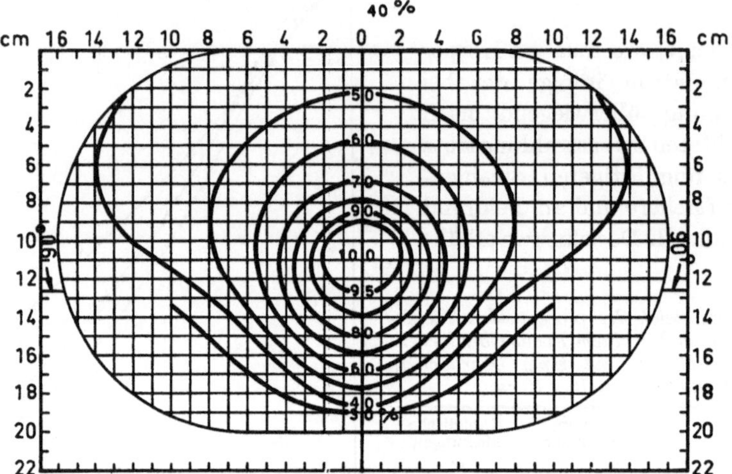

Fig. 10. Die Dosisverteilung bei Pendelbestrahlung mit Co⁶⁰ (HWS = 14,5 mm Cu), in der Pendelebene gemessen. Abstand Focus—Oberfläche = 50 cm. Feldgröße an der Oberfläche = 6 × 10 cm. Pendelwinkel = 180°. Tiefe der Pendelachse = 12,5 cm

fast punktförmige Dosisbestimmung. Die Kammern können unter sich gleichwertig ausgeführt werden. Bei den 30 Meßkammern, die verwendet wurden,

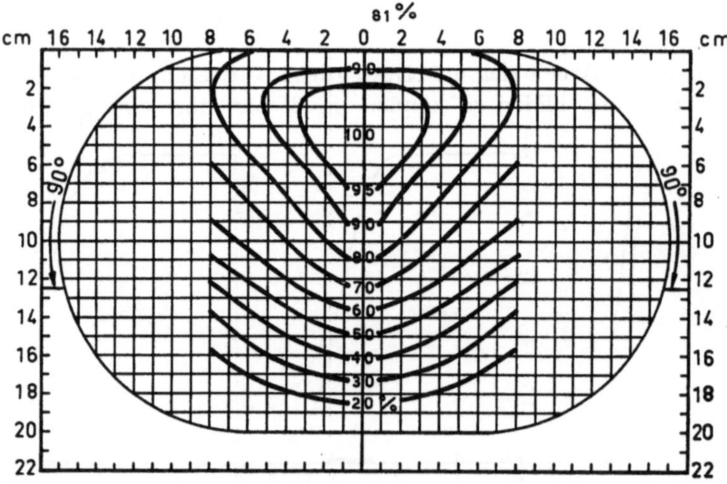

Fig. 11. Die Dosisverteilung bei Pendelbestrahlung mit 200 kV-Röntgenstrahlung (HWS = 1,1 mm Cu), in der Pendelebene gemessen. Bestrahlungsbedingungen wie die in Fig. 10 angegebenen

waren die Standardabweichungen geringer als ±2%. Um bei Co⁶⁰-Strahlung auch bei Messungen an der Oberfläche Elektronengleichgewicht zu erhalten, wurden die Kammern in Plexiglastuben (10 mm ⌀) gebracht.

Um bei Pendelbestrahlung eine Kenntnis der Dosisverteilung zu erhalten, genügt es im allgemeinen, die Verteilung in der Pendelebene zu bestimmen. Bei 200 kV haben wir deshalb an sieben „Mix D"-Phantomen verschiedener Größe etwa 400 Dosisverteilungen in der Pendelebene gemessen. Solche Verteilungen werden jetzt auch bei $Co^{60}$-Strahlung ermittelt.

Mit diesen Messungen an homogenen Phantomen haben wir, in Übereinstimmung mit der allgemeinen Praxis in der Strahlenheilkunde, danach gestrebt, Standardisodosen für die Pendelbestrahlung zu erhalten. Fig. 10 zeigt eine solche Isodose bei $Co^{60}$-Strahlung; als Vergleich wird die entsprechende für 200 kV-Röntgenstrahlung in Fig. 11 wiedergegeben. Fig. 12 zeigt, wie die Tiefe des Dosismaximums von der Zentrierungstiefe abhängt, teils bei $Co^{60}$-Strahlung, teils bei 200 kV-Röntgenstrahlung. Diese Messungen wurden auch an dem homogenen, 20 × 32 cm großen, annähernd elliptischen Phantom aus „Mix D" ausgeführt.

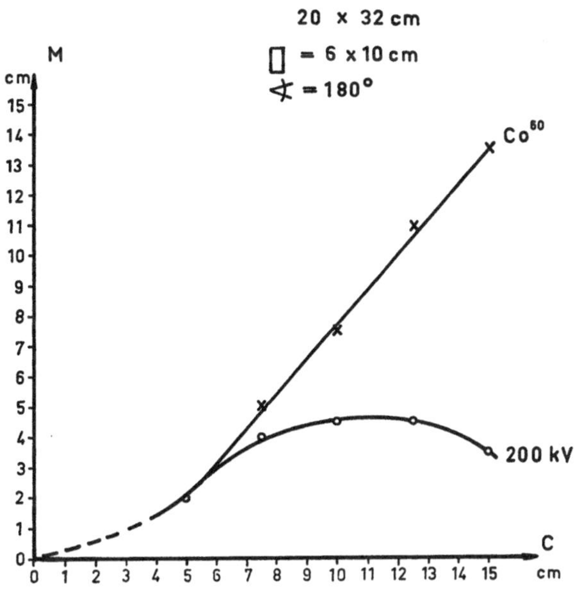


Fig. 12. Die Korrelation zwischen der Tiefe des Dosismaximums (*M*) und der Tiefe der Pendelachse (*C*) bei 180° Pendelbestrahlung

Zur Zeit werden auch Messungen in anatomischen Thorax- und Beckenphantomen gemacht. Fig. 13 zeigt die Isodosen in einem Beckenphantom bei

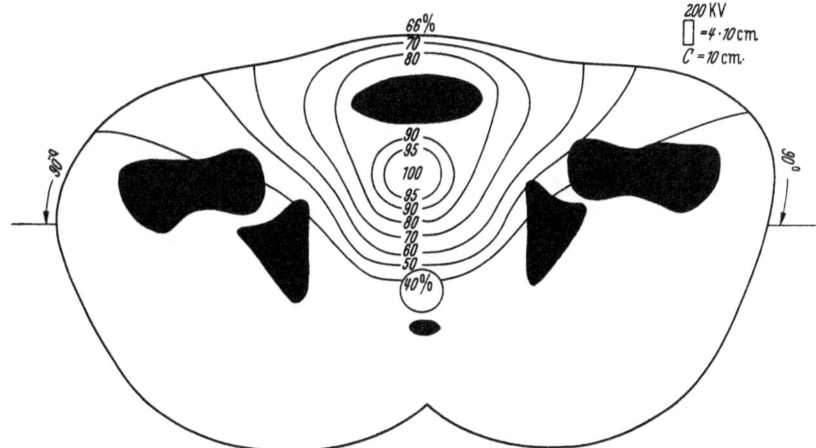


Fig. 13. Isodosen bei Pendelbestrahlung mit 200 kV-Röntgenstrahlung, an einem anatomischen Beckenphantom gemessen. Abstand Focus—Haut = 50 cm. Feldgröße an der Haut = 4 × 10 cm. Pendelwinkel = 180°. Tiefe der Pendelachse = 10,0 cm

200 kV-Röntgenstrahlung. Die Behandlungsbedingungen entsprechen ungefähr denen bei Bestrahlung von Blasentumoren. Fig. 14 zeigt die entsprechende Verteilung bei Co[60]. Die Hautbelastung vorn ist wesentlich niedriger als die bei 200 kV-Röntgenstrahlung.

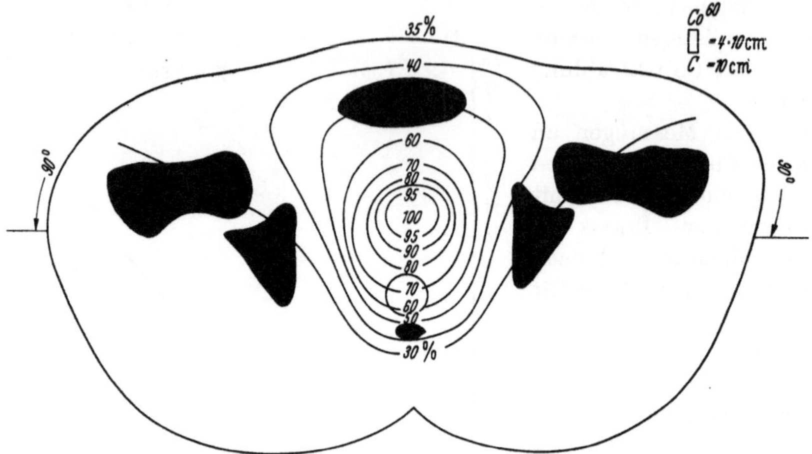


Fig. 14. Isodosen bei Pendelbestrahlung mit Co⁶⁰, an einem anatomischen Beckenphantom gemessen. Bestrahlungsbedingungen wie die in Fig. 13 angegebenen

In Fig. 15 schließlich ist die Dosisverteilung bei Vollrotation dargestellt. Die Belastung der benachbarten Organe wird beim Übergang von 200 kV-Röntgenstrahlung zu Co[60]-Strahlung erheblich reduziert, d. h. die Raumdosis wird kleiner und die relative Herdraumdosis wird größer.

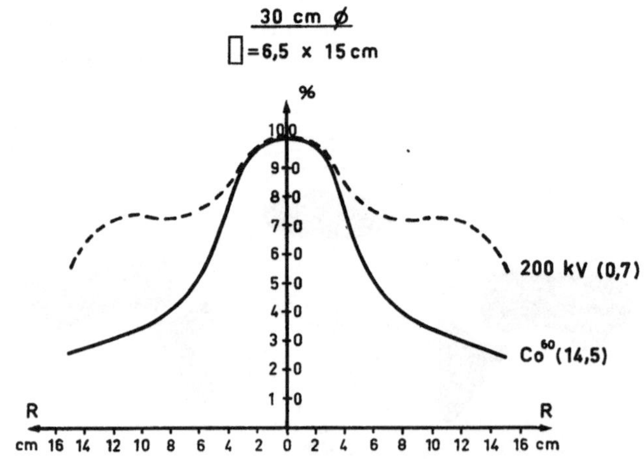


Fig. 15. Die Dosisverteilung bei Vollrotationsbestrahlung, an einem homogenen „Mix D"-Phantom (∅ = 30 cm) in der Rotationsebene gemessen

Gemäß den bisher durchgeführten Strahlungsmessungen und unseren praktisch-klinischen Früherfahrungen erfüllt das Gammatron sehr gut unsere anfangs aufgestellten Forderungen.

# Observations sur les phénomènes réactifs chez les malades soumis à la télécobalthérapie

Par

**A. Ratti,** Milano

La radiothérapie des tumeurs malignes exercée à l'aide des moyens, méthodes et techniques traditionnelles de la roentgenthérapie et de la radiumthérapie, n'est parvenue à se constituer une expérience large et approfondie qu'au bout de plusieur dizaines d'années.

Il est probable qu'avec les moyens, les méthodes et techniques introduites plus récemment, on parvienne à une expérience analogue plus rapidement, au moins pour les problèmes limités à une période d'observation pas trop longue et, en ligne de conduite, exception faite pour les résultats globals de longue durée, et pour les éventuels préjudices qui apparaitraient plus tard.

En fait, de nos jours, qui expérimente cliniquement les moyens actuels, peut disposer de connaissances d'ordre physique précises, de méthodes de dosimétrie satisfaisantes, de nombreuses donnés d'ordre radiobiologique et d'un riche matériel de comparaison constitué par les résultats des méthodes traditionnelles.

Par conséquent, les donnés d'une expérience relativement courte présentent déjà un intérêt notable, et permettent de dresser un bilan approximatif, différent peu de celui que fournirait une expérience plus longue.

Pour concourir à cette récolte d'éléments j'ai accepté avec plaisir l'invitation à rendre compte des donnés de notre expérience directe qui remonte à un peu plus de deux ans. Je me suis toutefois limité à considérer seulement les cas traités au cours des 19 premiers mois d'activité d'un service de t. c. t. institué dans une clinique privée, mais qui fonctionne en coopération directe au point de vue médical et physique avec le personnel des Universités de Milan et de Pavie.

Du 1-6-1955 au 31-12-1956, 329 patients ont été soignés, tous porteurs de tumeurs malignes vérifiés histologiquement. Nous avons disposé d'une unité de t. c. t. du type »Eldorado«, d'une puissance initiale d'un peu plus de 1000 curies, avec un débit d'environ 40 r/m à la distance foyer-surface irradiée de 80 cm et avec un taux de transmission d'à peu près 50% à 10 cm de profondeur.

Cette unité est construite pour la thérapie à champs fixes; avec une table de traitement tournante, la thérapie de mouvement est également possible. Cette dernière a fait l'objet d'une étude théorique déjà publiée.

Pour admettre les malades au traitement les indications ont été larges; nombre des cas étudiés sont donc particulièrement graves avec récidives à de préalables traitements chirurgicaux, ou traitements inachevés par suite de l'impossibilitè d'une intervention radicale.

Nous avons cependant veillé particulièrement aux conditions générales du malade, surtout lorsqu'il s'agissait de tumeurs de grande extension qu'il aurait fallu traiter en administrant des doses-volumes très fortes, incompatibles, justement, avec les conditions générales du malade.

Cette considération nous semble également claire parce que, comme nous pouvions le prévoir théoriquement et comme il en résultait d'après les communications de qui avait pu expérimenter la méthode précédemment, nous pouvions penser à priori qu'il était probable que la t. c. t. s'avère plus efficace que les méthode traditionnelles, mais c'était une erreur de croire q'on pouvait traiter avec succès formes et cas dépassant les limites d'un pouvoir d'action radiologique raisonnable; comme le sont, par exemple, certains cas de dissémination metastatique à foyers multiples, ou, pis encore, les cas de néoplasie généralisée de faible radiosensibilité, à moins qu'il y ait eu motifs pour recourir à quelque traitement palliatif limité.

Nous avons administré aux malades en traitement une dose de 5 a 6000 r, au foyer. Dans quelques cas, lorsqu'il s'agissait, par exemple, de grosses masses abdominales irradiées par de nombreux champs de moyennes et grandes amplitudes, nous nous sommes arrêtés à une dose moindre (3500 r); dans quelques cas de tumeurs bien délimités nous sommes arrivés jusqu'à 7000 r. Le traitement a duré de 3 à 5 semaines, au rythme habituel de 5 jours de traitement sur 7.

Chez les patients soumis à une irradiation par plusieurs champs nous avons traité 2 ou 3 portes dans la même séance, en administrant toujours au foyer, par principe, une dose quotidienne de 150 à 210 r. La localisation du foyer et l'étude physico-géométrique de la distribution des doses ont été faites avec le plus grand soin possible; mes collaborateurs ont même mis au point une méthode de contrôle radiographique du centrage dans l'irradiation du thorax, méthode fondée sur le principe d'une double impression et qui a déjà été publiée.

Les malades soignés peuvent, certains, être réunis en groupes relativement homogènes pour ce qui est des conditions du traitement. Si nous considérons seulement les tumeurs profondes nous pouvons les grouper de la manière suivante:

| | | | |
|---|---|---|---|
| Cancers du thorax | poumon | 36 | 49 |
| | oesophage | 13 | |
| de l'abdomen | ovaire | 19 | 32 |
| | divers | 13 | |
| du bassin | utérus | 26 | |
| | vessie | 30 | 68 |
| | rectum | 12 | |

L'observation des malades, durant et après le traitement, faite par examens de laboratoire, nous permet d'affirmer que le traitement n'a dû être interrompu en aucun cas. Pour quelques malades, nous avons jugé opportun, à un certain moment, de modifier le rythme de l'irradiation en réduisant les doses ou le nombre des séances hebdomadaires, ou encore en intercalant quelques jours de repos; cependant aucun phénomène d'intolérance locale ou générale n'a exigé la suspension définitive du traitement. Dans quelques cas plus graves, pour lesquels la cure avait été entreprise dans un but purement palliatif et limité, nous avons pu nous borner à des doses moins élevées que celles prévues.

L'importance qu'il faut attribuer à l'état du sang est unanimement reconnue. Nous avons donc suivi cette règle et nos expériences peuvent se résumer ainsi:

*Globules rouges.* Dans $^1/_3$ des cas environ, aucune diminution numérique appréciable. Il s'agissait spécialement de malades atteints de tumeurs de petite extension, quelquefois semiprofondes, irradiés par des champs d'amplitude limitée. Dans les autres $^2/_3$ des cas, nous avons eu, au contraire, une chute remarquable, avec une moyenne de 500000 globules rouges per $mm^3$ à la fin du traitement. Par rapport à la teneur initiale, cette diminution est de 10 à 15%. Mais presque toujours il y a eu récupération complète, ou presque, dans les deux mois suivants. Le comportement de la teneur en hémoglobine a évolué parallélement à celui des globules rouges.

*Globules blancs.* Dans 72% des cas, diminution notable, en moyenne 30% de la teneur initiale, sans différence significative dans le comportement des granulocytes, des lymphocytes et des monocytes. Contrairement à ce qu'on a observé pour les globules rouges, la récupération a été plus lente, quelquefois encore incomplète après quatre mois.

Pour ce qui est des globules blancs nous devons faire remarquer que chez certains patients subsistaient des complications inflammatoires et pour les globules rouges, que nous avons eu à faire plusieurs fois à tumeurs ulcérés avec hémorragies.

Nous devons enfin ajouter que, durant et après le traitement radiologique, nos malades ont été soignés contre l'anémie avec des extraits épathiques, vitamines, transfusions de sang grâce auxquels on a pu combattre cette anémie.

Toujours dans le cadre des phénomènes de caractère général, nous avons pu constater, spécialement pour l'irradiations de l'abdomen, que la t. c. t. provoque elle aussi le «mal des rayons», mais que celui-ci est moins accusé qu'avec la roentgenthérapie et cela particulièrement quand on traite des régions très sensibles comme les parties hautes de l'abdomen. Il nous est arrivé, ainsi qu'à d'autres auteurs, de soigner sans difficulté des cancers de l'estomac récidivés. D'autre part, alors qu'il en va du contraire avec la roentgenthérapie et la radium-thérapie traditionelles les altérations de la peau et du tissus souscutané sont minimes puisq'on sait que dans la t. c. t. le maximum des doses ne coïncide pas avec la surface cutanée, mais se trouve à 5 ou 6 mm au-dessous. Non observations confirment nettement ce fait et les altérations se sont limitées à des érythèmes, pigmentation et dans quelques cas à une épidermite sèche bénigne.

Nous avons eu peu d'accidents locaux au cours du traitement. Méritent d'être mentionnés:

1 hémorragie grave suivie de mort, dans une tumeur pulmonaire;
1 fistule oesophagotrachéale dans un cancer végétant de l'oesophage (5000 r au foyer);
1 fistule vésico-vaginale
1 fistule vagino-rectale } dans des cancers de l'utérus.

Les cancers du poumon, de l'oesophage et de la vessie sont, comme on le sait, parmi les plus aptes à être traités per la t. c. t. A propos de ces derniers et de quelques autres que nous avons traités, je crois de quelque intérêt les remarques suivantes:

*Cancer du poumon.* A côté des modifications subjectives (amélioration de la toux, diminution de la douleur et de la dyspnée) les modifications des images radiologiques offrent un intérêt considérable. Dans les cas d'obstruction d'une bronche importante, on peut voir disparaître l'atélectasie.

Chez d'autres malades, au contraire, l'opacité augmente à cause des phénomènes d'exudation réactive périfocale. On a constaté aussi l'apparition ou l'agrandissement d'images cavitaires prenant les caractères d'un véritable abcès, dans 4 cas sur 36. Ceux-ci sont entrés en régression et ont disparu, au moins sans laisser de symptomes cliniques, grâce aux antibiotiques,

Chez les malades qui ont eu une évolution favorable nous avons remarqué des images radiologiques qui nous paraissent correspondre à une fibrose résiduelle, du même aspect que celle vérifiée dans les traitements par roentgenthérapie rotatoire.

*Cancer de l'oesophage.* On y craint beaucoup la réaction locale ou oesophagite. Jusqu'à la dose de 3000 r, les altérations de l'oesophage ont été sans importance. Pour une dose de 3 à 4000 r, administrée dans les mêmes conditions on a eu disphagie sans gravité à la fin du traitement. Pour une dose de 4 à 5000 r et plus, la disphagie s'est aggravée devenant douloureuse au point de compromettre sérieusement l'alimentation à la fin du traitement. Toutefois ces phénomènes n'ont conservé un degré d'acuité que durant 4 ou 5 jours et aucun cas n'a nécessité une intervention chirurgicale.

*Cancer de la vessie.* Pour des doses allant jusqu'à 3500 à 4000 r, les réactions de cystite ont été bien tolérées. Au delà de 4000 r on a eu au contraire une cystite un peu grave (18 cas) qui a duré de 10 à 20 jours; au delà de 5000 r la cystite s'est résolument aggravée et a duré plus longtemps (de 20 à 40 jours).

Chez ces malades, lorsque l'examen cystoscopique a été possible, il a révélé une congestion des vaissaux avec des phénomènes de muscosite bulleuse.

Dans le traitement des *tumeurs du rectum* dans la moitié des cas environ, on a eu des phénomènes de proctite notable, avec douleurs, émission de sang et mucus qui ont duré 2 ou 3 semaines tout au plus après la fin de traitement  Chez ces malades, chez certains desquels nous avons obtenu des résultats satisfaisants qui durent plus d'un an, on a noté un rétrécissement des parois de l'ampoule, qui a pris une forme tubulaire, avec une diminution de l'élasticité des parois, sans aller jusqu'à une rigidité capable d'occasioner des troubles de la canalisation.

Le traitement de volumineuses *tumeurs de l'abdomen* a été suivi des phénomènes d'intolérance intestinale du type de l'entérocolite. Cependant ceux-si ont été moins accusés que les phénomènes constatés par nous directement avec la roentgenthérapie, à égalité des doses et de volume irradiés.

Les connaissances préliminaires d'ordre physique, biologique et technique et les résultats obtenus jusqu'à présent, permettent d'affirmer que, ainsi qu'il est nécessaire dans un bon nombre de cas, la t. c. t. permet d'administrer de hautes doses au foyer sans provoquer de réactions d'une gravité telle à empêcher de continuer le traitement, ou des réactions qui échappent au contrôle médical. On peut donc accepter ces réactions dans le traitement des tumeurs malignes.

Au point ou sont actuellement nos connaissances, il me semble possible d'établir les termes d'une comparaison entre la t. c. t. et les autres méthodes de traitement, de cette manière:

vis-à-vis des méthodes traditionnelles de roentgenthérapie à 200 kV, la t. c. t. constitue une amélioration décisive dans le traitement des tumeurs profondes et semiprofondes;

par rapport à la roentgenthérapie de mouvement, on peut la mettre sur le même plan que la thérapie rotatoire dans le traitement des certaines tumeurs parmi lesquelles se trouvent celles du thorax à siège central et les tumeurs du tiers moyen de l'oesophage et certaines tumeurs pulmonaires;

par rapport à la roentgenthérapie de convergence, la t. c. t. n'offre pas d'avantages importants pour les tumeurs du massif facial et les tumeurs oro-pharyngolaryngées. On note toutefois que l'emploi correcte de la thérapie convergente exige une considérable habileté; comme il en va pour toutes les techniques modernes.

En face du bétatron et des autres sources de haute énergie, cette comparaison ne pourra être faite que sur les donnés de résultats contrôlés dans plusieurs années.

En ce qui concerne le danger de porter préjudice, nous signalons qu'il est opportun de rappeler qu'il est possible de nuire à des structures particulièrement délicates comme le sont celles de l'axe cérébro-spinal. Celles-ci, en effet, par suite du haut pouvoir de pénétration des radiations émises par le cobalt quand elles traversent les structures osseuses, peuvent être exposées à recevoir des doses rarement administrées jusqu'à présent. D'autre part les caractéristiques du système nerveux rendent possible des lésions qui peuvent se manifester très tardivement.

Cette considération est surtout valable pour l'irradiation appliquée à des sujets jeunes et il faut l'incorporer au problème général des bienfaits à espérer et des préjudices à craindre dans tout cas de maladie, en union au sens que le traitement a pour la vie et pour la validité de l'individu.

# Klinische Erfahrungen mit Theratron Junior*

Von

O. Hubacher, Thun

**Allgemeines.** Über die Vorteile, welche die Bewegungsbestrahlung besonders in Kombination mit ultraharter Strahlung bietet, ist andernorts berichtet worden. Mit dieser Kombination wird eine maximale Belegung des tiefliegenden Tumorherdes unter weitgehender Schonung der Umgebung gewährleistet. Es entstehen keine ungünstigen Dosisüberschneidungen. Bei der kontinuierlichen Rotationsbestrahlung z. B. erfolgt eine Halonierung im Randgebiet des Tumors, welche die Bekämpfung der Randzone sicherstellt.

**Apparaturen.** Seit Mai 1956 arbeiten wir mit der Kobaltanlage Theratron Junior der Atomic Energy of Canada. Der Bestrahlungskopf mit einem Gewicht von 500 kg enthält im Zentrum das radioaktive Kobalt. Dieses hat eine Aktivität von 60 C/g. Solche Aktivitäten können bisher in der westlichen Welt nur im Reaktor Chalk River, Canada, erzeugt werden. Im Bestrahlungskopf sind 600 C eingelagert. Die initiale r/min Leistung im Abstand von 55 cm betrug 40 r.

Das radioaktive Kobalt ist im Zentrum des Bestrahlungskopfes angeordnet und verschiebt sich nicht. Vor dem Bestrahlungskanal liegen zwei Verschlußblöcke. Der regelmäßig in Betrieb stehende Verschlußblock wird mit Preßluft betätigt, der zweite Verschlußblock dient als Notverschluß. Bestrahlungstubusse erübrigen sich, da ein zweckmäßiges Lichtvisier, welches als Tiefenblende ausgebildet ist, montiert ist. Ein ähnliches Verschlußsystem wie die kanadische Kobaltbombe, besitzt die Siemens-Anlage Gammatron.

Die Auswechslung der Strahlenquelle ist mit dem international standardisierten Container einfach. Es wird nicht nur das Kobalt gewechselt, sondern auch die Wolframschublade. Bei der Auswechslung können Bestrahlungskopf und Container gut adaptiert werden, so daß keine, die Toleranzgrenze übersteigende Strahlenmenge austreten kann.

Wegen des als primärer Strahlenschutz ausgebildeten Gegengewichtes sind die Strahlenschutzmaßnahmen beim Theratron Junior relativ einfach. Es sind Betonmauern von 35 cm Dicke notwendig. Die Transmission durch den Bestrahlungskopf unterschreitet, wie eigene Messungen ergeben haben, die Toleranzgrenze bei weitem. Am Äquator des Bestrahlungskopfes messen wir kontakt 7 mr/Std. und 0,6 mr in 1 m Abstand von der Strahlenquelle.

Der Bestrahlungstisch besteht aus Aluminium, welches nur 5% der Strahlung absorbiert, also praktisch nicht ins Gewicht fällt. Er ist in allen Richtungen verschieblich.

---

* Röntgeninstitut Dr. O. Hubacher, Thun

**Erfahrungen bei der Bestrahlung mit $CO^{60}$.** Aus der Literatur ist eine neue Bestrahlungsmethode nicht zu erlernen, da alle Publikationen lückenhaft sind. Unsere ersten Erfahrungen wurden anläßlich eines Englandaufenthaltes vor $1^1/_2$ Jahren gesammelt (Hammersmith Hospital, Mount Vernon Hospital, London). Nicht nur bestrahlen die Engländer in Europa schon seit einigen Jahren mit der Kobaltbombe (Theratron Senior der Atomic Energy), sondern sie besitzen auch große Erfahrung mit dem Linearaccelerator, wie Sie heute vormittag gehört haben. Der Accelerator mit 4 MeV und beweglicher Strahlenquelle entspricht physikalisch weitgehend der Kobaltbombe, allerdings mit dem Unterschied der verschiedenen spektralen Energieverteilung. Wir möchten jedem Kollegen, der mit der Kobaltbombe zu arbeiten beginnt, raten, sich in England oder den USA auszubilden.

Wir selbst haben seit mehr als 1 Jahr 150 Fälle mit malignen Tumoren mit der Kobaltbombe bestrahlt. Über 133 Fälle möchte ich berichten.

Unsere Bestrahlungsmethode geht aus dem anschließenden Film hervor. Von den verschiedenen Methoden zur Konstruktion der Isodosenkurven bedienen wir uns meistens derjenigen von BRAESTRUP, New York, indem wir von den bestehenden Isodosen für Stehfeldkobaltbestrahlungen ausgehen und bei der Drehung der Isodosenkarte auf dem Rotationszentrum alle 30° das Integral bilden und somit endgültige Isodosenfiguren erhalten.

Der optische Modellversuch dient dazu, uns Anhaltspunkte zu geben, wie unsere graphische Methode sich praktisch auf das Objekt übertragen läßt. Einzelheiten gehen ebenfalls aus dem Film hervor.

Im folgenden zeige ich 5 prägnante Fälle, welche besonders interessieren mögen.

1. KG Nr. 23803. Nachbestrahlung einer Struma maligna, welche nicht radikal hat operiert werden können. Gesamtdosis 4000 r $Co^{60}$ in $4^1/_2$ Wochen mit stehender Bombe. Es ist, wie die Fig. 1 zeigt, lediglich eine minimale Rötung auf einem Feld von $10 \times 10$ cm aufgetreten. Eindrucksvoll war für den praktisch tätigen Radiologen, daß keinerlei Zeichen von Larynxödem, Schleimhautreizung, Perichondritis, entstanden sind. Die Patientin ist nach 10 Monaten symptomfrei.

2. KG Nr. 23856. Axilläre Bestrahlung bei einem Mammacarcinom Steinthal II. Es handelt sich um eine Vorbestrahlung mit 4000 r $Co^{60}$ in $4^1/_2$ Wochen. Es ist keine Rötung entstanden, aber eine vollständige Epilation. Die baumnußgroßen, derben axillären Drüsen waren bei Abschluß der Bestrahlung nicht mehr palpabel und bei der nachfolgenden Operation histologisch in Serienschnitten negativ.

3. KG Nr. 24290. Vergleich der Hautreaktion bei Nachbestrahlung eines Mammacarcinomes. Das Mammafeld wurde in $4^1/_2$ Wochen mit 2000 r/o mit konventionellen Bedingungen bestrahlt. Es ist eine ziemlich starke exsudative Reaktion aufgetreten. Dagegen haben wir axillär, supraclaviculär und sternal mit der Kobaltbombe bestrahlt und zwar wurden in derselben Zeit 4500 r $Co^{60}$ gegeben. Es hat sich im Bereiche der Kobalt-bestrahlten Felder nur eine minimale Rötung gebildet, trotzdem diese mehr als die doppelte Dosis des Mammafeldes erhalten haben.

4. KG Nr. 24492. Auffallend rasche Rückbildung eines ausgedehnten, tiefsitzenden Oesophaguscarcinomes bei einer 80jährigen Patientin. Vor der Bestrahlung zeigt sich ein ausgedehnter Defekt. 5 Wochen später, d. h. nach Applikation einer Gesamtdosis von 4000 r $Co^{60}$ in 4 Wochen und einem Rotationswinkel von 360° war der Defekt vollständig verschwunden. Die Patientin ist symptomfrei (nach 10 Monaten). Es ist keine Hautreaktion entstanden. Namentlich zeigte sich aber auch bei dieser sehr alten und gebrechlichen Patientin keine Allgemeinreaktion.

5. KG Nr. 23405. Bestrahlungs- und Isodosenschema einer Rezidivbestrahlung eines tumorösen Infiltrates im re. Parametrium. Dabei wurde ein Pendelwinkel von 150° gewählt,

damit Rectum und Blase ausgespart bleiben. Gesamtdosis 5000 r $Co^{60}$ auf der 90%igen Isodose in 5 Wochen. Bei diesem Fall ist sehr gut ersichtlich, daß die Modellierung der Isodosenkurven in praktisch beliebigen Grenzen ermöglicht wird.

**Provisorische Zusammenstellung der Frühresultate.** Die Darstellung (Fig.2) stützt sich nur auf ein Kriterium, die Überlebenszeit. Dadurch wird sie einfacher und objektiver. Uns scheinen Angaben mit „gebessert, oder teilweise gebessert" unzuverlässig.

Großes Gewicht wird auf die Vergleichbarkeit der Kurven gelegt. Mit jeder der 4 Bestrahlungsmethoden wurden über 100 Carcinompatienten behandelt, auf

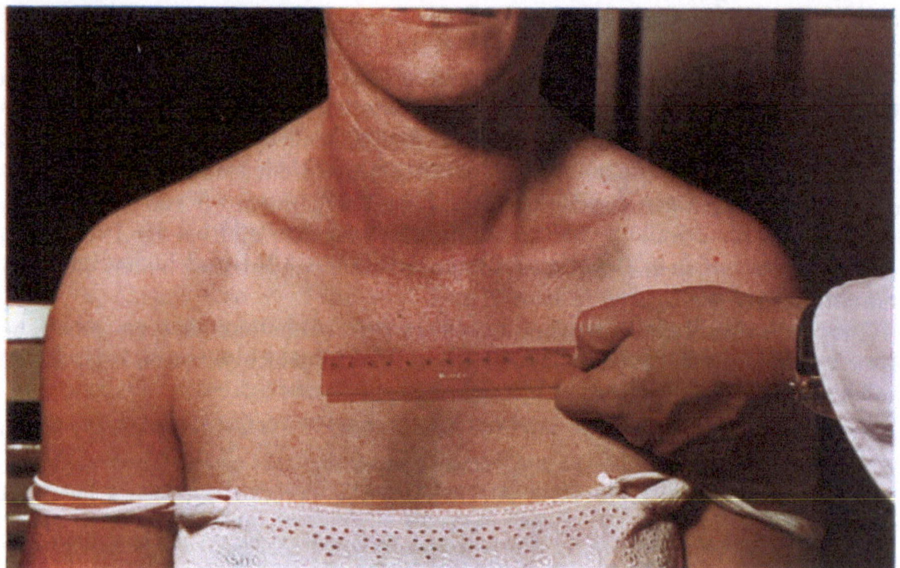

Fig. 1. Hautreaktion bei Nachbestrahlung einer Struma maligna nach 4000 r $Co^{60}$ in 4 Wochen.
Nur geringe Rötung

dieselben Lokalisationsgruppen entfielen ungefähr die gleiche Anzahl Fälle. Auch die Schwere der Carcinomfälle war dieselbe. Bei den 150 Kreuzfeuer-, 107 Konvergenz- und 133 Kobaltbomben-Bestrahlungspatienten handelt es sich um eigene Fälle; über die 112 Patienten, welche mit dem 31 MeV Stehfeld-Betatron bestrahlt wurden, berichtete B. RAVNIHAR, Ljubljana. Die sehr aufschlußreiche Statistik der erwähnten Autorin ist, wie eine persönliche Besprechung ergab, mit unserer Statistik zwanglos vergleichbar.

Als Resultat zeigt sich ein signifikanter Unterschied zugunsten der Kobaltbestrahlungsfälle. Nach 10 Monaten leben 82% der Kobaltfälle, während noch 60—70% der Konvergenz- und Betatronfälle, und nur 56% der Kreuzfeuerbestrahlungsfälle am Leben sind.

Es ist uns bewußt, daß eine Erfolgsstatistik im jetzigen Zeitpunkt noch verfrüht ist, und daß jede Statistik Fehler hat. Die Zusammenstellung soll lediglich als Zusammenfassung unseres ersten, wirklich guten Eindruckes von dieser Therapie aufgefaßt werden.

Ich kann nicht erklären, wie die besseren Resultate der Kobaltbestrahlung zustandegekommen sind. Ich könnte mir aber vorstellen, daß diese mit der verbesserten relativen Herdraumdosis zusammenhängen, welche für die Kombination von Bewegungsbestrahlung mit ultraharter Strahlung charakteristisch ist.

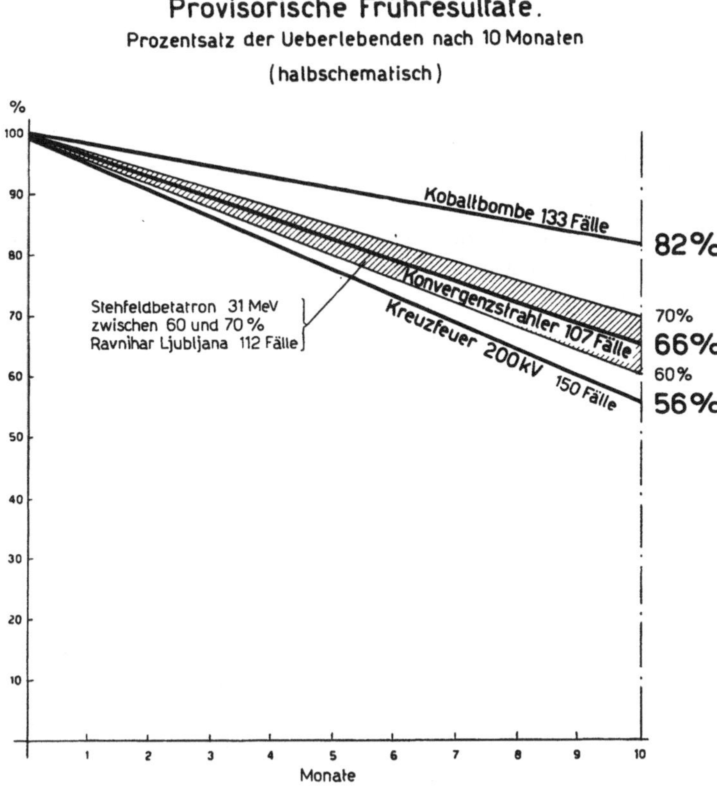


Fig. 2

**Zusammenfassung.** Die Vorteile, welche die Telekobalttherapie bietet, besonders wenn dieselbe als Bewegungsbestrahlung durchgeführt wird, sind folgende:

1. Die Hauttoleranz wird gegenüber den konventionellen Bestrahlungsmethoden auf mehr als das Doppelte erhöht.

2. Man kann mit der Kobaltbombe auf einfache Weise auch Halbtiefentherapie betreiben. Sie eignet sich besonders zur Bestrahlung von etwas tiefer liegenden Drüsen.

3. Man arbeitet mit einer vom Radium her bekannten homogenen Strahlung. Man begibt sich also nicht auf therapeutisches Neuland.

4. Die Apparaturen sind relativ einfach aufgebaut und erfordern praktisch keinen Unterhalt, da die Hochspannung wegfällt. Die Therapie mit Kobalteinheiten ist nicht nur in Universitätsinstituten, sondern auch in Spitälern und privaten Instituten möglich.

5. Dosierung: Wegen der praktisch wegfallenden Allgemeinreaktion kann man große Einzeldosen verabreichen. Die Carcinombestrahlungen sind demzufolge wie in England sehr gut ambulant durchführbar.

Einzeldosis: 150—350 r CO$^{60}$.

Gesamtdosis: Wir haben darauf geachtet, daß wir eine approximative Wochendosis von 1000 r Co$^{60}$ applizieren und zu einer Gesamtdosis von 3500—6500 r gelangen. Weil die Späterscheinungen noch nicht beurteilt werden können, ist Vorsicht geboten.

Wöchentliche Bestrahlungssitzungen 4—6.

Wir haben uns mit dieser Dosierung den Dosen der Engländer und Amerikaner (Mount Vernon Hospital London, Monte Fiori Hospital New York) angeschlossen und waren erfreut, daß auch G. J. L. Thurgar und E. T. Farmer, Newcastle, über eine ganz ähnliche Dosierung berichtet haben.

6. Die Isodosen können in praktisch beliebiger Weise modelliert werden. Es ist deshalb möglich, strahlensensible Organe auszusparen.

7. Die Allgemeinreaktion ist praktisch Null. Dies hängt unter anderem wahrscheinlich mit der kleinen relativen Herdraumdosis zusammen. Dieser letzte Vorteil der Telekobalttherapie ist für den Strahlentherapeuten vielleicht der eindrucksvollste und für den Patienten der wichtigste.

## Literatur

Hubacher, O.: Strahlenther. **102**, 2, 332 (1957).
— Kongreßbericht dos Radiologistas e Electrogistas de Cultura latina IV, Lisboa, 2, 360 (1957).
— Kongreßbericht, Rad. Austriaca, Wien (1957).
— Radiol. clin. (Basel) **26**, 5, 254 (1957).

# Die biologische Wirkung energiereicher Strahlung

## Die unterschiedliche biologische Wirkung von 30 MeV-Elektronen, 31 MeV-Röntgenstrahlen und 180 keV-Röntgenstrahlen*

Von

H. FRITZ-NIGGLI, Zürich

Die Strahlenbiologie beschäftigt sich mit der Wirkung rein physikalischer Elemente auf das Lebewesen und auf Lebenseinheiten. Es bestand nun — lange Zeit — die begreifliche Hoffnung, die Strahleneffekte in mathematischer Formulierung auf gemeinsame Nenner bringen zu können. Die Hoffnung hat sich nicht erfüllt, im Gegenteil, es zeigte sich, daß die lebendige Materie in überaus vielfältiger Weise den Strahlen antwortet, die ihrerseits in eine keineswegs starre Wechselwirkung treten. Das Reaktionssystem und sein physiologischer und physikalischer Zustand spielen die entscheidende Rolle. Bei den vergleichenden Experimenten mit Strahlen verschiedener Art tritt dieses Grundphänomen klar zutage.

Die vergleichenden Betrachtungen stoßen nun auf zahlreiche Schwierigkeiten, welche oft eine klare Auswertung hemmen oder gar unmöglich gestalten. Die hauptsächlichsten Fehlerquellen sind:

*1. Dosismessung.* Die Dosen werden zumeist als Ionendosen gemessen. Es ist nun durchaus möglich, daß die Dosismessung hoher Energien, wie sie 31 MeV-Photonen und 30 MeV-Elektronen darstellen, der tatsächlich vom Gewebe absorbierten Dosis nicht adäquat ist.

*2. Versuchsanordnung.* Die Versuchsanordnung sollte so gewählt werden, daß die Objekte sich in einem flachen Teil der Tiefenabsorptionskurve befinden. Anderenfalls bewirken geringe Tiefenverschiebungen große Unterschiede der Wirkung.

*3. Wahl der Dosis.* Es hat sich als tunlich gezeigt, die Dosis niedrig zu wählen. Große Dosen gestatten es nicht, die oft feinen Wirkungsunterschiede verschiedenartiger Strahlung zu differenzieren. So reagieren die Zellen der Vicia Faba-Wurzelspitze auf 500 r 31 MeV-Photonen gleich wie auf 180 keV-Photonen, während sich bei der Bestrahlung mit 250 r klare Wirkungsunterschiede zeigen.

*4. Wahl des Objektes.* Da die Strahlen verschiedener Energie in Lebewesen unterschiedlicher Größe und Bau nicht das gleiche Verteilungsmuster aufweisen, sind die Resultate aus Versuchen mit größerem Testobjekt schwer zu analysieren. Kleinste Objekte erweisen sich als zweckmäßig.

*5. Inhomogenität des Materials.* Es zeigte sich in unseren Versuchen, daß die Wirkungsunterschiede abhängig sind vom Entwicklungsstadium der Objekte.

---

* Aus dem Strahlenbiologischen Laboratorium (Leiterin: Privatdozentin Dr. HEDI FRITZ-NIGGLI) der Radiotherapeutischen Klinik des Kantonsspitals (Direktor: Prof. Dr. H. R. SCHINZ) Zürich

Eine relative biologische Wirksamkeit, gewonnen an einem Objekt, das biologisch
inhomogen ist, sagt überhaupt nichts aus.

Es sei nun von unseren Experimenten berichtet, zunächst summarisch, um dann
an einem Beispiel die Schwierigkeiten vergleichender Experimente zu ermessen
und zu deuten.

**Methode.** Bestrahlt wurde mit 180 keV-Photonen in einem Plexiglasphantom,
mit 31 MeV-Photonen wiederum in einem Plexiglasphantom im Maximum der
Transitionskurve und mit 30 MeV-Elektronen in Luft. Gemessen wurde an
Objektsstelle von uns mit dem Victoreen-r-Dosimeter 31 MeV-Photonen und 180

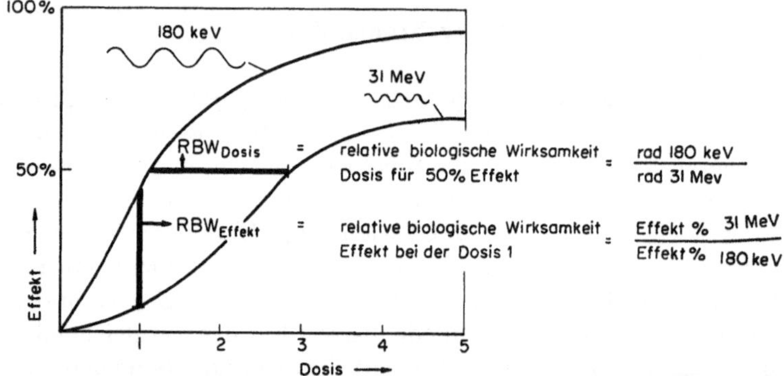


Fig. 1. Berechnung der relativen biologischen Wirksamkeit (RBW) nach 2 verschiedenen Kriterien. Die RBW-
Dosis besteht aus dem Quotienten zweier Dosen, die den gleichen Effekt, in diesem Fall 50% Effekt, hervorrufen.
Der RBW-Effekt vergleicht die Effekte der beiden Strahlenarten bei gleicher Dosis

keV-Photonen, während die 30 MeV-Elektronen von Dr. M. Sempert, dem wir
herzlich danken möchten, gemessen wurden. Die energiereiche Strahlung stammt
aus dem Brown-Boveri-Betatron.

Eine relative biologische Wirksamkeit (RBW) kann man messen, indem man
die Dosen zweier Strahlungen vergleicht, die einen bestimmten, gleichen biologischen Effekt hervorrufen. Auf Fig. 1 ist die RBW dargestellt, die aus dem
Quotient der Dosen 180 keV und 31 MeV besteht, die einen bestimmten Effekt,
nämlich 50%ige Tötung hervorrufen. Ich nenne diese RBW in der Folge RBW-
Dosis. Ebenso kann eine RBW errechnet werden, die aus dem Quotienten der
Effekte bei gleichen Dosen besteht, ich nenne diese relative biologische Wirksamkeit RBW-Effekt. Je nach der Wahl der Vergleichspunkte entstehen nun ganz
verschiedene RBW. 2 Kurven z. B. liefern verschiedene Vergleichswerte und
damit unterschiedliche RBW. Die festgestellte RBW kann uns also nur als Richt-
linie dienen. Ihr zahlenmäßiger Wert darf niemals überschätzt werden.

Unsere Versuche umfaßten die Strahlenreaktionen: Tod, Entwicklungsstörung,
Mitosehemmung, Mutation, Chromosomenbruch.

**Chemische Reaktion.** Zunächst sei eine anscheinend einfache chemische Reaktion zitiert, nämlich die Oxydation von Ferrosulfat zu Ferrisulfat [10]. Es stellt
sich beim Vergleich der Wirkung der 31 MeV- zu 180 keV-Photonen kein Unterschied heraus, wenn 50 $\gamma$ Ferrosulfat bestrahlt wurden (Tab. 1). Die Strahlenreaktion erfordert allerdings sehr hohe r-Dosen (10000—80000 r), so daß feinere
Wirkungsunterschiede nicht erfaßt werden können.

Tabelle 1. *Vergleich der Wirkung von 31 MeV-Photonen und 180 keV-Photonen*
*180 keV-Effekt = 1* (n. B. = nach Bestrahlung)

| Objekt | Reaktion | Dosisbereich r | Vergleichsbasis | $RBW_E$ | $RBW_D$ |
|---|---|---|---|---|---|
| | | *Chemische Reaktion* | | | |
| Ferrosulfat | Oxydation zu Ferrisulfat | 10000—80000 | Umsetzung der Ferromoleküle | 1 | 1 |
| | | *Entwicklungsstörungen und Tod* | | | |
| Drosophila-Embryonen 1 Std. alt | Nichtausschlüpfen aus Eihülle | 30— 600 | Dosis für 30% Nichtgeschlüpfte | | 1 |
| $1^3/_4$ Std. alt | Nichtausschlüpfen aus Eihülle | 30— 400 | Dosis für 30% Nichtgeschlüpfte | | 1 |
| 3 Std. alt | Nichtausschlüpfen aus Eihülle | 30—1400 | Dosis für 30% Nichtgeschlüpfte | | 0,76 |
| 4 Std. alt | Nichtausschlüpfen aus Eihülle | 30—1200 | Dosis für 30% Nichtgeschlüpfte | | 0,74 |
| $5^1/_2$ Std. alt | Nichtausschlüpfen aus Eihülle | 30—1400 | Dosis für 30% Nichtgeschlüpfte | | 0,86 |
| 7 Std. alt | Nichtausschlüpfen aus Eihülle | 30—1800 | Dosis für 30% Nichtgeschlüpfte | | 0,75 |
| Drosophila-Vorpuppen (5 Std.) | Nichtausschlüpfen aus d. Puppenh. | 1500—4500 | Effekt in % bei 4000 r | 0,22 | |
| Drosophila-Vorpuppen (5 Std.) | Phänokopie, gespreizte Flügel | 1500—4500 | Dosis für 50% gespreizte Flügel | | 0,8 |
| Drosophila-Vorpuppen (5 Std.) | Ausfall von Borsten | 300—4000 | Effekt bei 4000 r | 0,8 | |
| Embryonen Xenopus | Tod | 1000—2000 | Dosis für gleiche Überlebenskurve | | 0,6 |
| Escherichia coli-Bakterien | Unfähigkeit zur Vermehrung | 360—5150 | Dosis für 50% Geschädigte | | 0,9 |
| Maus | Tod | 800 | Überlebensdauer für 50% | 0,8 | |
| | | *Mitosehemmung* | | | |
| Vicia Faba-Wurzelspitzen | Mitosehemmung | 250—500 | Dosis, die ein Ausbleiben der Mitos. 12 Std. n. B. bewirkt | | 0,5 |
| Ehrlich-Carcinom der Maus (solid) | Mitosehemmung | 3000 | Mitoserate in $^0/_{00}$ (8 Std. n. B.) (16 Std. n. B.) | 0,1 0,4 | |
| Ehrlich-Carcinom (Ascites) | Mitosehemmung | 150 | Mitosen in $^0/_{00}$ (5 Std. n. B.) | 0,6 | |
| | | *Genetische Änderungen* | | | |
| Drosophila ♂ adult | Recessive sichtbare Mutationen | 2000 | Mutationsrate in % (5—7 Tage n. B.) | 1 | |
| Drosophila ♂ adult | Recessive geschl.-geb. Letalfaktoren | 3000 | Mutationsrate in % (1—7 Tage n. B.) | 0,7 | |
| Drosophila ♂ adult | Dominante Letalfaktoren | 1000 | Mutationsrate in % (3 u. 4 Tage n. B.) | 1,8 | |
| Drosophila ♂ adult | Dominante Letalfaktoren | 1000 | Mutationsrate in % (6 u. 7 Tage n. B.) | 0,7 | |
| Drosophila ♂ adult | Dominante Letalfaktoren | 1000 | Mutationsrate in % (8 u. 9 Tage n. B.) | 0,3 | |

**Strahlentod und Entwicklungsstörungen.** Eine komplizierte Strahlenreaktion stellt der *Strahlentod der Säugetiere* dar. 800 r wirken unterschiedlich auf adulte weiße Mäuse wenn sie einerseits vom 31 Millionenvolt-Betatron und andererseits

                                   H. Fritz-Niggli:

von der konventionellen Röntgenapparatur abgegeben werden, und zwar sind die Betatronstrahlen weniger effektiv (Fig. 2). Sowohl 30 MeV-Elektronen als auch 31 MeV-Photonen sind in ihrer Wirkung 180 keV unterlegen [6, 16]. Die unterschiedliche Dosisverteilung der verschiedenen harten Strahlung im Objekt gestattet keine weitere Analyse dieses Phänomens. Sehr viel kleiner sind die Objekte, die Lindenmann [13] bestrahlte, nämlich *Escherichia coli-Bakterien*. Hier zeigte sich, wobei als Kriterium Nichtvermehrung gewählt wurde, ein geringer Wirkungsunterschied zugunsten von 180 keV gegenüber 31 MeV-Photonen, wobei allerdings wiederum mit relativ hohen Dosen gearbeitet werden mußte.

Embryonen sind viel strahlensensibler als adulte Tiere, so daß hier mit sehr viel kleineren Dosen eine deutliche Strahlenreaktion hervorgerufen werden kann. Ein be-

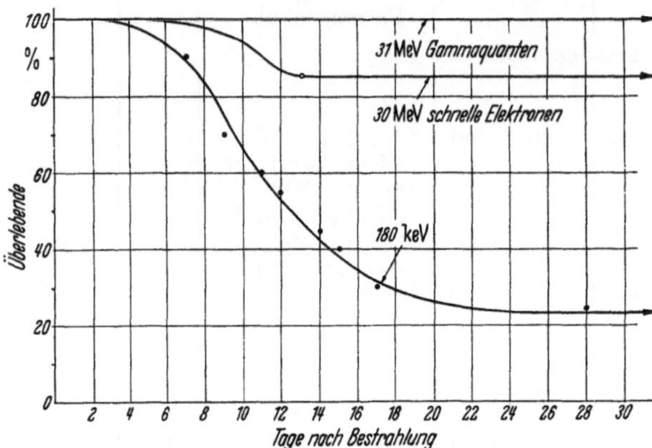


Fig. 2. Strahlentod weißer, männlicher Mäuse (21—24 g) mit 31 MeV-Photonen (hinter 33 mm Plexiglas), 180 keV-Photonen (hinter 5 mm Plexiglas, 1 mm Cu, 6 mA) und 30 MeV-Elektronen in Luft. Dosis in allen 3 Versuchen: 800 r Ionendosis

kanntes Testobjekt ist das *Ei von Drosophila*, das allerdings nicht besonders günstig ist, weil es ein Lebewesen darstellt, das in stürmischster Entwicklung begriffen ist. Die beobachtete Strahlenreaktion ist das Nichtausschlüpfen aus der Eihülle. Die Embryonen vermögen nun nach der Bestrahlung z. T. noch lange zu

Tabelle 2. *Vergleich der Wirkung von 30 MeV-Elektronen und 180 keV-Photonen*
*180 keV-Effekt = 1*

| Objekt | Reaktion | Dosisbereich r | Vergleichsbasis | RBW$_E$ | RBW$_D$ |
|---|---|---|---|---|---|
| Drosophila ♂ adult | Recessive geschl.-geb. Faktoren | 2000 | Mutationsrate in % (5—7 Tage n. B.) | 0,5 | |
| Drosophila ♂ adult | Chromosomenbrüche | 2000 | Mutationsrate in % (5—7 Tage n. B.) | 0,4 | |
| Drosophila ♂ adult | Translokation | 2000 | Mutationsrate in % (5—7 Tage n. B.) | 0,4 | |
| Drosophila ♂ adult | Dominante Letalfaktoren | 1000 | Mutationsrate in % (3—4 Tage n. B.) | 1,9 | |
| Drosophila ♂ adult | Dominante Letalfaktoren | 1000 | Mutationsrate in % (6—7 Tage n. B.) | 0,6 | |
| Drosophila ♂ adult | Dominante Letalfaktoren | 1000 | Mutationsrate in % (8 Tage n. B.) | 0,8 | |

leben, indem sie sich zu schlüpfbereiten aber schlüpfunfähigen Tieren entwickeln. In Fig. 3 ist ein normaler schlüpfbereiter Embryo zu sehen und ein früher bestrahlter, der differenziertes Gewebe, aber beispielsweise einen defekten Kauapparat und Muskulatur aufweist. Der Strahlentod scheint die Folge von enzymatischen Störungen zu sein, indem bestimmte Gewebe, deren Differenzierungsstoffe z. Z. der Bestrahlung noch nicht ausgebildet waren, zugrunde gehen. Sind diese Gewebe oder Organe lebenswichtig, so ist Tod die Folge. Die verschiedenen Stadien sind verschieden strahlenempfindlich, und zwar wechselt die Sensibilität von Viertelstunde auf Viertelstunde. Bis zum Alter von 3 Std. sind die Embryonen sehr strahlensensibel mit einem Sensibilitätsmaximum im Alter von $1^3/_4$ Std. Der Gastrulationsbeginn (3 Std.) scheint noch sehr empfindlich zu sein, dann steigt die Strahlen-Resistenz, um mit dem Alter von 7 Std. schlagartig zuzunehmen [3]. Zu

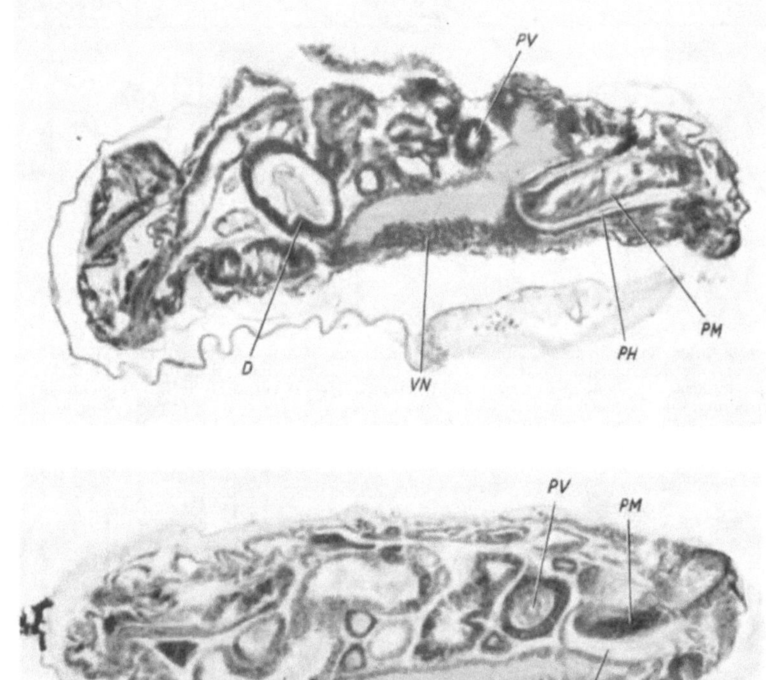


Fig. 3. Sagittalschnitt durch schlüpfreife Embryonen, *a* unbestrahlt, *b* bestrahlt im Alter von 4 Std. mit 1100 r 180 keV. Es zeigen sich deutliche Schädigungen in der Mundregion; Pharynxmuskulatur nicht ausgebildet; chitinige Ablagerungen. *D* Darm, *PH* Pharynx, *PM* Pharynxmuskulatur, *PV* Proventrikel, *VN* Ventrales Nervensystem, *a* und *b* gleich vergrößert (Größenunterschied nicht auf Bestrahlung zurückzuführen)

dieser Zeit beginnen sich die Eingeweide und Skeletmuskeln zu differenzieren, so daß angenommen werden kann, der sog. Spättod hänge mit einer Differenzierungs-Störung der Muskulatur und der Hypodermis zusammen. Bereits differenziertes Gewebe ist strahlenresistent. Die gleiche Beobachtung konnten wir übrigens auch an bestrahlten Puppen machen [4].

Bei der Bestrahlung von 1 Std.-Eiern mit der notwendigen Alterseinschränkung von $\pm$ $^1/_4$ Std. zeigte sich zwischen 31 MeV-Photonen und 180 keV kein Wirkungsunterschied, (Fig. 4) ebenso nach der Bestrahlung von $1^3/_4$ Std.-Eiern. Die Schädigungskurve der $1^3/_4$ Std.-Embryonen verläuft deutlich anders, und die Tiere sterben, wie auch die 1 Std.-Embryonen an Frühtod. Mit dem Alter von

3 Std. stellen sich Wirkungsunterschiede ein, ebenso ändert die Todesart, indem lediglich Spättod vorkommt. Das gleiche gilt für 4 Std.- und 5¹/₂ Std.-Embryonen, und mit 7 Std. wird der Unterschied erheblich. Aus Fig. 4 wird deutlich, daß sich

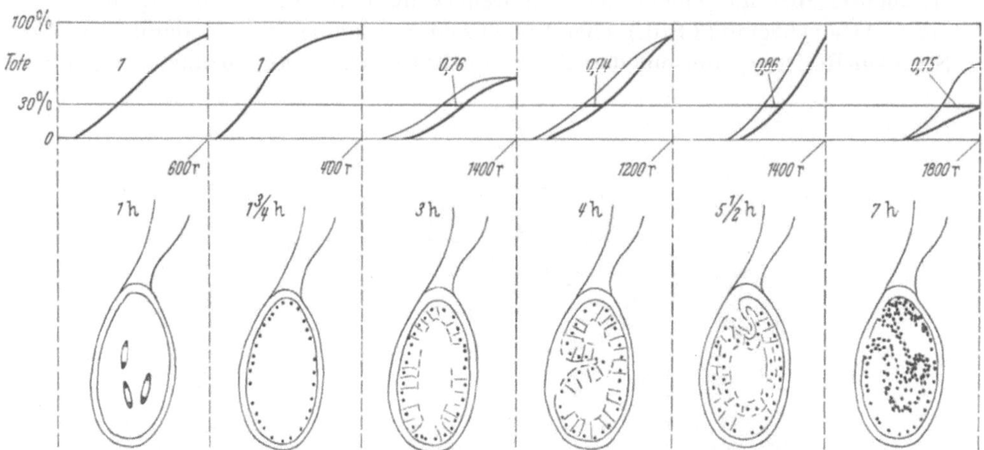


Fig. 4. Dosis-Effekt-Kurven von Drosophila-Embryonen verschiedenen Alters, mit 180 keV- und 31 MeV-Photonen Darunter sind die verschiedenen bestrahlten Entwicklungsstadien schematisch dargestellt. Zunächst Auswander der Furchungskerne aus dem Ei-Innern (1 Std.), Bildung eines Blastems (3 Std.) und nachfolgende Gastrulation Differenzierung von Geweben (5¹/₂—7 Std.). Oben Dosis-Effektkurven. Ordinate: Tote in Prozent, Abszisse Dosisbereich. Dünne Linie : 180 keV-Photonen. Dicke Linie: 31 MeV-Photonen, beide in Plexiglasphantom Eingetragen die RBW-Dosis für 30% Tote

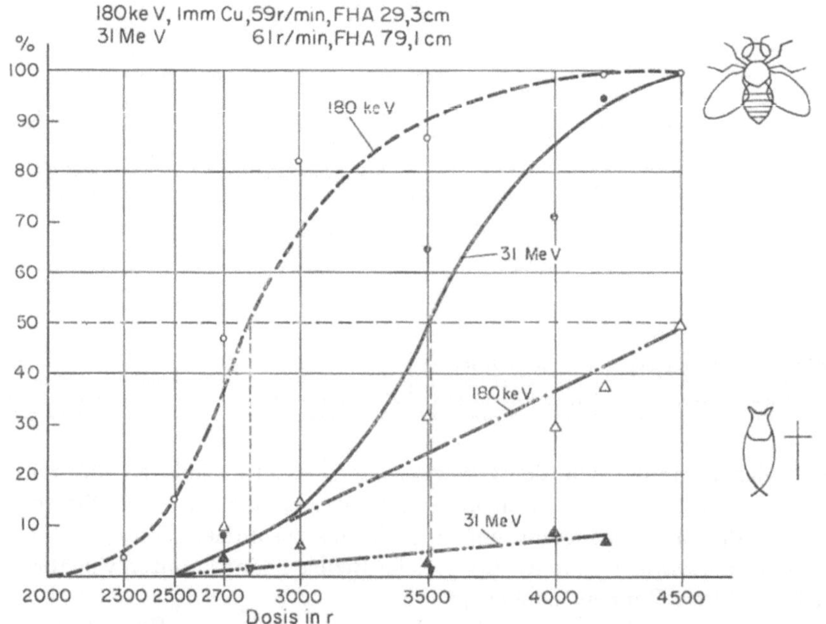


Fig. 5. Auftreten der Modifikation „gespreizte Flügel" nach Bestrahlung fünfstündiger Drosophila-Vorpuppe mit 180 keV und 31 MeV. Kreise: Modifikation. Dreiecke: Nichtgeschlüpfte Tiere

je nach dem Entwicklungsalter die Sensibilität und die Kurvenform ändern Ebenso variieren die Werte der RBW-Dosis. Mit 1 und 1³/₄ Std. ist die RBW$_D$ 1 und zwar zu dem Zeitpunkt, da noch keine Gestaltungsbewegungen und Differenzierungen eingeleitet sind. Die Kerne befinden sich im Innern und wandern nach außen, um ein Blastem zu bilden. Das biologische Objekt bleibt in allen Entwicklungsstadien gleich groß, so daß wir hier für die verschiedenen relativen biologischen Wirksamkeiten keinen Meßfehler verantwortlich machen können. Die unterschiedliche biologische Wirksamkeit muß vielmehr biologisch begründet sein [7].

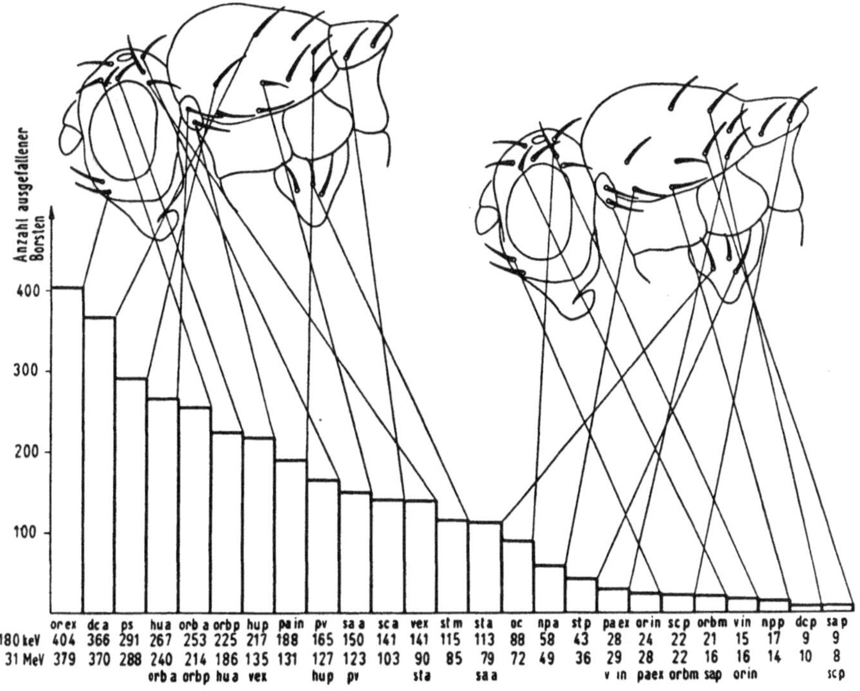



| | or ex | dc a | ps | hu a | orb a | orb p | hup | pa in | pv | sa a | sc a | vex | st m | st a | oc | npa | st p | pa ex | or in | sc p | orbm | v in | npp | dcp | sa p |
|---|---|---|---|---|---|---|---|---|---|---|---|---|---|---|---|---|---|---|---|---|---|---|---|---|---|
| 180 keV | 404 | 366 | 291 | 267 | 253 | 225 | 217 | 188 | 165 | 150 | 141 | 141 | 115 | 113 | 88 | 58 | 43 | 28 | 24 | 22 | 21 | 15 | 17 | 9 | 9 |
| 31 MeV | 379 | 370 | 288 | 240 | 214 | 186 | 135 | 131 | 127 | 123 | 103 | 90 | 85 | 79 | 72 | 49 | 36 | 29 | 28 | 22 | 16 | 16 | 14 | 10 | 8 |
| | | | | orb a | orb p | hu a | vex | | hup | pv | | sta | | sa a | | | v in | paex | orbm | sap | orin | | | | scp |

Fig. 6. Strahlensensibilität verschiedener Borsten von Drosophila nach Bestrahlung der Vorpuppe im Alter von 5 Std. mit 180 keV- und 31 MeV-Photonen (total 1200 Tiere bestrahlt). Angegeben die Anzahl der Fehlenden für jeden einzelnen Borstentyp. Alle Dosen zusammengefaßt. Ordinate: Summe der fehlenden Borsten. Abszisse: Anordnung der Borsten nach abfallender Empfindlichkeit. Man beachte, daß die Dorsocentrales anteriores (dca) außerordentlich sensibel sind, die unmittelbar dahinter liegenden Dorsocentrales posteriores aber strahlenresistent (aus NAVILLE [14])

Bei der Bestrahlung von Drosophila-Vorpuppen stellt sich beim erwachsenen Tier die Modifikation „gespreizte Flügel" ein [2]. In Fig. 5 sehen wir die Dosis-Effektkurven für 31 MeV-Photonen und 180 keV-Photonen. Die 31 MeV-Strahlen sind deutlich unterlegen [5]. Ebenso töten beispielsweise 3500 r 31 MeV weniger als 5% der Tiere, während dies nach 180 keV-Bestrahlung zu 30% geschieht. Eine andere interessante Modifikation ist der *Ausfall von Borsten* im streng determinierten Borstenmuster der adulten Fliege, wiederum nach Bestrahlung der Vorpuppe. Es zeigte sich [14], daß gesetzmäßig gewisse Borsten fehlen. Fig. 6 demonstriert die verschiedene Strahlenempfindlichkeit der Borsten, wobei es sich zeigt, daß sehr strahlensensible Borsten sich in unmittelbarer Nähe der unempfindlichsten befinden; dies beweisen die beiden dorsozentralen Thoraxborsten. In Fig. 7 wird ersichtlich, daß die 31 MeV-Photonenstrahlung nach einem gewissen Schwellenwert wiederum 180 keV unterlegen ist. Diese Strahlenreaktion stellt vermutlich eine Störung in der Bildung von Enzymen oder Differenzierungsstoffen dar.

**Mitosehemmung.** Eingehende Untersuchungen wurden angestellt, um vergleichend die Mitosehemmung in verschiedenen Geweben zu prüfen. Ein gutes Untersuchungsobjekt, das in Wasser bestrahlt werden kann, sind die *Wurzelspitzen von Vicia Faba*. 31 MeV-Photonen und schnelle Elektronen sind 180 keV-Photonen deutlich unterlegen [8, 9]. Dasselbe stellte sich bei der Bestrahlung des soliden Ehrlich-Carcinoms der weißen Maus heraus [18]. Wie Fig. 8 demonstriert, welche die Zahl Mitosen in Promille der Gesamtzellzahl nach Bestrahlung mit 3000 r darstellt, setzt die Mitosehemmung nach Betatronbestrahlung später und weniger ausgeprägt ein. Beim Ehrlich-

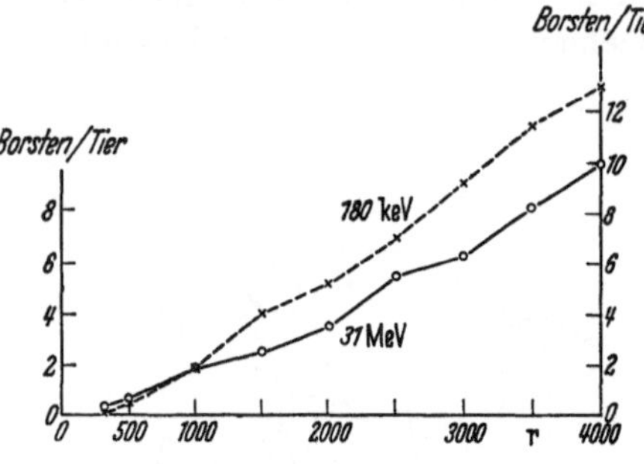


Fig. 7. Durchschnittliche Zahl der fehlenden Borsten pro Tier nach Bestrahlung der Vorpuppen mit 180 keV und 31 MeV (aus Naville [14])

Ascites-Carcinom, das sich als Testobjekt weniger eignet, erwies sich nach den Untersuchungen von Reiche [15] die 31 MeV-Strahlung ebenfalls als weniger wirksam.

**Mutationen.** *a) Sichtbare Mutationen.* Am aufschlußreichsten sind nun die vergleichenden Untersuchungen über die strahlenbedingte Auslösung des Mutationsvorganges, können sie doch erstens mit verhältnismäßig großen Zahlen angestellt werden und sind sie 2. relativ einfache fundamentale biologische Prozesse, bei denen der Weg physikalischer Effekt zum biologischen Effekt kurz ist. Zum dritten werden kleine Objekte, die Taufliege, bestrahlt, ein Umstand, der einer genauen Dosisbestimmung zugute kommt, und nicht zuletzt gestatten sie vielleicht, den Mutationsvorgang zu klären. Es hat sich leider gezeigt, daß die meisten strahlengenetischen Vorstellungen revisionsbedürftig sind.

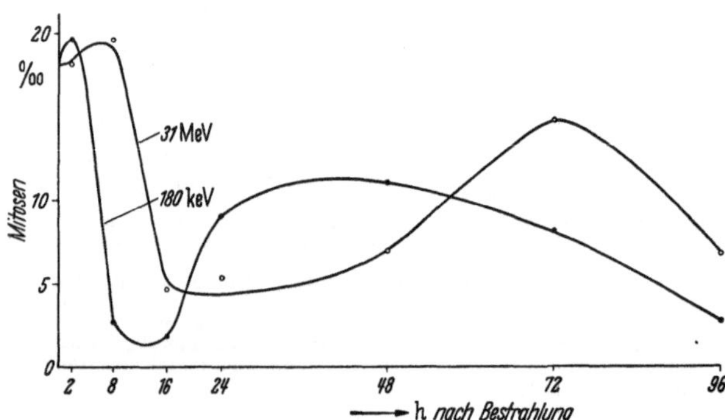


Fig. 8. Wirkung von 180 keV-Photonen und 31 MeV-Photonen auf die Mitosen des soliden Ehrlich-Carcinoms (3000 r Ionendosis, Lokalbestrahlung, hinter Plexiglas). Ordinate: Mitosen in Promille der Gesamtzellzahl. Abszisse: Std. nach Bestrahlung. Man beachte die später und weniger stark einsetzende Mitosehemmung der 31 MeV-Strahlen. Die Mitoserate steigt nach 31 MeV-Photonenbestrahlung später an als nach 180 keV-Bestrahlung (Zahlen aus Würmli [18])

Wir untersuchten die strahleninduzierte Mutationsrate des einzelnen Gens und zwar den Mutationsschritt eines normalen Gens zu seinem recessiven Allel. Geprüft wurden 5 Gene im Geschlechtschromosom [17]. Die Versuchsanordnung gestattete gleichzeitig die Aufdeckung von teilweisem Chromosomenverlust in der Entstehung von Gynandern. Fig. 9 zeigt die Anordnung. Bestrahlt werden normale Männchen, gekennzeichnet durch ihren Geschlechtskamm, die mit für die recessiven Gene homozygoten Weibchen gepaart werden (Gene: white, forked, yellow, crossveinless und vermilion). Auf der Figur das Gen für Gabelborsten (forked). Normalerweise entstehen heterozygote Bastarde, die einen normalen Phänotyp aufweisen, da ja das Gen für Gabelborsten recessiv ist. Mutiert im bestrahlten Männchen aber das normale Gen zu "forked", dann demonstriert das Bastardtier (bestrahltes Chromosom mal markiertes Chromosom) die forked-Eigenschaft. Beide Gene sind da. Ebenso wird in der ersten Generation eine dominante Mutation sichtbar. Wenn nun in den ersten Furchungsteilungen Zellen entstehen, die z. T. das bestrahlte X-Chromosom verloren haben, resultieren Gynander, die

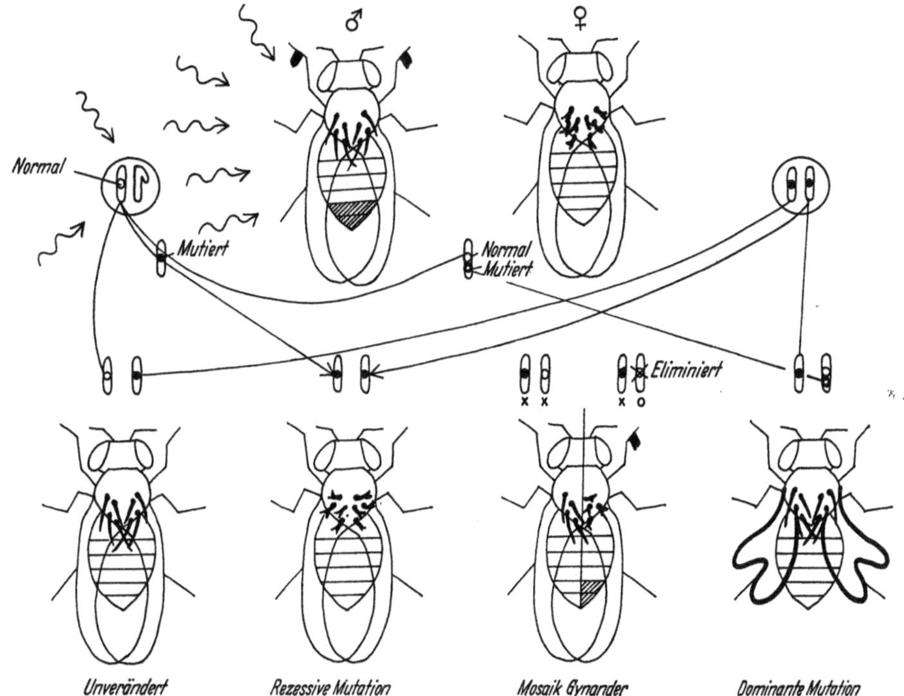


Fig. 9. Feststellung der Mutationsrate eines einzelnen, recessiven, sichtbaren Gens, ferner von Gynandern, dominanten, sichtbaren Mutationen und Mosaikmutationen. Bestrahlt werden normale Männchen, die gekreuzt werden mit Weibchen, die homozygot für die zu findenden recessiven Faktoren sind, in diesem Falle für forked = Gabelborsten. Ohne Mutation ist die $F_1$ normal. Jede Mutation des forked-Gens wirkt sich aber phänotypisch aus und kann festgestellt werden, ebenso der Verlust des X-Chromosoms in einem Teil des Körpers: Entstehung von XO-Gewebe, das männlich ist. Ebenso können dominante sichtbare Mutationen festgestellt werden

auf der einen Seite beide X-Chromosomen tragen, also Weibchen sind, auf der andern Hälfte aber nur ein X-Chromosom aufweisen, also Männchen, charakterisiert durch Geschlechtskamm und äußere Geschlechtsmerkmale, sind. Ebenso kann mosaikartig ein Teil des Körpers mutierte Gene aufweisen.

Beim Vergleich 180 keV zu 31 MeV stellte sich bei der Berechnung der durchschnittlichen Mutationsrate des Einzelgens kein Unterschied heraus (Experimente von Schmid [17]). Sie beträgt pro r etwa $9 \cdot 10^{-8}$ für Betatron und 180 keV. Total 40000 Chromosomen wurden getestet. Gynander traten nach Betatronbestrahlung weniger häufig auf.

*b) Recessive geschlechtsgebundene Letalfaktoren* (Tab. 1 u. 2). Recessive geschlechtsgebundene Letalfaktoren werden nach folgender Methode getestet: Männchen werden bestrahlt (Fig. 10) und mit markierten Weibchen (Muller-5-

Chromosom) gekreuzt. Bastardweibchen mit einem bestrahlten und einem unbestrahlten Geschlechts-Chromosom werden weitergekreuzt. Liegt nun im unmarkierten bestrahlten Chromosom ein Letalfaktor, so ist dieser männliche Genträger lebensunfähig. Sowohl 31 MeV-Photonen als auch 30 MeV-Elektronen erzeugen nun deutlich weniger recessive letale Mutationen als 180 keV-Strahlen.

Mit 31 MeV-Strahlen entstanden 1—7 Tage nach Bestrahlung mit 3000 r unter 1 859 geprüften Chromosomen 59 Mutationen = 3,2%, gegenüber 73 Mutationen unter 1 556 Chromosomen = 4,7% mit 180 keV. Die $RBW_E$ beträgt demnach 0,7. Nach Bestrahlung des empfindlichsten Keimzellstadiums (Spermien, die 5—7 Tage nach Bestrahlung zur Befruchtung gelangen) mit 30 MeV-Elektronen ergab sich eine Mutationsrate von 4,5% (73 Letalfaktoren unter 1 610 getesteten Chromosomen) gegenüber 8,9% = 30 Mutationen unter 336 getesteten Chromosomen nach 180 keV-Bestrahlung. Die Ionendosis war in beiden Versuchen 2000 r*.

*c) Chromosomenfragmentation.* Getestet wurde die Chromosomenfragmentation und Verlust (Fig. 11). Bestrahlt wurden Männchen mit dem recessiven Faktor yellow = Honigfarbe, im X-Chromosom. Der recessive Faktor kann sich aber nicht manifestieren, weil er durch den normalen Faktor, der sich in einem dem Y-Chromosom angeklebten Stück befindet, verdeckt wird[1]. Bricht dieses Stück durch Bestrahlung ab oder fehlt das ganze X- oder Y-Chromosom, so manifestiert das betroffene Tier die Eigenschaft honigfarben.

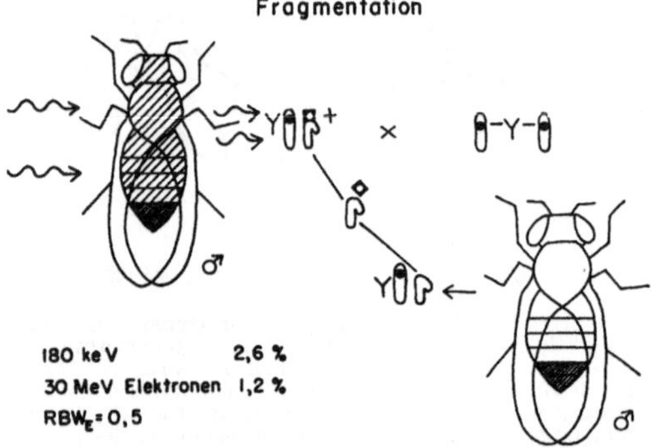


Fig. 10. Feststellung recessiver geschlechtsgebundener Letalfaktoren. Bestrahlt werden normale Männchen, die mit markierten Weibchen gekreuzt werden (X-Chromosom markiert mit dem dominanten Faktor Bar = Bandauge). Markiertes X-Chromosom = schraffiert. Die $F_1$-Weibchen werden mit gewöhnlichen Männchen gekreuzt. Ist im bestrahlten X-Chromosom ein recessiver Letalfaktor entstanden, fällt in der $F_2$ die Klasse der normalen Männchen aus

Fig. 11. Feststellung von Fragmentationen. Bestrahlt werden Männchen, die im X-Chromosom den recessiven Faktor y = yellow = gelbe Körperfarbe tragen. Dem Y-Chromosom der bestrahlten Männchen ist ein Stück X-Chromosom mit dem Normalallel von y angeklebt. Die Männchen sind demnach normalfarben. Fehlt in der $F_1$ das X-Stück des Y-Chromosoms, so kann sich y in den Söhnen hemizygot manifestieren

Unter 1767 mit Elektronen (2000 r Ionendosis) bestrahlten Chromosomen ereigneten sich 22 = 1,2% Brüche gegenüber 27 unter 1001 mit 180 keV, 2000 r bestrahlten Chromosomen = 2,7%.

---

* Vorversuche mit allerdings abweichenden Versuchsbedingungen ergaben andere Resultate, was auf eine nicht adaequate Dosismessung zurückzuführen ist

[1] An dieser Stelle möchte ich Herrn Dr. I. I. Oster, der uns die Stämme zum Bestimmen der Translokationen und Chromosomenbrüche zur Verfügung stellte, herzlich danken

Die Elektronen sind mit einer RBW von 0,4 deutlich unterlegen.

*d) Translokationen.* Dasselbe stellte sich bei der Erzeugung von Translokationen, dem Austausch von Chromosomenstücken heraus, nämlich eine $RBW_E$ von 0,4 (Fig. 12). Unter 471 mit 30 MeV-Elektronen (2000 r Ionendosis) bestrahlten Keimzellen fanden sich 10 Translokationen zwischen dem II. und III. Chromosom = 2,1%, während unter 184 mit 180 keV bestrahlten Keimzellen 10 Translokationen = 5,4% auftraten.

*e) Dominante Letalfaktoren.* Am interessantesten erwiesen sich die Experimente mit dominanten Letalfaktoren. Bestrahlt werden hier (Fig. 13) Männchen, die mit unbefruchteten Weibchen gepaart werden. Die von den bestrahlten Spermien befruchteten Eier werden auf ihre Fähigkeit zur normalen Weiterentwicklung geprüft.

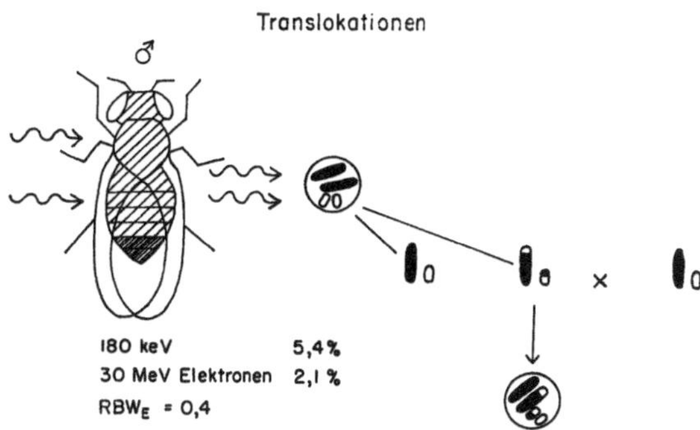


Fig. 12. Entstehung von Translokationen = Chromosomenstückaustausch. Dargestellt reziproker Stückaustausch zwischen dem schwarzen großen und dem kleinen hellen Chromosom

Indem zu verschiedenen Tagen die Weibchen gewechselt wurden, konnte das ganze Sensibilitätsmuster der Spermatogenese erfaßt werden. Fig. 14 zeigt den Versuch mit drei verschiedenen Strahlenqualitäten und der gleichen Dosis (etwa 910—930 rad). Auf der Abszisse sind die Tage nach Bestrahlung eingetragen, auf der Ordinate die embryonale Sterblichkeit in %. Die einzelnen Kurvenstücke umfassen die Zahlen aus den verschiedenen Zuchten. Wir sehen zunächst, daß die einzelnen Stadien der Spermatogenese sehr verschieden strahlenempfindlich sind. 8—10 Tage nach Bestrahlung sind die Keimzellen außerordentlich sensibel. Ich schließe aus andern Versuchen an, daß hier Gameten zur Befruchtung gelangten, die im Stadium der Reifeteilung bestrahlt werden. Unempfindlich sind die Spermatogonien, empfindlicher die reifen Spermien, während reifende Spermien bereits eine erhöhte Strahlenempfindlichkeit aufweisen. Die dicke Linie repräsentiert die embryonale Sterblichkeit nach 180 keV- Bestrahlung.

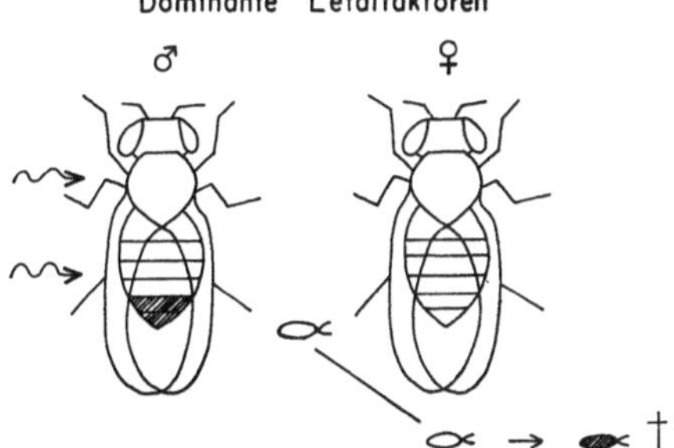


Fig. 13. Feststellung dominanter Letalfaktoren. Bestrahlt werden Männchen, die mit normalen Weibchen gekreuzt werden. Es wird die Entwicklung der Embryonen überprüft, die mit bestrahlten Spermien befruchtet worden waren. Die Zahl der Nichtweiterentwickelten gibt einen (allerdings nicht absoluten) Maßstab für die Zahl der strahleninduzierten dominanten Letalfaktoren

Jeder Mittelwert setzt sich aus 6 Versuchen mit je 400-500 untersuchten Embryonen zusammen, insgesamt 2400-3000 geprüften Embryonen pro Punkt. Die beiden kleinen Kreise zeigen den Versuch mit dem höchsten und dem niedrigsten Wert. Die Experimentestreuung ist außerordentlich gering, an manchen Punkten weniger als ±2% und kommt damit der Genauigkeit physikalischer Messungen gleich. (Zwischen den einzelnen Versuchen liegen oft Intervalle von einem Monat.) Dieser überraschend gut reproduzierbare, genaue strahlenbiologische Test erlaubte erstmals, ohne zweifelnde Hemmung, eine Analyse der verschiedenen biologischen Wirksamkeiten.

**Abhängigkeit der RBW vom Entwicklungsstadium.** Im gesamten zeigt sich zunächst, daß die verschiedenen Keimzellstadien ganz unterschiedlich auf die Strahlen ansprechen. Aus der Darstellung wird ersichtlich, daß energiereiche Photonen und Elektronen bei reifen Spermien mit derselben Ionendosis wie

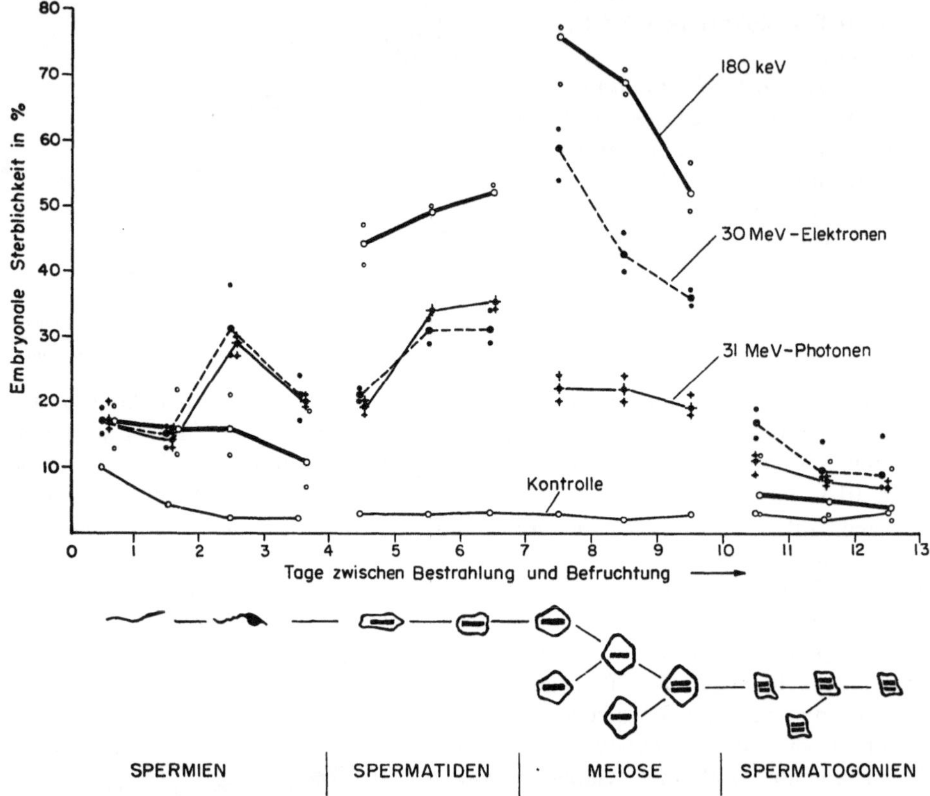



Fig. 14. Erzeugung dominanter Letalfaktoren durch 180 keV-Photonen (hinter 12,5 mm Plexiglas), 31 MeV-Photonen (hinter 40,5 mm Plexiglas) und 30 MeV-Elektronen in Luft (Ionendosis 1000 r = 80—110 r/min). Jeder Durchschnittswert stellt die Werte aus sechs Experimenten mit je 400—500 Embryonen, also total etwa 3000 Embryonen dar. Angegeben sind ferner die niedrigsten und die höchsten Werte. Es wird ersichtlich, daß die Werte nur wenig streuen. Für gewisse Punkte ist die Schwankung nicht mehr als ±2% und erreicht damit die Genauigkeit physikalischer Arbeiten. Bestrahlt wurden Männchen, die zu verschiedenen Tagen wieder mit neuen unbefruchteten Weibchen gekreuzt wurden. Die verschiedenen Zuchten stellen Keimzellen dar, die zu verschiedenen Stadien der Spermatogenese bestrahlt worden waren. Man sieht, daß zunächst in der 1. Zucht die energiereichen Strahlen den 180 keV-Strahlen überlegen sind, dann in der 2. Zucht gemeinsam unterlegen. In der Zucht 3 separieren sich die 3 Kurven, um wieder in Zucht 4 zusammenzufallen

180 keV die gleiche Wirkung haben, dann aber den weicheren Strahlen deutlich überlegen sind. In der Zucht 2 (Bestrahlung reifender Spermien) ändert sich das Bild. 180 keV-Strahlung ist deutlich überlegen, Photonen und Elektronen hoher Energie gemeinsam unterlegen. Nach der Bestrahlung meiotischer Stadien aber teilen sich alle drei Kurven und wir stellen zu unserer Überraschung fest, daß die 31 MeV-Photonen den 30 MeV-Elektronen unterlegen sind. Dosismeßfehler können ausgeschlossen werden, da sich in früheren Stadien die Wirkungen decken. Die Unterschiede müssen demnach im biologischen Reaktionssystem begründet sein.

**Qualitative Unterschiede.** Gleichzeitig zeigt die Darstellung, daß die Strahlen qualitativ verschieden wirken, eine Annahme, die durch weitere Untersuchungen mit verschiedenen Strahlendosen gestützt werden konnte. Fig. 15 stellt das Sensibilitätsmuster der einzelnen Keimzellstadien dar. Weiß bezeichnet die resistenten Stadien, hier die Spermatogonien und schwarz die sensiblen Stadien. Es sind die Spermatocyten in Reifeteilung dargestellt, die reifenden Spermien und die reifen Keimzellen. Nach Bestrahlung mit 30 MeV-Elektronen (1000 und 2000 r Ionendosis) erweisen sich stets die Zellen in Meiose am empfindlichsten, nach

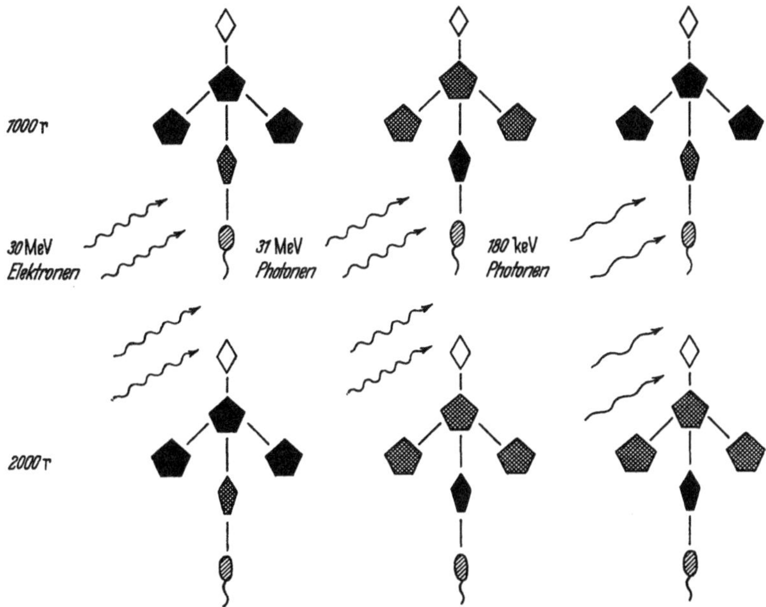



Fig. 15. Sensibilitätsmuster verschiedener Stadien der Spermatogenese von Drosophila. Angegeben zuoberst Spermatogonien, dann meiotische Stadien, dann reifende Spermatiden und zuunterst reife Spermien. Schwarz bedeutet: strahlensensibel, hell: relativ resistent. Es zeigt sich, daß mit 30 MeV stets, auch bei 2000 r, die meiotischen Stadien am empfindlichsten sind, während mit 31 MeV-Photonen die reifenden Spermatiden am meisten geschädigt werden. Mit 180 keV ändert die Sensibilität je nach Dosis

Bestrahlung mit Photonen aber die reifenden Spermien. Nach 180 keV-Bestrahlung wechselt das Sensibilitätsmuster mit der Dosis, 2000 r produzieren am meisten Letalfaktoren bei den reifenden Stadien, während 1000 r die meiotischen Stadien am meisten beeinflussen.

Bei der Betrachtung der verschiedenen RBW stellt sich das Faktum ein, das beim selben Objekt die RBW je nach Stadium ändert (Fig. 16). Eingezeichnet ist die RBW gegenüber 180 keV. 1 bedeutet gleiche Wirksamkeit der energiereichen Strahlung und der 180 keV-Photonen. In den ersten Stadien steigt die RBW bis auf 2 an, um dann unter 1 zu sinken. Die RBW ist ebenfalls dosisabhängig. Vergleichen wir die Wirkungen von 2000 r Ionendosis schnelle Elektronen mit 180 keV, so liegt die RBW zunächst unter 1, um später mehr als 1 zu betragen.

**Zusammenfassung der vergleichenden biologischen Versuche.** Zusammenfassend zeigt sich:

1. Die relative biologische Wirksamkeit der Betatronstrahlung gegenüber der konventionellen Röntgenstrahlung ist je nach Objekt und Entwicklungsstadium

verschieden. Sie ist ebenfalls dosisabhängig. Die verschiedene Wirkung muß also zum Teil wenigstens in einer differentiellen Wechselwirkung der Materie mit den verschiedenartigen Strahlungen begründet sein.

2. Es zeigten sich Unterschiede im Strahleneffekt von 30 MeV-schnellen Elektronen und 31 MeV-Photonen. Ebenso fanden sich qualitative Unterschiede, indem die Sensibilitätsmuster anders gestaltet waren.

3. Die weichere energieärmere Strahlung ist meistens, aber nicht immer den Betatronstrahlen überlegen. In einigen wenigen Fällen sind Betatronstrahlen wirksamer. Diese Ausnahmen interessieren uns.

**Diskussion.** Die Deutung dieser Phänomene ist schwierig. Es sei mir gestattet, den Versuch einer Deutung zu unternehmen und zwar auf Grund der Ergebnisse

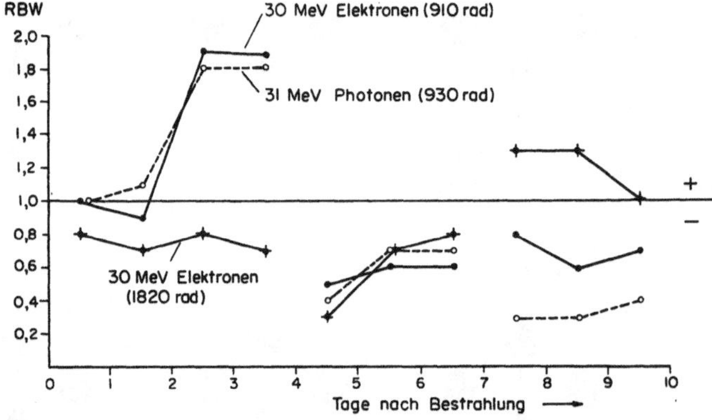


Fig. 16. Relative biologische Wirksamkeit in der Produktion dominanter Letalfaktoren zu verschiedenen Zeiten der Spermatogenese. 1 = Effekt für 180 keV Photonen. Zunächst liegt die RBW über 1, um wieder abzusinken, während in der 3. Phase die RBW der 31 MeV- und 30 MeV-Strahlen voneinander abweichen. Angegeben ist ebenfalls die RBW für 2000 r Ionendosis (rad umgerechnet mit Faktor 0,93 für 180 keV und 31 MeV-Photonen und 0,91 für 30 MeV-Elektronen)

mit dominanten Letalfaktoren. Prinzipiell können Wirkungsunterschiede der Betatronstrahlung und 180 keV-Strahlen folgendermaßen begründet sein:

1. Die verschiedene Emissionsform. Betatronstrahlen, sowohl Elektronen als auch Photonen werden in Blitzen von $10-15\mu$ 'sec abgegeben, während konventionelle Röntgenstrahlung kontinuierlicher erfolgt. Wirkungsunterschiede wie wir zwischen Elektronen und Betatron-Photonen beobachtet haben, lassen sich aber nicht erklären, es sei denn die minimale Änderung des Röntgenblitzes bei Elektronenstrahlung, die verschiedene Impulsform spiele eine Rolle. Wir sind daran diese Frage zu lösen.

2. Verschiedene Energieübertragung und Diffusion.

3. Der verschiedene Energieverlust der primären und sekundären Elektronen pro Weglänge, also der lineare Energieverlust spielt eine Rolle. Je geringer die Energie ist, um so größer ist der lineare Energieverlust. Entsprechend dem Energieverlust verhält sich die Anzahl der erzeugten Ionenpaare pro Weglänge. Man bezeichnet die lineare Verteilung der Ionenpaare als spezifische Ionisation. 180 keV-Photonen geringer Energie ionisieren dicht, Betatronstrahlen mehr als 10mal weniger dicht. 180 keV-Strahlen erzeugen etwa 100 Ionenpaare pro Mikron. Mit steigender Energie sinkt nun die Ionisationsdichte in Wasser auf den minimalen Wert von 5,7 bei 1 MeV, um aber nachher wieder anzusteigen. Die 31 MeV-

Photonen mit einer mittleren Energie von 10 MeV und ihrem großen Energie-spektrum haben vermutlich eine geringere Ionisationsdichte (6—7 pro $\mu$) als die homogenen 30 MeV-Elektronen, deren Ionisationsdichte über 8 beträgt (nach Ta-bellen von JOHNS [12]). Wir hätten demnach die 180 keV-Photonen als die am dichtesten ionisierenden Strahlen, dann die 30 MeV-Elektronen und ferner als die Strahlen mit geringstem linearen Energieverlust die 30 MeV-Photonen. Was hat nun die verschiedene Ionisationsdichte zur Folge? Es sei zur Erklärung eine chemische Wirkung der Ionisationen in Erinnerung gerufen, die vermutlich die

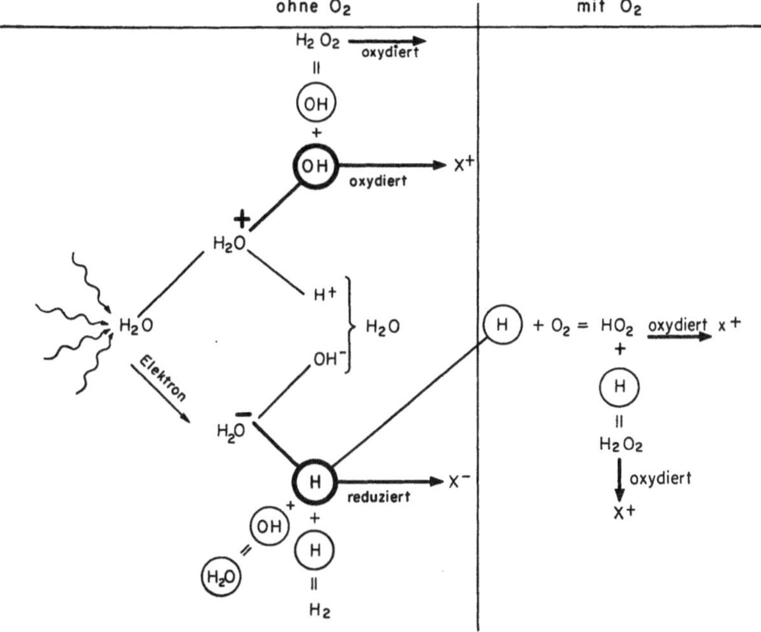


Fig. 17. Darstellung der Entstehung von biologisch aktiven Stoffen nach Bestrahlung von Wasser. Primärvorgang = Ionisation der Wassermoleküle. Zerfall der positiven und negativen Ionen in freie OH- und H-Radikale die oxydierend und reduzierend wirken. Links der Vorgang ohne Anwesenheit von O₂, rechts mit O₂. O₂-Anwesenheit vermehrt H₂O₂-Produktion, indem H-Radikale sich mit O₂ kombinieren können. X = Reaktionssystem

meisten strahlenbiologischen Reaktionen einleitet oder unterstützt. Ein bestrahltes Wassermolekül liefert OH- und (vermutlich) H-Radikale, die primär oxydie-rend und reduzierend wirken (Fig. 17). Bei Abwesenheit von $O_2$ können sie untereinander zu wiederum biologisch aktiven Verbindungen, wie $H_2O_2$, rekombi-nieren. Ist Sauerstoff anwesend, dann wird die $H_2O_2$-Produktion verstärkt. Die $H_2O_2$-Produktion ist stark milieuabhängig und ebenso variiert sie mit der Ioni-sationsdichte, wie dies EBERT [1] demonstrierte.

In Fig. 18 ist nach den Angaben von GRAY [11] die Entstehung der positiven Ionen und der negativen Partner durch den Durchgang eines dicht ionisierenden Partikels und eines Elektrons von 60 keV durch die Materie dargestellt. Die positiven Ionen sind beim Durchgang des $\alpha$-Partikels in einer Säule von $8 \cdot 10^{-4}\mu$ Radius angeordnet. Darum herum liegen die negativen Ionen. Nach der Diffusion, geringe Zeit später, haben sich die Verteilungen unmerklich verschoben. Anders beim Durchgang eines Elektrons. Negative und positive Ionen liegen distanziert voneinander und weichen nach der Diffusion noch mehr auseinander.

Bei der Verteilung der freien Wasserradikale ergibt sich nun ein gleiches Bild (Fig. 19). Bei dicht ionisierender Strahlung rekombinieren die OH-Radikale als

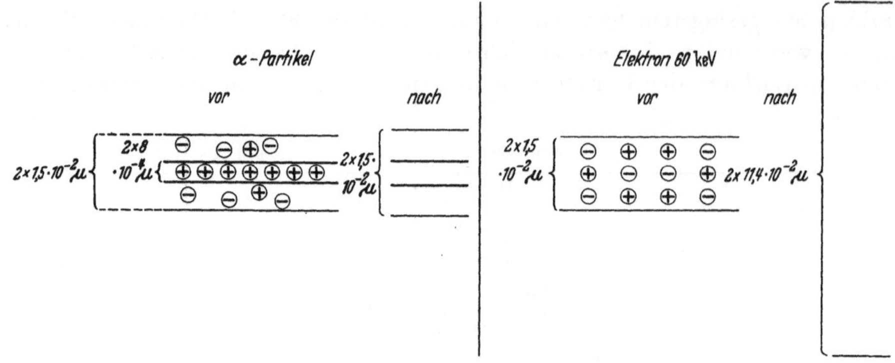


Fig. 18. Diagramm der räumlichen Verteilung der positiven und negativen Ionen im Augenblick ihrer Bildung durch ein α-Partikel und ein 60 keV-Elektron. Rechts die Ausdehnung der Säulen nach der Diffusion, bzw. nach der Zeit, in der 50% der Radikale einen Schock erleiden, (α-Partikel = $10^{-9}$—$10^{-11}$ sec, 60 keV-Elektron: $2 \times 10^{-8}$ sec Maße nach Gray 1955)

Abkömmlinge der positiven Wasserteilchen zu $H_2O_2$. Sauerstoffmoleküle können mit dieser Rekombination nicht konkurrieren. Anders die Reaktionen dünner ionisierender Strahlung. Die freien Radikale leben länger und Sauerstoff übt eine verstärkende Wirkung auf die $H_2O_2$-Produktion aus.

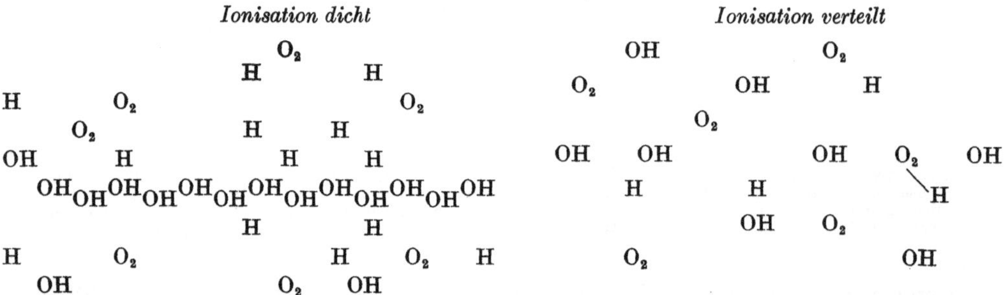


Fig. 19. Verteilung der OH- und H-Radikale nach dem Durchgang dicht ionisierender (links) und verteilt ionisierender Strahlung (rechts). Die dichte Anordnung der OH-Radikale (links) erleichtert ihre Rekombination zu dem biologisch aktiven $H_2O_2$: Die freien Radikale sind kurzlebig. Anwesendes $O_2$ übt kaum eine verstärkende Wirkung aus. Bei verteilten Ionisationen (rechts), Radikale langlebiger; Anwesenheit von $O_2$ verstärkt $H_2O_2$-Bildung.

Beides nun, $H_2O_2$ und die freien Radikale können biologisch aktiv sein. Ich nehme nun an, daß bei der Entstehung dominanter Letalfaktoren je nach Keimzellstadium einmal die freien Radikale als solche für die Reaktion verantwortlich sind, dann aber auch die $H_2O_2$-Moleküle. In der ersten Phase reagieren vornehmlich die Radikale. Die freien Radikale leben länger bei der energiereicheren Strahlung und tatsächlich ist, wie wir aus Fig. 14 sehen, die Betatronstrahlung wirksamer als 180 keV. Die primäre Wirkung der freien Radikale muß milieuunabhängig sein. Tatsächlich stellten wir fest, daß reife Spermien durch Bestrahlung in Stickstoffatmosphäre nicht geschützt wurden. Bei der Bestrahlung meiotischer Stadien gesellt sich nun ein anderer Wirkungsfaktor hinzu, das Wasserstoffsuperoxyd oder andere sekundär entstandene biologisch aktive Teilchen. Je verteilter nun die Ionenpaare sind, um so weniger $H_2O_2$ entsteht durch Rekombination. Die Entstehung dieser Sekundärprodukte muß ebenfalls milieuabhängig sein. Tatsächlich tritt hier nach Sauerstoffentzug ein starker Milieueffekt auf. In Fig. 14 sehen wir, daß in diesen Stadien die Strahlung mit der geringsten spezifischen Ionisation die unwirksamste ist. Allerdings müssen hier schon geringste Abweichungen der Ionisationsdichten von den Chromosomen registriert werden, verstärkt vielleicht durch die Anwesenheit von $O_2$.

Anschließend stellen wir fest, daß scheinbar einfache fundamentale strahlenbiologische Reaktionen eine Fülle von Problemen aufgeben. Vermutlich spielen Verteilung und damit Lebensdauer der primärwirkenden Radikale eine Rolle, und die Entstehung von Sekundärprodukten ist von entscheidender Bedeutung. Milieu, Strahlenbeschaffenheit, und vor allem das Reaktionssystem erzeugen gemeinsam die komplizierte Strahlenreaktion, die wir einmal zu erkennen hoffen[1].

## Zusammenfassung

1. Es wurde die biologische Wirkung gleicher Ionendosen von 31 MeV-Photonen, von schnellen 30 MeV-Elektronen, sowie von 180 keV-Photonen vergleichend geprüft.

2. Die relative biologische Wirksamkeit (RBW) der Betatronstrahlung gegenüber der konventionellen Röntgenstrahlung ist je nach Objekt und Entwicklungsstadium verschieden. Sie ist ebenfalls dosisabhängig. Untersucht wurde:

3. *Strahlentod.* Sowohl 30 MeV-Elektronen, als auch 31 MeV-Photonen sind in bezug auf den Strahlentod der weißen Maus den konventionellen Röntgenstrahlen unterlegen. Dasselbe gilt für die Tötung von 3-, 4-, $5^1/_2$- und 7 stündigen Drosophila-Embryonen. Hingegen sind bei der Tötung von 1 und $1^3/_4$ stündigen Drosophila-Embryonen Betatron- und 180 keV-Strahlen gleich wirksam. Bei der Inaktivierung von Escherichia coli-Bakterien stellte sich kaum ein Wirkungsunterschied heraus.

4. *Phänokopien.* In der Erzeugung von Modifikationen an Drosophila, welche Flügelstellung und Borstenmuster betreffen, sind die 180 keV-Photonen den Betatronstrahlen überlegen.

5. *Mitose.* 31 MeV-Photonen und schnelle Elektronen sind in ihrer Wirkung auf Mitosen in der Wurzelspitze von Vicia Faba, sowie im soliden Ascites-Ehrlich-Carcinom der Maus deutlich unterlegen.

6. *Mutationen.* Betatronstrahlen sind in der Produktion von recessiv geschlechtsgebundenen Letalfaktoren, Chromosomenfragmentation, Translokationen den 180 keV-Strahlen unterlegen. Je nach Entwicklungsstadium sind bei der Produktion von dominanten Letalfaktoren 180 keV-Photonen effektiver oder den Betatronstrahlen unterlegen.

---

[1] Die Experimente konnten durchgeführt werden dank der Mitarbeit von Fräulein J. Berger, Fräulein H. Inauen und Frau I. Misani sowie dank der Unterstützung durch die Schweiz. Nationalliga für Krebsbekämpfung und Krebsforschung und einer Zuwendung der Firma Brown-Boveri, Baden, denen allen herzlich gedankt sei.

7. Bei der Bestrahlung männlicher Keimzellen (Drosophila) zeigten sich qualitative und quantitative Unterschiede im Strahleneffekt von 30 MeV-schnellen Elektronen und 31 MeV-Photonen.

8. Die unterschiedlichen biologischen Wirkungen werden diskutiert. Zur Deutung wird der unterschiedliche lineare Energieverlust und eine verschiedene biologische Bedeutsamkeit von Sekundärreaktionen und Sekundärprodukten herbeigezogen. Vermutlich werden die Schädigungen bei reifen Spermien vornehmlich durch freie OH- und H-Radikale erzeugt, während unreife Spermien durch die $H_2O_2$-Moleküle geschädigt werden.

## Literatur


1. Ebert, M.: Angew. Chem. **67**, 169 (1955).
2. Fritz-Niggli, H.: Schweiz. med. Wschr. **1951**, 1218.
3. — Naturwiss. **39**, 485 (1952).
4. — Fortschr. Röntgenstr. **76**, 217 (1952).
5. — Fortschr. Röntgenstr. **80**, 28 (1954).
6. — Experientia (Basel) **10**, 209 (1054).
7. — Fortschr. Röntgenstr. **83**, 178 (1955).
8. — Naturwiss. **43**, 112 (1956).
9. — Oncologia (Basel) **9**, 269 (1956).
10. — u. J. Schmidlin-Mészaros: Strahlenther. **95**, 551 (1954).
11. Gray, L. H.: Aus «Actions chimiques et biologiques des radiations». Paris: Masson et Cie. 1955.
12. Johns, H. E.: Aus "Radiation Dosimetry". P. 531—596. New York: Academic Press Inc. 1956.
13. Lindenmann, J.: Oncologia (Basel) **6**, 1 (1953).
14. Naville, B.: Oncologia (Basel) **8**, 55 (1955).
15. Reiche, K.: Strahlentherapie **97**, 549 (1955).
16. Schinz, H. R., u. H. Fritz-Niggli: Radiol. clin. (Basel) **25**, 371 (1956).
17. Schmid, W.: In Veröffentlichung.
18. Würmli, H.: Oncologia (Basel) **7**, 306 (1954).

Die Einwirkung schneller Elektronen
auf Gewebekulturen des HeLa-Stammes*

Von

H. GÄRTNER, Tübingen

Die Gewebezüchtung hat sich nach den bisher gesammelten Erfahrungen zum Nachweis strahlenbedingter Effekte und vornehmlich auch zur Klärung der Wirkungsweise verschiedener Strahlenqualitäten als geeignet erwiesen.

Als Testobjekte wurden von mir bisher Hühnerherzfibroblastenkulturen, also normales embryonales Gewebe mesenchymaler Herkunft verwendet. Sowohl aus rein theoretischem Interesse als auch von praktisch-therapeutischen Gesichtspunkten aus erschien es folgerichtig, nunmehr die strahlenbiologische Reaktion an Tumorzellen selbst, und zwar möglichst an menschlichem Carcinomgewebe in vitro zu überprüfen.

Seit Beginn meiner Tätigkeit im Strahleninstitut der Universität Tübingen habe ich daher mit der dankenswerten Förderung durch Herrn Professor BAUER, der krankheitshalber heute nicht anwesend sein kann, Bestrahlungsversuche an den sogenannten HeLa-Kulturen vorgenommen. Die ersten Ergebnisse möchte ich Ihnen heute vorlegen. Ich darf vorausschicken, daß es sich bei den HeLa-Zellen um Explantate eines sehr unreifen, nicht verhornenden Plattenepithel-Carcinoms der Portio uteri handelt, das GEY und Mitarbeiter 1951 anzüchten konnten und das seitdem als Stamm HeLa fortlaufend in Roller-Tubes kultiviert wird.

Das epitheliale Wachstum dieses Tumors erfolgt in vitro vorwiegend einschichtig und erleichtert somit die cytologische Auswertung. Der maligne Charakter kommt in einer erheblichen Teilungsaktivität und zahlreichen Mitosestörungen in Form von mehrpolaren Teilungen und Riesenzellbildung zum Ausdruck. Die pathologischen Mitosen treten in einer Frequenz von etwa 25% auf, eine Tatsache, die bei der Beurteilung von Strahleneffekten an HeLa-Zellen von vornherein Berücksichtigung finden muß.

Unter Einhaltung der erforderlichen biologischen und physikalischen Versuchsbedingungen erhielten die verschiedenen Versuchsserien Einzeldosen von 250 und 500 r schneller Elektronen des 15 MeV-Betatrons mit einer Dosisleistung von 40 r/min. Zum Vergleich kamen in Parallelversuchen 250 und 500 r Röntgenstrahlen von 180 kV einer angeglichenen Dosisleistung zur Anwendung.

Zur *cytologischen* Auswertung wurden durch entsprechende, insgesamt 12 verschiedene Fixierungszeiten jeweils der Primär- und Sekundäreffekt getrennt erfaßt. Die Gesamtbeobachtungszeit erstreckte sich über 24 Std. In jeder

---

* Aus dem Medizinischen Strahleninstitut der Universität Tübingen (Direktor: Prof. Dr. R. BAUER).

9*

Versuchsreihe wurden pro Fixierungszeitpunkt etwa 2000 Zellen ausgewertet und dabei folgende Kriterien berücksichtigt.

1. Mitosehäufigkeit
2. Mitosephasenverteilung
3. geschädigte Mitosen
4. mehrkernige Mitosen und solche mit Nebenkernen
5. nekrobiotische Zellvorgänge, also Pyknosen, Chromatolysen und Cytolysen.

Die *statistische Aufarbeitung* der Auswertungsergebnisse erfolgte durch Mittelbildung unter Berücksichtigung der Zeitintervalle als Gewichte und anschließende statistische Sicherung der ermittelten Häufigkeitsdifferenzen.

Beim Vergleich der verwendeten Strahlenarten fällt zunächst auf, daß sowohl nach Einstrahlung von 250 r als auch 500 r Röntgenstrahlen üblicher Härte, einer Dosisleistung von 40 r/min, die *Mitosefrequenz* stärker absinkt als nach gleichen Dosen schneller Elektronen angeglichener Dosisleistung. Dieser Effekt tritt sowohl während der Ablaufzeit des Primäreffektes als auch der Späteffekte und damit auch während der Gesamtbeobachtungszeit in Erscheinung.

Tabelle 1. *Vergleich der Mitosenfrequenz*
(HeLa-Kulturen)

| | Grad der statistischen Sicherung der Differenz 0—24 Std. | Dosisäquivalentfaktor El/Rö |
|---|---|---|
| 250 r El/250 r Rö 40 r/min | +++ | 0,541 |
| 500 r El/500 r Rö 40 r/min | +++ | 0,567 |

+++ = sehr gut gesichert.

Die ermittelten Häufigkeitsdifferenzen sind statistisch signifikant. *Die errechneten Wirkungsfaktoren betragen für beide Versuchsdosen etwa 0,5. Schnelle Elektronen sind demnach in den untersuchten Dosis- und Dosisleistungsbereichen bezüglich der Mitosehemmung von HeLa-Zellen nur halb so wirksam wie entsprechende Dosen Röntgenstrahlen klassischer Härte bei gleicher Dosisleistung.*

Tabelle 2. *Vergleich der Schädigungsraten*
(HeLa-Kulturen)

| | geschädigte Mitosen 0—24 Std. | Mehrkernige Zellen Nebenkerne 0—24 Std. |
|---|---|---|
| 250 r El/250 r Rö 40 r/min | 0 | 0 |
| 500 r El/500 r Rö 40 r/min | +++ | ++ |

0 = nicht gesicherte     ⎫ Frequenz-
++ = gut gesicherte     ⎬ unter-
+++ = sehr gut gesicherte ⎭ schiede.

Wie Sie sehen, liegt dabei *die Rate der morphologisch sichtbar geschädigten mitotischen Zellen* nach Einstrahlung der geringeren Versuchsdosis von 250 r Röntgenstrahlen üblicher Härte und den schnellen Elektronen des Betatrons praktisch in der gleichen Größenordnung. Nach Anwendung der doppelt so hohen Dosis von 500 r steigt die Schädigungsrate nach Einwirkung von Röntgenstrahlen jedoch stärker an. Während des Primär- und Sekundäreffektes und damit auch während der Gesamtbeobachtungszeit läßt sich die stärker schädigende Wirkung der klassischen Röntgenstrahlen als statistisch signifikant nachweisen. Dieser Befund wird noch durch die Tatsache gestützt, daß nach der höheren Versuchsdosis auch die Zahl der mehrkernigen und derjenigen Zellen mit Nebenkernen als Ausdruck der Mitoseschädigung signifikant erhöht ist.

Da die Untersuchung der *nekrobiotischen Zellvorgänge* seinerzeit an Hühnerherzfibroblasten auf die Möglichkeit qualitativer Wirkungsunterschiede aufmerksam
gemacht hatte, verdient die jetzt an HeLa-Kulturen festgestellte Frequenz an
Chromatolysen, Cytolysen und Pyknosen besonderes Interesse.

Es läßt sich feststel-
len, daß nach Ein-
strahlung der Versuchs-
dosis von 250 r beider
Strahlenarten kein nen-
nenswerter Frequenz-
unterschied an tödlich
geschädigten Zellen be-
steht. Erst nach Ein-
wirkung von 500 r
nimmt die Letalitätsrate

Tabelle 3. *Vergleich der nekrobiotischen Zellformen*
(HeLa-Kulturen)

| | Cytolysen 0—24 Std. | Chromatolysen 0—24 Std. | Pyknosen 0—24 Std. | zusammen 0—24 Std. |
|---|---|---|---|---|
| 250 r El/250 r Rö 40 r/min | 0 | 0 | 0 | 0 |
| 500 r El/500 r Rö 40 r/min | + + | 0 | 0 | + + |

0 = nicht gesicherte Frequenzunterschiede.
+ + = gut gesicherte Frequenzunterschiede.

nach Röntgenbestrahlung signifikant zu. Diese Zunahme kommt ausschließlich durch stärkere cytolytische Vorgänge während der Ablaufzeit des Primäreffektes zustande und tritt somit auch während der Gesamtbeobachtungszeit
statistisch in Erscheinung. Während des Sekundäreffektes ist dieser Frequenz-
unterschied nicht mehr vorhanden. Nach der vorläufigen Analyse handelt es sich
um einen rein quantitativen Unterschied, der nach der vorhin genannten stärkeren
Zunahme an geschädigten Mitosen, mehrkernigen und Zellen mit Nebenkernen
auch durchaus zu erwarten war.

*Zusammenfassend kann also gesagt werden, daß an HeLa-Kulturen quanti-
tative Wirkungsunterschiede zwischen schnellen Elektronen und Röntgenstrahlen
üblicher Härte eindeutig in Erscheinung treten, während qualitative Unterschiede
an HeLa-Zellen bisher nicht nachweisbar sind.*

Es liegt nunmehr nahe, abschließend die Strahlenreaktion an Hühnerherzfibroblasten und HeLa-Zellen, die je nach Ursprung und Art ganz verschieden
sind, einem kritischen Vergleich zu unterziehen.

Sie sehen zunächst, daß 250 r und
500 r schneller Elektronen und 180 kV-
Röntgenstrahlen angeglichener Dosisleistung bei Hühnerherzfibroblasten zu einer
signifikant stärkeren Mitosehemmung
führen als bei Carcinomzellen des HeLa-
Stammes. Es muß weiteren Versuchen

*Tabelle 4*

| Strahlenart | Mitosenhemmung |
|---|---|
| Rö  40 r/min (180 kV) | Fibroblasten > HeLa |
| El  40 r/min (15 MeV) | Fibroblasten > HeLa |

vorbehalten bleiben zu klären, ob bei höheren Dosen oder Veränderungen der
zeitlichen Dosisverteilung eine Angleichung bezüglich der Herabsetzung der
Mitoserate in beiden Gewebsarten stattfinden wird. Jedenfalls scheinen zur
anhaltenden Abbremsung der Teilungsaktivität bei den äußerst malignen HeLa-
Zellen höhere Dosen notwendig zu sein.

Festzuhalten ist jedoch die Tatsache, daß die ermittelten Dosisäquivalent-
faktoren in dem untersuchten Dosis- und Dosisleistungsbereich für beide Gewebsarten mit etwa 0,5 von gleicher Größenordnung sind.

Was aber demgegenüber die tödliche Schädigung beider Zellarten anbelangt,
so läßt sich folgendes sagen (s. Tabelle 5).

Bei röntgenbestrahlten HeLa-Zellen besteht die gleiche Zellschädigungsfrequenz wie bei Hühnerherzfibroblasten.

Nach Einwirkung schneller Elektronen läßt sich demgegenüber aber bei HeLa-Zellen eine stärkere Nekrobiosefrequenz als bei Hühnerherzfibroblasten nachweisen. Das bedeutet, *daß schnelle Elektronen in bezug auf die tödliche Schädigung von Carcinomzellen des HeLa-Stammes einen stärkeren Effekt hervorzurufen vermögen als bei Hühnerherzfibroblasten.* Es läßt sich daraus folgern, daß gegenüber der Röntgenbestrahlung zumindest in vitro die irreversiblen Zellvorgänge im Carcinomgewebe nach Elektronenbestrahlung häufiger zu beobachten sind als im normalen Gewebe.

*Tabelle 5*

| Strahlenart | Nekrobiotische Zellvorgänge |
|---|---|
| Rö  40 r/min<br>(180 kV) | HeLa = Fibroblasten |
| El  40 r/min<br>(15 MeV) | HeLa > Fibroblasten |

Wenn man davon ausgeht, daß die irreversiblen Zellschädigungen einen entscheidenderen Faktor zur Beurteilung der Strahlenwirksamkeit darstellen als die Mitosehemmung, dann deutet der Vergleich der Versuchsergebnisse zwischen elektronenbestrahlten HeLa-Kulturen und elektronenbestrahlten Hühnerherzfibroblastenkulturen erneut auf die schon anhand der früheren Ergebnisse diskutierte Möglichkeit einer stärkeren elektiven Wirksamkeit der schnellen Elektronen hin.

## Diskussionsbemerkungen

H. R. SCHINZ (Zürich):

Zur Ergänzung des Vortrages meiner Mitarbeiterin Frau FRITZ-NIGGLI möchte ich noch folgende Ausführungen machen:

1. Die Strahlenempfindlichkeit verschiedener biologischer Systeme ist eine ganz unterschiedliche: so stirbt der Mensch nach 600 r Allgemeinbestrahlung, das 1 Std. alte Drosophilaei nach 100 r, die Drosophilafliege erst nach 60000—100000 r.

2. Die Drosophila-Versuche sind mit großen Zahlen durchgeführt. Die einzelnen Versuchspunkte liegen so nahe beieinander, daß auch nach dem berühmten Mathematiker VAN DER WAERDEN in Zürich eine Signifikanz-Berechnung überflüssig wird.

3. Die letzte biologische Einheit, die wir kennen, ist das Gen. Deshalb die Versuche mit Punktmutationen. Die höhere biologische Einheit ist das Chromosom, dann folgt die Zelle.

4. Die Strahlenbiologie versucht die Lebensvorgänge aufzuklären. Sie kann noch nicht Grundlage der Strahlentherapie sein. Diese geht immer noch rein empirisch vor und muß an der klinischen Statistik geprüft werden. Mitosenstopp und Mitosenuntergang genügt nicht; die ganze Zelle muß cancerostatisch oder cancericid geschädigt werden. Bei der großen Verschiedenheit zwischen den Carcinomtypen hinsichtlich ihrer Strahlenempfindlichkeit dachte man früher an verschiedenen Gehalt von sensiblem Eiweiß; es ist aber auch möglich, daß der Unterschied im Vorhandensein von zelleigenen Schutzstoffen liegt.

B. RAJEWSKY (Frankfurt a. M.):

An welchem Testobjekt halten Sie es für zweckmäßig, die RBW zu prüfen, an Einzelzellen, Zellkolonien, Geweben oder Organismen? Oder wollen Sie noch kleinere Einheiten wählen, also Chromosomen, Moleküle, Ribonucleinsäure, Desoxyribonucleinsäure.

R. KEPP (Gießen):

Der Unterschied in der Strahlenempfindlichkeit verschiedener Gene, auf den Frau FRITZ-NIGGLI hingewiesen hat, dürfte auf die unterschiedliche Strahlenempfindlichkeit von verschiedenen Aminosäurekombinationen zurückzuführen sein, der neuerdings in der physiologischen Chemie nachgewiesen wurde.

Mit der Frage, ob die Ultrafraktionierung, d. h. der Impulsbetrieb bei Apparaten zur Erzeugung schneller Elektronen oder ultraharter Röntgenstrahlen, wie Elektronenschleuder, Linearbeschleuniger und Synchrotron, für die biologische Wirkung der Strahlung eine Rolle spielt, habe ich mich mit meinen Mitarbeitern beschäftigt. Herr D. HOFMANN wird in seinem Vortrag näher darauf eingehen. Ich möchte hier nur darauf hinweisen, daß sich im Modellversuch eindeutige Unterschiede bei ultrafraktionierter Bestrahlung zwischen klassischen Röntgenstrahlen und Elektronen finden. Um strahlenbiologische Untersuchungen vergleichen zu können, müßte stets angegeben werden, mit welcher Dosisleistung gearbeitet wurde und ob die energiereiche Strahlung ultrafraktioniert war bzw. welche Impulsfrequenz pro sec sie aufwies, oder ob es sich um eine kontinuierliche energiereiche Strahlung wie die des Radiokobalts handelte.

### G. HÖHNE (Hamburg):

Wir haben vor 7 Jahren an der Göttinger 6 MeV-Elektronenschleuder mit Elektronen die Letalfaktorenrate bestimmt und dabei gefunden, daß die relative biologische Wirksamkeit ungefähr 1 ist. Wir haben die Versuche mit Elektronen von 15 MeV jetzt wieder aufgenommen. Sie sind noch im Laufen, aber es scheint auch hier wieder ungefähr der Faktor 1 herauszukommen. Die Streuungsbreite ist jedenfalls so groß, daß man keinen sicheren Unterschied etwa zwischen 0,8 und 1 machen kann. Wir haben weiter Chromosomen-Translokationsversuche an Drosophila-Eiern gemacht und auch hierbei ergab sich das Wirkungsverhältnis von ungefähr 1.

### R. WIDERÖE (Baden/Schweiz):

Was ergaben die Experimente mit 10 MeV- und 30 MeV-Elektronen ?

### O. EICHLER (Heidelberg):

Die Reichweite der Radikale errechnet sich aus ihrer Lebensdauer von etwa $10^{-8}$ sec im wäßrigen Milieu nach der statistischen Mechanik, wie sie von WEISS ausgeführt wurde. Die Rechnung gilt aber nicht mehr im biologischen Milieu, bei der Hemmungen auftreten und Regulationen (dargelegt in „Prinzipien des Lebendigen") wirksam werden. Um diese auszuschalten, müssen 2 benachbarte Ionisierungen (ohne Energieübertragung) nicht zu weit auseinander liegen.

Deshalb müssen Strahlungen mit dichteren Ionisierungen wirksam sein. Diese Differenzen werden eines Tages zur Prüfung periodischer Strukturen des Zellplasmas Bedeutung erlangen.

### S. RÖSINGER (Höchst a. M.):

Die verschiedene Wirksamkeit der Strahlungen von 180 kV und 31 MeV sowie die Unterschiede nach verschiedenen Meßzeiten kann auch eine rein chemische Ursache haben. Wir haben den Einfluß definierter org. Substanzen auf das System Fe II bis Fe III untersucht und z. B. bei den Molekülen, die 2 C-Atome enthalten, gefunden, daß Essigsäure praktisch keinen Einfluß auf die Oxydationsrate hat. Alkohol, Aldehyd und Kohlenwasserstoff dagegen steigern die Ausbeute an Fe III.

Die Abhängigkeit einer strahlenchemischen oder strahlenbiologischen Reaktion vom $p_H$ kann auch zu größeren Unterschieden in den Meßwerten führen. Gerade bei der Oxydation von Fe II steigt die Ausbeute mit abnehmendem $p_H$.

Da bei verschiedenen strahlenchemischen Reaktionen keine Abhängigkeit von der Dosisleistung in einem Bereich von 0,2—$10^5$ r/min gefunden wurde, ist nur mit einer momentanen Radikalverteilung zu rechnen. Es treten also, wie z. B. RAJEWSKY andeutete, keine verschiedenen Radikalkonzentrationen nach verschiedenen Bestrahlungszeiten auf. Es ist nur mit einer unterschiedlichen Diffusionslänge der gebildeten Radikale mit der Ionisationsdichte zu rechnen.

### B. MARKUS (Göttingen):

Eine scheinbar noch so genaue Arbeit zur Bestimmung relativer biologischer Wirksamkeit verschiedener Strahlenarten ist wertlos, wenn sie nicht angibt, wie die Dosis jeweils gemessen wurde. Darüber hinaus ist es wesentlich, daß die verwendeten räumlichen Dosisverteilungen gleich oder wenigstens mit guter Näherung gleich sind. Bei schnellen Elektronen gegenüber Quantenstrahlen sind prinzipiell die Dosisverteilungen verschieden, daher Dosisäquivalente besonders schwierig zu erhalten. Bisher hat man es so gemacht, daß entweder nur solche biologischen Testobjekte verwendet wurden, bei denen infolge ihrer Kleinheit angenähert gleichmäßige Dosisverteilungen vorlagen, oder aber, man hat die Verschiedenheit der räumlichen Dosisverteilungen ignoriert. — Da für die praktische Strahlenanwendung in der Medizin ausgedehnte Testobjekte, z. B. die menschliche Haut im weiten Sinne, besonders interessant sind, wurde versucht, die räumliche Dosisverteilung schneller Elektronen an geeignet gewählte von konventionellen Röntgenstrahlen für Wasser als Phantomsubstanz angenähert anzugleichen. Dies ist uns in einigen Fällen gelungen. Nähere Angaben werden in einer im Druck befindlichen Mitteilung gemacht.

H. Fritz-Niggli (Zürich):

Als Elementareinheit im biologischen System erscheint das Chromosom im lebenden Organismus zur Forschung des Strahleneffektes am zweckmäßigsten. Isolierte Systeme, wie z. B. Mitochondrien, zeigen in vitro eine andere Reaktion als in vivo. Die RBW als Bezugswert für den Kliniker muß empirisch erprobt werden.

Ultrafraktionierung ist im makroskopischen Bereich wichtig. Chromosomenbrüche und nachfolgende Rekombination werden durch konzentrierte Strahlen gefördert.

Die RBW für rezessive Letalfaktoren ändert sich mit dem Entwicklungsstadium. Reife Spermien sind milieuunabhängig und besitzen für 31 MeV Photonen und Elektronen/180 kV Photonen eine RBW von 1. Die RBW für Translokationen von reifen Spermien ist vermutlich ebenfalls 1.

Bei Experimenten mit herabgesetzter Energie von 30 MeV auf 10 MeV entsprechen sich die Ergebnisse, lediglich bei reifen Spermien ergaben sich geringe Unterschiede.

M. Sempert (Baden/Schweiz):

Zur Frage nach der Art der Dosismessung bei den Experimenten von Frau Fritz-Niggli möchte ich sagen, daß wir mit einer dünnwandigen Ionisationskammer (0,01 mm Aluminium) vor dem Objekt gemessen haben. Dabei ist eine Genauigkeit von 0,8% bezüglich der Victoreenkammer von 250 r erreichbar.

Bei der Dosismessung mittels Fingerhutkammer ist die Anordnung der Meßkammer vor oder hinter dem Objekt ungünstig wegen der Störung durch Vielfachstreuung. Es sind Fehler bis zu 50% möglich. Günstiger ist die Anordnung der Ionisationskammer neben dem Objekt, vorausgesetzt, daß ein homogenes Elektronenfeld vorliegt.

Der zeitliche Verlauf der Strahlung des 31 MeV-Betatron von BBC ist so, daß die Halbwertsbreite für Gammaimpulse 7 $\mu$/sec und für Elektronen 5 $\mu$/sec beträgt.

H. Lettré (Heidelberg):

1. Der HeLa-Stamm wird in vielen Laboratorien insbesondere für Virusforschung verwandt und ist leicht erhältlich. Es ist offen, ob bei der Züchtung in verschiedenen Laboratorien der Erde eine Aufspaltung in Linien verschiedener Qualität eingetreten ist. Eine Standardisierung wäre gerade mit ionisierender Strahlung möglich.

2. Ergebnisse an in vitro gezüchteten normalen und malignen Zellen müssen nicht unmittelbar auf die Verhältnisse in vivo übertragbar sein. Gerade Fibroblasten werden in vitro zu starkem Wachstum angetrieben und müssen daher in vitro strahlenempfindlicher sein als in vivo.

H. R. Schinz (Zürich):

Daß auch in vitro Tumorgewebe empfindlicher ist als Normalgewebe, ist nicht erstaunlich, basiert doch unsere ganze Strahlentherapie auf der elektiven Wirkung auf Tumorgewebe im Vergleich zum Muttergewebe. Hingegen würde mich noch interessieren zu erfahren, was der Vergleich 15 MeV Elektronen zu 15 MeV Photonen ergeben hat.

R. Wideröe: (Baden/Schweiz):

Wie wurde die Dosis gemessen?

B. RAJEWSKY (Frankfurt a. M.):

Eine wesentliche Feststellung der Experimente von Frau Dr. GÄRTNER ist die, daß die RBW dosisabhängig ist. Bei kleinen Dosen treten keine Unterschiede auf, wohl aber bei großen. Ich möchte fragen, warum haben Sie die Experimente nicht mit Dosen über 500 r fortgesetzt?

H. GÄRTNER (Tübingen):

Bei den hier mitgeteilten Versuchen handelt es sich um die ersten Ergebnisse. Eine Steigerung der Dosis wie in früheren Versuchen mit Hühnerherzfibroblasten bis zu einer Höchstdosis von etwa 2000 r ist für die nächste Zeit vorgesehen. Dabei wird sich zeigen, ob die Wirkungsfaktoren sich mit Steigerung der Dosis, ebenso wie bei Hühnerherzfibroblasten, wiederum dem Faktor 1 nähern. — Ebenso ist für weitere Experimente auch eine Veränderung der Dosisleistung vorgesehen.

G. BREITLING (Tübingen):

Die Dosismessung bei den Experimenten von Frau Dr. GÄRTNER habe ich mit einem Victoreendosimeter durchgeführt, und zwar derart, daß ein Monitor mit Hilfe der am Bestrahlungsort angeordneten Victoreenkammer geeicht wurde.

# Die relative biologische Wirksamkeit ultraharter Strahlungen und aus ihr sich ergebende Folgerungen für die praktische Strahlentherapie*

Von

F. WACHSMANN und G. BARTH, Erlangen

Der Hauptgrund dafür, daß die ultraharten Strahlungen in die Therapie ein-
geführt wurden, ist selbstverständlich die Erhöhung der relativen Tiefendosis.
Neben der Verbesserung der ,,makroskopischen Dosisverteilung" tritt bei diesen
Strahlungen aber auch eine räumlich gänzlich andere ,,mikroskopische Dosis-
verteilung" auf. Diese ist auf die verschiedene Ionisationsdichte der ,,klassischen"
Röntgenstrahlen und der neuen Strahlungen zurückzuführen.

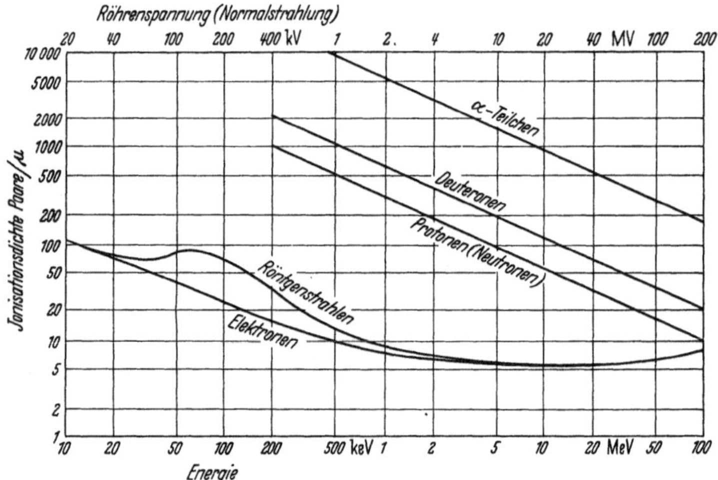


Fig. 1. Mittlere integrale Ionisation bei verschiedenen Strahlungen

Man kann damit rechnen, daß die mittlere integrale Ionisation bei der bisher
verwendeten Röntgenstrahlung im Bereich von 20–200 kV 75 Paare/$\mu$ beträgt
[9, 7]. Bei den ultraharten Strahlungen von 2 MeV und mehr sinkt sie dagegen
auf etwa den zehnten Teil dieses Wertes ab. Bei Neutronen, Protonen und Deute-
ronen liegt sie umgekehrt etwa 10 mal und bei $\alpha$-Teilchen sogar 100 mal höher [11]
(Fig. 1). Rein schematisch ergeben sich dabei die in Fig. 2 gezeigten Verhältnisse.

Die Frage der biologischen Wirksamkeit – heute sagen wir die relative biologische Wirksamkeit (RBW) – verschiedener Röntgenstrahlenqualitäten des
klassischen Bereiches wurde seinerzeit viel diskutiert. Angeblich beobachtete
Unterschiede – besonders bezüglich der erythemerzeugenden Wirkung – haben

---

* Aus der Medizinischen Universitätsklinik Erlangen (Direktor: Prof. Dr. N. HENNING)

sich als nicht vorhanden erwiesen. Dies wundert uns heute nicht mehr, wissen wir doch, daß die mittlere integrale Ionisation im Gebiet von 20—200 kV, des Überganges von Photo- zu Comptonprozessen wegen, praktisch gleich ist. Umgekehrt wissen wir aber auch, daß die RBW von dicht ionisierenden Strahlungen wie Neutronen, Protonen usw. in der Regel etwa um den Faktor 5—10 oder auch mehr größer ist als die der klassischen Strahlungen (z. B. [4, 6]). Es ist also nicht verwunderlich, daß die RBW der weniger dicht ionisierenden ultraharten Röntgen- und Elektronenstrahlungen auch nicht gleich 1, d. h. verschieden von der RBW der bisher verwendeten Strahlungen ist.

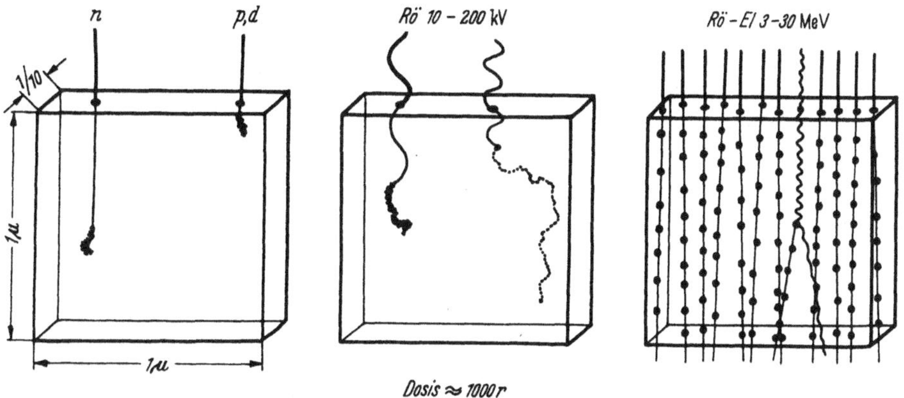


Fig. 2. Verteilung der Ionen in Gewebe (schematisch)

Eine Zusammenstellung von RBW-Werten, die wir selbst und andere Autoren gefunden hatten, haben wir bereits 1950 veröffentlicht [8]. Zu dieser Zeit glaubte man noch nicht allgemein daran, daß ultraharte Strahlungen, verglichen mit den bis dahin verwendeten Strahlungen, quantitativ in verschiedenem Maße unterschiedlich wirken könnten.

Besonders die Amerikaner haben diese Werte häufig angezweifelt und angenommen, die RBW ultraharter Strahlungen sei für alle Objekte und Reaktionen einheitlich gleich etwa 0,7—0,8.

Heute können wir die RBW-Werte durch eine lange Reihe neuer Beobachtungen ergänzen. Anläßlich des Heidelberger Betatron-Symposions wurden vor allem von FRITZ-NIGGLI neue interessante RBW-Werte genannt. Nach diesen steht fest, daß die RBW ultraharter Strahlungen etwa in den Grenzen zwischen 0,4—1,2 variiert, unter Umständen ausnahmsweise aber noch tiefer bzw. höher liegen kann.

Wichtig ist dabei, daß die RBW nicht nur nach Objekt und Reaktion in verschiedenem Grade von 1 abweicht. Sie ist vielmehr auch von der Dosishöhe und dem Zeitpunkt der Beobachtung abhängig. Als Beispiel hierfür erwähnen wir die Reaktionen von Rattenhoden, bei denen die RBW ultraharter Strahlungen (schnelle Elektronen von 15 MeV) bei Beobachtung am 30. Tag nach der Bestrahlung deutlich größer, am 60. Tag nach der Bestrahlung jedoch kleiner als 1 ist [2] (vgl. Fig. 3).

Dies Ergebnis ist von Wichtigkeit. Wäre die RBW ultraharter Strahlungen bei allen Objekten und Reaktionen stets gleich, so würde dies nur bedeuten, daß stets im gleichen Verhältnis anders dosiert werden muß als bei den konventionellen Strahlungen, um dann aber die gleichen biologischen Wirkungen zu erhalten. Da nun aber die RBW nicht immer gleich ist, bedeutet dies — bei der Bestrahlung komplexer Objekte — zwangsweise eine Verschiebung der Elektivitätsverhältnisse.

Wir möchten jedoch vermeiden, allgemein von einer „Erhöhung der Tumorelektivität" schlechthin zu sprechen. Diese erhoffen wir auf Grund gewisser experimenteller Ergebnisse, die durch klinische Beobachtungen gestützt werden.

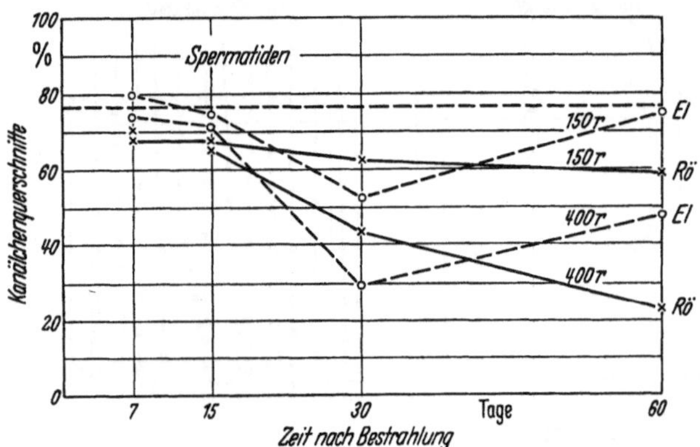


Fig. 3. Ab- und Wiederzunahme der Spermatiden im bestrahlten Rattenhoden bei Bestrahlung mit 180 kV-Röntgenstrahlen und 15 MeV Elektronen (nach Börner, Neff, Riemann und Wachsmann)

Wir glauben jedoch keinesfalls, daß sie unbedingt und immer — d. h. bei allen Tumorarten gleich — vorhanden sein müßte!

Für die praktische Tumortherapie ergeben sich aus der Verschiedenheit der RBW im übrigen zunächst folgende zwei Fragen:

1. Wie groß muß bei ultraharten Strahlungen die Tumordosis gewählt werden, um gleiche Wirkungen zu erzielen wie bei den bisher verwendeten Strahlungen?

Die Kanadier Burkell und Watson [3] halten 10%, der Amerikaner Fletcher [5] 10—20% höhere Dosen für erforderlich, um mit ultraharten Strahlungen gleiche Tumorwirkungen zu erzielen wie mit 200 kV. Die Italiener Bonomini und Valdagni [1] glauben dagegen mit etwa gleichen Dosen auszukommen. Der gleichen Auffassung ist — soweit wir informiert sind — Becker, Heidelberg.

2. Die zweite Frage betrifft die Verträglichkeit des Gesunden. Sie zu beantworten ist schwer. Einmal weil das „gesunde Gewebe" ja selbst in hohem Grade komplex ist, zum anderen aber, weil auch die räumliche Dosisverteilung bei klassischen Bestrahlungsbedingungen und ultraharten Strahlungen so verschieden ist, daß direkte Vergleiche unmöglich werden. Immerhin glauben wir, auf Grund der Ergebnisse experimenteller Untersuchungen, annehmen zu können, daß hier allgemein mit RBW-Werten von 0,7—0,8 gerechnet werden kann.

Für die Praxis von besonderer Bedeutung ist, daß die Toleranz der Haut bei ultraharten Strahlungen — unabhängig von dem Aufbaueffekt — offenbar deutlich größer ist als bei klassischen Strahlungen (vgl. auch [10]). Soeben teilte uns Herr Degener, Berlin, mit, daß die Erfahrungen an der Kobaltbombe ergeben haben, daß, was die Hautreaktion anbetrifft, etwa mit einer RBW von 0,5 gerechnet werden muß.

Der Unterschied gegenüber neoplastischem Gewebe ist aber keineswegs so groß, daß eindeutig von einer *biologischen* Überlegenheit der ultraharten Strahlungen gesprochen werden kann. Das Vorhandensein eines allgemein günstigen klinischen Eindruckes bei der Anwendung ultraharter Strahlungen soll damit jedoch nicht bestritten werden. Es kann aber unserer Auffassung nach zur Zeit noch nicht entschieden werden, inwieweit die Überlegenheit der ultraharten Strahlungen auf ihre zweifellos vorhandene günstigere makroskopische oder verschiedene mikroskopische Dosisverteilung zurückzuführen ist. Nur klinische Beobachtungen und Erfahrungen werden, wie anläßlich des Heidelberger Betatron-Symposions WEITZEL, Heidelberg, ausgeführt hat, weitere Aufschlüsse in dieser Beziehung ermöglichen.

**Zusammenfassung.** Die auf der Verschiedenheit der mittleren integralen Ionisationsdichte von „klassischen" und „ultraharten" Strahlungen beruhenden Unterschiede in der RBW werden heute allgemein anerkannt. Die von verschiedenen Autoren gemessenen RBW-Werte liegen normalerweise etwa zwischen den Grenzen 0,4—1,2. Wichtig ist jedoch, daß die RBW nicht nur nach Objekt und Reaktion verschieden ist, sondern auch von der Dosishöhe und dem Zeitpunkt der Beobachtung abhängt. Eine Steigerung der „Tumorelektivität" bei ultraharten Strahlungen läßt sich wohl erhoffen, heute aber noch nicht beweisen. Erst klinische Beobachtungen werden endgültige Aufschlüsse geben.

## Literatur


1. BONOMINI, B., u. C. VALDAGNI: Vortrag anl. der 33. Zusammenkunft der Radiologen von Triveneto, Roncegno, 15.—16. 6. 57.
2. BÖRNER, W., V. NEFF, H. RIEMANN u. F. WACHSMANN: Strahlenther. **101**, 101 (1956).
3. BURKELL, C. C., and T. A. WATSON: Amer. J. Roentgenol. **76**, 865 (1956).
4. Empfehlungen des Internat. Radiologen-Kongresses, Kopenhagen 1953. Brit. J. Radiol. Suppl. 6 (1955).
5. FLETCHER, G. A.: Amer. J. Roentgenol. **76**, 866 (1956).
6. RILEY, E. F., R. C. EVANS, R. B. RHODY, P. J. LEINFELDER and R. D. RICHTADS: Radiology **67**, 673 (1956).
7. SCHINZ, H. R., H. FRITZ-NIGGLI u. K. SCHARER: Radiol. clin. (Basel) **24**, 317 (1955).
8. WACHSMANN, F.: Strahlenther. **81**, 273 (1950).
9. —: Strahlenther. **89**, 128 (1952).
10. —: Strahlenther. **90**, 438 (1953).
11. —, u. A. DIMOTSIS: Kurven und Tabellen für die Strahlentherapie. Stuttgart: S. Hirzel 1957.

# Die relative biologische Wirksamkeit der Betatronstrahlung*

Von

**W. Dittrich**, Hamburg

Unter relativer biologischer Wirksamkeit (RBW) sei das Verhältnis der in rad gemessenen, biologisch gleich wirksamen Dosen verschiedener Strahlenarten verstanden. Im Nenner dieses Bruches steht die Dosis der Standardstrahlung.

Die RBW-Werte der Betatronstrahlung, welche für die verschiedenartigsten Strahlenreaktionen experimentell ermittelt wurden, und zwar mit harter (180- oder 200 kV-) Röntgenstrahlung als Standardstrahlung, lassen sich in umfangreichen Tabellen zusammenstellen. Hier mag der Hinweis genügen, daß die bisher gefundenen Werte zumeist wenig von 1 abweichen, daß aber trotz aller statistischen Unsicherheit im einzelnen, von einer einheitlichen, für *alle* Strahlenreaktionen und für alle strahlenbiologischen Objekte gültigen RBW sicherlich nicht die Rede sein kann. Die gefundenen Unterschiede sind vielmehr in einigen Fällen zweifellos reell.

Wohl eine der ersten Untersuchungen, aus der dies eindeutig hervorging, war die Strahlentötung von Drosophilaeiern verschiedener Altersklassen [2]. Bei verschieden alten Eiern ergaben sich unter Versuchsbedingungen, deren Einheitlichkeit gewährleistet war, deutliche Unterschiede der RBW, die reell sein mußten, selbst wenn damals die absoluten Dosisangaben noch erheblich korrekturbedürftig gewesen wären. Es stand damit schon zu Beginn der Betatronära fest, daß es eine für alle Strahlenreaktionen gleiche RBW der Betatronstrahlung verglichen mit harter Röntgenstrahlung nicht geben kann.

Übrigens haben Schinz u. Mitarb. [7] später an der 31 MeV-Schleuder Untersuchungen mit ähnlichem Ergebnis durchgeführt.

Die Betatronstrahlung wird biologisch wirksam hauptsächlich über primäre und, wenn es sich um ultraharte Röntgenstrahlen handelt, über sekundäre schnelle Elektronen, bzw. Positronen. Es ist daher vielfach möglich und auch zweckmäßig, $\beta$- und $\gamma$-Strahleneffekte gemeinsam zu betrachten.

Zu Beginn der Betatronära war es für den Strahlentherapeuten sehr beruhigend zu hören, daß die Betatronstrahlung je Einheit absorbierter Energie biologisch in der Regel etwas weniger wirksam ist als harte Röntgenstrahlung. Man brauchte nur die Dosen der üblichen Röntgentherapie in die Betatrontherapie zu übernehmen, um der Gefahr einer Überdosierung zu entgehen – ein Gedanke, der meines Wissens erstmals von Schubert [9] ausgesprochen wurde und der sich in der Praxis bewährt hat. Dabei ist freilich zu beachten, daß selbst bei geringen Abweichungen der RBW-Werte von 1 in umgekehrter Richtung, also nach oben hin,

---

* Aus der Universitäts-Frauenklinik Hamburg-Eppendorf (Direktor: Prof. Dr. G. Schubert)

die Gefahr einer Überdosierung doch auch dann im allgemeinen noch nicht gegeben sein dürfte. Denn in der Betatrontherapie werden für gewöhnlich höhere relative Herdraumdosen erreicht. Pauschalurteile dieser Art, denen natürlich stets eine gewisse Willkür anhaftet, werden mit wachsender therapeutischer Erfahrung entbehrlich.

Mit der Entwicklung der Betatrontherapie hat die strahlenbiologische Forschung auf dem Gebiet der ultraharten Strahlungen Schritt gehalten. Überblickt man die Ergebnisse der letzten Jahre, dann fällt auf, daß sich doch bei einigen, z. T. sehr sorgfältig untersuchten Strahlenreaktionen RBW-Werte von nahezu 1 und manchmal auch RBW-Werte über 1 ergeben haben. Ohne auf Einzelheiten einzugehen, möchte ich lediglich ein Ergebnis herausgreifen. Bei Untersuchungen über geschlechtsgebundene recessive Letalfaktoren bei Drosophila fand MOSSIGE [6] für die weniger strahlenempfindlichen Entwicklungsstadien der Keimzellen RBW-Werte, die über 1 lagen. Untersuchungen an der Göttinger 6 MeV-Schleuder, bei denen solche feineren Unterscheidungen nicht gemacht worden waren, hatten gleiche Wirksamkeit von schnellen Elektronen und Röntgenstrahlen ergeben [3]. Diese Feststellungen sind im Hinblick auf Strahlenschutzfragen von besonderem Interesse. Im praktischen Strahlenschutz wird man nämlich keinesfalls mit RBW-Werten kleiner als 1 rechnen dürfen.

Bei vielen Strahlenreaktionen besteht eine einsinnige Abhängigkeit der RBW von der differentialen Ionisation, bzw. linearen Energieübertragung (LET) der verwendeten Strahlenart. Man ist daher in solchen Fällen in der Lage, Voraussagen zumindest über die Richtung zu machen, in der die RBW der Betatronstrahlung, bezogen auf harte Röntgenstrahlung als Standard, vom Wert 1 abweichen wird, allerdings unter der Voraussetzung, daß der Impulscharakter der Betatronstrahlung die biologische Wirksamkeit nicht wesentlich beeinflußt.

So nimmt z. B. bei gewissen Pflanzenzellen die Häufigkeit von Chromosomenaberrationen des Sekundäreffektes mit der differentialen Ionisation zu. Diese Chromosomenaberrationen werden, wie schon Untersuchungen an der Göttinger 6 MeV-Schleuder an Vicia faba, vor allem von HEINRICH und SCHMERMUND [8] gezeigt haben, nach Elektronenbestrahlung relativ seltener beobachtet als nach Röntgenbestrahlung.

Bei tierischen Zellen können strahleninduzierte Chromosomenaberrationen des Sekundäreffektes anscheinend gleichfalls seltener auftreten.

Dies zeigen Untersuchungen am Yoshidasarkom der Ratte (Fig. 1). 24 Std. nach Bestrahlung mit Dosen von 200 r, bzw. 186 rad liegen die Elektronenwerte für Brückenbildung und Chromosomenfragmentierung durchwegs unter den Röntgenwerten. Die angegebenen Häufigkeiten beziehen sich jeweils auf 800 (Elektronenversuch), bzw. 1300 (Röntgenversuch) Mitosezellen. Trotzdem finden sich bei beiden Strahlenarten annähernd gleich große Sauerstoffeffekte — in Fig. 1 gekennzeichnet durch den Höhenunterschied der schwarzen und schraffierten Säulen. Die eine Gruppe der Versuchstiere atmete während der Ganzbestrahlung Sauerstoff unter Überdruck, die andere Luft. Unter den gewählten Versuchsbedingungen ist der Einfluß der Sauerstoffspannung auf die Strahlenwirkung erheblich deutlicher als der Einfluß der differentialen Ionisation.

Das Beispiel lehrt, daß die Strahlenwirkung durch Nebenbedingungen modifiziert werden kann, deren Einfluß offenbar auch beim Übergang von harter

Röntgenstrahlung zur Betatronstrahlung der allgemeinen Regel genügt, mit der differentialen Ionisation abzunehmen.

Im Zusammenhang damit verdient ein anderer Gesichtspunkt Beachtung. Anscheinend gleichfalls erstmalig konnte an der Göttinger 6 MeV-Schleuder der Nachweis einer relativ hohen Variabilität eines strahlenbiologischen Effektes erbracht werden. Nach Elektronenbestrahlung ergab sich eine vergleichsweise höhere Streuung der Blattlänge von Gerstenkeimlingen als nach Röntgenbestrahlung [4]. Ein mit der differentialen Ionisation zunehmendes Eintönigwerden der Strahlenreaktion ist den Pflanzenzüchtern, soweit sie sich mit der Erzeugung neuer Mutanten durch Strahlung befassen, geläufig geworden [1, 5]. So finden sich z. B. nach Neutroneneinwirkung unter den überlebenden Pflanzen erheblich mehr

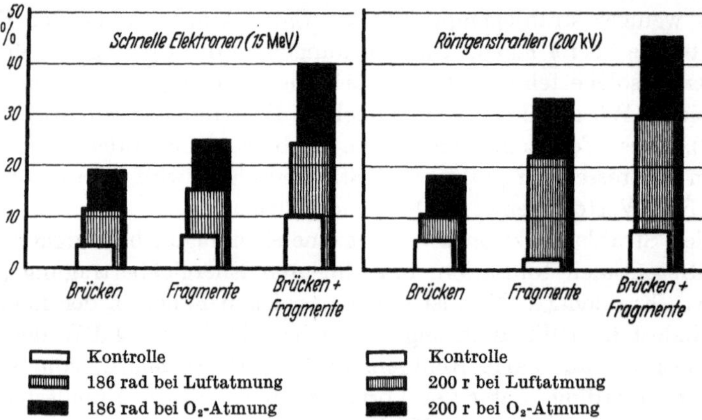


Fig. 1. Relative Häufigkeit von Chromosomenaberrationen in Mitosezellen (späte Ana- und frühe Telophase) des Yoshida-Sarkom-Ascites bei Luft- und Sauerstoffatmung (2 at)

kranke, in ihrer Vitalität beeinträchtigte Individuen als nach Röntgenbestrahlung, deren Wirkung sich in Richtung auf einen „Alles-oder-Nichts"-Typ hin verschiebt. Anscheinend liegen mit der Betatronstrahlung zu diesem Punkt weitere Untersuchungen, neben der erwähnten, nicht vor.

Eine Erklärungsmöglichkeit der besonders großen Variabilität der Betatroneffekte liegt in ihrer besonders großen Beeinflußbarkeit durch Nebenbedingungen — wirksam sind neben der Sauerstoffspannung auch Wassergehalt, Schutzstoffe, Temperatur usw. — und eine andere in einer größeren Reversibilität „leichter" Treffer, die bei der Betatronstrahlung bevorzugt auftreten.

Und nun abschließend zur Frage des Einflusses einer pulsierenden Dosisleistung auf die biologische Wirkung:

Die Untersuchungen von KEPP und WITTE weisen darauf hin, daß bei ionisierenden Strahlenarten ein Einfluß der Dosisflimmerung vorhanden sein kann. Von vornherein ist ja zu erwarten, daß bei erwiesenem Verdünnungs- bzw. Konzentrierungseffekt auch ein gleichsinnig gerichteter Flimmereffekt auftritt. Ja, man könnte von solchen Untersuchungen geradezu Aufklärung über die Lebensdauer von instabilen Zuständen, Intermediärprodukten oder von strahleninduzierten Funktionsstörungen erwarten. In Fällen aber, wo einfache Protrahierung oder Konzentrierung der Dosis unwirksam bleibt oder gar die Strahlenwirkung gegensinnig beeinflußt, wie von der Ultrafraktionierung behauptet wird, dürfte vorläufig noch große Zurückhaltung bei der Beurteilung der Ergebnisse angebracht und ihre Bestätigung durch andere, unabhängige Untersucher abzuwarten sein.

Zusammenfassend darf ich feststellen, daß die RBW der Betatronstrahlung, bezogen auf harte Röntgenstrahlung als Standard bei verschiedenen Strahlenreaktionen, unterschiedliche Werte annehmen kann. Maßnahmen des praktischen Strahlenschutzes wird man mit Rücksicht auf mögliche genetische Schäden RBW-Werte von mindestens 1 zugrunde legen müssen. Durch äußere Faktoren, insbesondere durch die Sauerstoffspannung, läßt sich die biologische Wirkung der Schleuderstrahlung mindestens ebenso deutlich beeinflussen wie die Wirkung harter Röntgenstrahlen. Die hieraus resultierenden Unterschiede können die durch einen Wechsel der Strahlenart bedingten sogar um ein Vielfaches übertreffen.

## Literatur


1. CALDECOTT, R. S.: Ann. N. Y. Acad. Sci. **59**, 514 (1955).
2. DITTRICH, W., H. FASS, G. HÖHNE u. G. SCHUBERT: Strahlenther. **81**, 223 (1950).
3. — G. HÖHNE, W. PAUL u. G. SCHUBERT: Naturwiss. **37**, 545 (1950).
4. — W. PAUL, M. RIEDEL u. G. SCHUBERT: Strahlenther. **80**, 19 (1949).
5. MACKEY, J.: Hereditas (Lund) **40**, 65 (1954).
6. MOSSIGE, J.: Progress in Radiobiology. Herausgeber: J. S. MITCHELL et al., London 137 (1956).
7. SCHINZ, H., H. FRITZ-NIGGLI u. E. FREY: Experientia (Basel) 8, 16 (1952).
8. SCHMERMUND, H. J., u. H. L. HEINRICH: Strahlenther. **86**, 227 (1952).
9. SCHUBERT, G.: Schweiz. med. Wschr. **1950**, 193.

# Strahlenbiologische Untersuchungen
## über die Wirkung schneller Elektronen*

Von

**D. Hofmann**, Gießen

Ein unbestrittener Vorteil der in die Therapie eingeführten energiereichen
Strahlungen besteht — jedenfalls, was die schnellen Elektronen betrifft — in der
günstigen Dosisverteilung derselben im Gewebe. Was die Voraussetzung einer
wenigstens angenäherten biologischen Gleichwertigkeit der energiereichen Strahlungen gegenüber der herkömmlichen Röntgenstrahlung betrifft, so ist eine
quantitative Klärung bis jetzt in bezug auf eine große Anzahl von Dosisäquivalen-
ten erfolgt, jedoch nur an niederen biologischen Objekten bzw. durch Ganzkörperbestrahlung von Versuchstieren. Auch andere Faktoren, welche die letzten Endes
interessierenden Elektivitätsverhältnisse beeinflussen, wie z. B. der der zeitlichen
Dosisverteilung, sind an überwiegend niederen Objekten hinsichtlich ihrer
biologischen Auswirkung untersucht. Im ganzen steht jedoch die Beantwortung
der Frage nach der therapeutischen Gleichwertigkeit der neuen energiereichen
Strahlungen, d. h. also nach einer mindestens gleichwertigen Tumorselektivität
noch aus, indem wohl immer wieder unterschiedliche Wirkungen der neuen Strahlungen gegenüber der herkömmlichen festgestellt werden konnten, Schlüsse auf
das komplexe Endresultat, die Elektivität, jedoch nicht gezogen werden konnten.

Nach den Ergebnissen gewisser Tierversuche und auf Grund klinischer Beobachtungen kann wohl gesagt werden, daß wesentliche biologische Wirkungsunterschiede, welche die therapeutische Applikationsweise der neuen Strahlungen in
besonderer Weise beeinflussen könnten, nicht bestehen, vorausgesetzt, daß man
ein für die Tumortherapie gültiges Dosisäquivalent beachtet.

Was letzteres betrifft, so darf wohl angenommen werden, daß es für verschiedene
Objekte unterschiedliche Werte besitzt und daß die Wirkung schneller Elektronen
das 0,4—0,7 fache der Wirkung herkömmlicher Röntgenstrahlen beträgt. Der
Wirkungsfaktor 0,7 ist nun derjenige, der von amerikanischen Autoren auf Grund
von Ergebnissen der Ganzkörperbestrahlung von Mäusen eingeführt wurde und
nach allgemeiner Erfahrung in etwa richtig sein dürfte. Die Schwierigkeit hinsichtlich der exakten Bestimmung eines Dosisäquivalentes für Tumoren besteht ja in
der Auffindung eines quantitativ faßbaren Kriteriums für die Tumorwirkung.

Wir stellten uns die Aufgabe, anhand eines größeren Materials weißer Mäuse
die Gültigkeit des soeben bezeichneten Wirkungsfaktors für den Fall von Tumoren
zu überprüfen und tasteten uns zu diesem Zweck mit den verglichenen Strahlungen — einer Röntgenstrahlung von 180 kV und schnellen Elektronen von

* Aus der Frauenklinik der Justus-Liebig-Universität Gießen (Direktor: Prof. Dr.
R. K. Kepp)

Strontium-Yttrium[90] — an diejenige Dosis heran, welche in der Lage ist, die verwendeten Ascitestumoren lokal restlos zum Verschwinden zu bringen. Das Ergebnis der so gewonnenen Dosiseffektkurven ist aus der Fig. 1 ersichtlich. Aus ihr läßt sich feststellen, daß der Dosisquotient für einen Bezugseffekt von 50% 0,41 beträgt, ein Wert, welcher an mehreren niederen Objekten bereits gefunden worden war.

Fragt man sich, für welchen Effekt der klinisch verwendete Quotient 0,7 zustande kommt, so ergibt sich aus der Fig. 1 der Effekt etwa 80%. Nimmt man diesen Effekt in Anbetracht der Reaktion des gesunden Gewebes als wünschenswert an, so ergibt sich daraus in der Tat die Richtigkeit des bisher für die Therapie als gültig angenommenen Dosisäquivalentes.

Aus einer Reihe von Untersuchungen, die wir zur Frage der indirekten Strahlenwirkung anstellten, kann in Anbetracht der zur Verfügung stehenden Zeit nur kurz auf eine solche eingegangen werden, bei welcher wir den Wirkungsradius zweier Röntgenstrahlungen einerseits und einer Elektronenstrahlung andererseits auf biologische Weise bestimmten.

Wir gingen ursprünglich von dem Umstand aus, daß bei Bestrahlung von Wasser mit $\alpha$-Teilchen leicht die Bildung von $H_2O_2$ und auch von $H_2$ nachzuweisen ist, kaum jedoch bei Bestrahlung mit $\gamma$- oder Röntgenstrahlen. Ein Deutungsversuch hierfür besteht bekanntlich darin, daß es bei Bestrahlung von Wasser zu

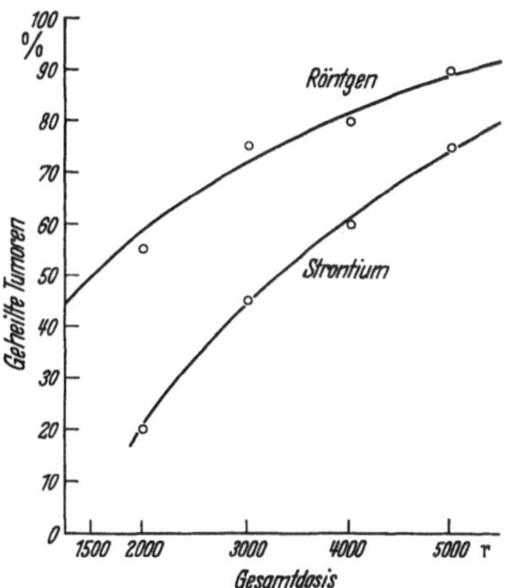


Fig. 1. Dosiseffektkurven für Röntgenstrahlen und schnelle Elektronen, ermittelt an Ascitestumoren

einer unterschiedlichen Anordnung von Hydroxyl und Wasserstoff kommt. Durch den relativ kurzen Abstand der primären Ionisationen voneinander, wie er bei $\alpha$-Teilchen vorliegt, soll daher die Bildung von $H_2O_2$ begünstigt sein. Neben anderen Erklärungen sei auch eine solche von ALLEN erwähnt, derzufolge die Rückreaktion $OH + H_2 \rightarrow H_2O + H$ bei $\alpha$-Teilchen unwahrscheinlich ist.

Es liegen nun keine festen Anhaltspunkte über die Lebensdauer von $H_2O_2$ in wäßrigen Systemen vor, doch bestehen u. a. auch biologische Hinweise dafür, daß die Lebensdauer des $H_2O_2$ und dementsprechend sein Diffusionsweg diejenigen der anderen eigentlich kurzlebigen Radikale weit übertrifft. Wir erwähnen hier die Versuche von EVANS und BARRON. Es scheint, als müsse zwischen die beiden bekannten Gruppen der kurzreichenden indirekten Strahlenwirkung, basierend auf aktiven Radikalen, und der weitreichenden indirekten Strahlenwirkung eine weitere Gruppe eingeschaltet werden, die den allgemeinbiologisch naheliegenden Übergang darstellt.

Ohne aus dem Gesagten bereits eine überzeugende Deutung geben zu wollen, zeigen wir in Fig. 2 einige Versuchsergebnisse. Über das technische Vorgehen muß soviel gesagt werden, daß 1,5 Std. alte Drosophila-Eier in Wasser liegend zunächst durch eine Röntgenstrahlung vorsensibilisiert wurden und dann einer irgendwie gearteten indirekten Wirkung ausgesetzt wurden, welche von einem zentralen Wasservolumen bei hochdosierter Bestrahlung desselben in die Peripherie ging. Besonderer Wert wurde bei der zweiten Bestrahlung darauf gelegt, daß die Objekte selbst keinerlei Strahlung oder Streustrahlung erhielten.

Die Fig. 2 läßt signifikante Unterschiede der Kurvenverläufe erkennen, welche gewissermaßen die relative Tiefendosis der indirekten Strahlenwirkung darstellen.

Die drei verwendeten Strahlungen unterscheiden sich nun mikrophysikalisch lediglich durch ihre unterschiedliche primäre spezifische Ionisation, was unter

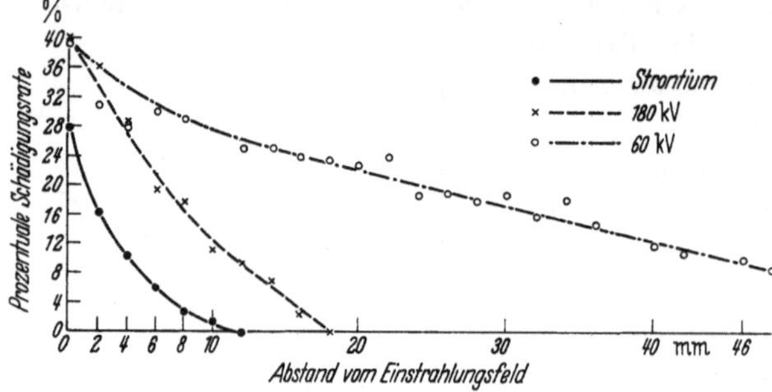


Fig. 2. Zusammenfassende Darstellung der Kurven für die Schädigung von Drosophilaeiern im Wasser durch indirekte Strahlenwirkung bei 60 kV, 180 kV und Strontium 90-Bestrahlung mit 7900 r

Berücksichtigung der von Burton angegebenen Reaktionsmechanismen die Möglichkeit einer Erklärung in Form einer mehr oder weniger großen $H_2O_2$-Bildung gibt.

Die vorliegenden Untersuchungen sind noch nicht abgeschlossen, doch möchten wir noch einmal auf die bereits behandelte Frage des Dosisäquivalentes zurückkommen und auf die eventuelle Möglichkeit einer Erklärung für die unterschiedlichen Wirkungen schneller Elektronen und harter Röntgenstrahlen hinweisen, die darin bestehen könnte, daß von vornherein eine unterschiedliche Konzentration und Verteilung indirekt wirkender Radikale bzw. Substanzen entsteht.

Ein weiterer Zusammenhang mit den angeschnittenen Fragen bietet sich bei Betrachtung der qualitativen Wirkungsunterschiede der verglichenen Strahlungen,

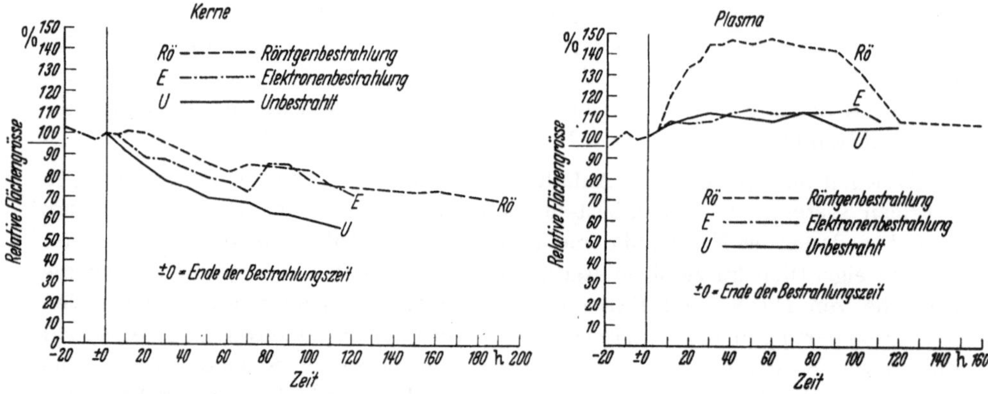


Fig. 3. Gemittelte Ergebnisse der Kernplanimetrie     Fig. 4. Gemittelte Ergebnisse der Plasmaplanimetrie

deren Existenz wir in Übereinstimmung mit Gärtner für wahrscheinlich, von der Frage der Tumorselektivität im übrigen aber unabhängig erachten. Der Grund für die Existenz qualitativer Wirkungsunterschiede kann letzten Endes auch nur in der Ionisationstopographie der Strahlungen zu suchen sein, welche ja ihrerseits Unterschiede bezüglich Ausbeute und Verteilung der Substrate der indirekten Strahlenwirkung bedingt. Die oft gebrauchten Begriffe etwa des „schweren" und des „leichten Treffereignisses" könnten demnach gleichbedeutend mit unterschiedlichen Konzentrationsverhältnissen besagter Substrate sein. Es liegt in der Tat nahe, unterschiedliche Verläufe des Zellunterganges bei verschiedenen Strahlenarten weniger durch die schwerpunktartige Schädigung empfindlicher Bereiche in Form von Treffern, als vielmehr durch eine relativ homogene Verteilung indirekter Noxen durch die genannten Substanzen anzunehmen.

Zur Verifizierung qualitativer Wirkungsunterschiede führten wir kinematographische Untersuchungen an röntgen- und elektronenbestrahlten Ascitestumorzellen aus. Als Ergebnis der quantitativen Auswertung des soeben fertiggestellten Filmes seien hier die Resultate der Kern- und der Plasmaplanimetrie wiedergegeben. Entgegen oft gewonnenen Eindrücken ist in der Fig. 3 kein unterschiedliches Verhalten der Kerngrößen röntgen- und elektronenbestrahlter Zellen festzustellen. Dagegen zeigt die Fig. 4 einen deutlichen Unterschied der Strahlenwirkungen in bezug auf das zeitliche Verhalten der Plasmavolumina. Wir müssen uns mit der Feststellung begnügen, daß die Betrachtung des Filmes auch mancherlei andere Unterschiede erkennen läßt, wie zum Beispiel unterschiedliche Kernschwärzung, deren objektive Erfassung naturgemäß technische Schwierigkeiten bietet. Alles in allem bestätigt sich jedoch das Vorliegen qualitativer Wirkungsunterschiede, welche indes mit keinem der quantitativ erfaßbaren und therapeutisch zu berücksichtigenden Wirkungsunterschiede in unmittelbaren Zusammenhang gebracht werden können.

## Literatur

ALLEN, A. O.: J. Phys. Colloid Chem. **52**, 479 (1948).
BURTON, M.: J. Phys. Colloid Chem. **51**, 611 (1947).
— J. Phys. Colloid Chem. **51**, 786 (1947); **52**, 564 (1948); **52**, 810 (1948).
— Brit. J. Radiol. **24**, 416 (1951).
— Sympos. on Radiobiology, p. 117, New York-London 1952.
— and J. L. MAGGEE: Naturwissenschaften **19**, 433 (1956).
BARRON, E. S. G., et. al.: J. gen. Physiol. **32**, 537 (1949); **32**, 595 (1949).
CLARK, R. V., and H. QUASTLER: Amer. J. Roentgenol. **54**, 723 (1945).
DALE, W. M.: Biochem. J. **34**, 1367 (1940).
— Biochem. J. **36**, 80 (1940).
— Brit. J. Radiol. **24**, 433 (1951).
EVANS, T. C.: Physiol. Zool. **8**, 521 (1935).
— Biol. Bull. **92**, 99 (1947).
GÄRTNER, H.: Strahlenther. **86**, 217 (1952); **89**, 26 (1952); **96**, 201 (1955); **96**, 378 (1955).
HOFMANN, D.: Strahlenther. **95**, 209 (1954); **97**, 231 (1955).
— Strahlenther., Sonderbd. **35**, 183, 189, 194 (1956).
— Filmvortrag, Deutscher Gynäkologen-Kongreß, Heidelberg 1956.
— u. F. WACHSMANN: Strahlenther. **86**, 2, 282 (1952).
— R. K. KEPP, G. OEHLERT u. H. W. VASTERLING: Strahlenther. **96**, 1 (1955).
MAGGEE, J. L., and M. BURTON: J. Amer. chem. Soc. **72**, 1965 (1950); **73**, 523 (1951).
NÖDL, F.: Strahlenther. **92**, 576 (1953).
PATT, H. M., D. E. SMITH, E. B. TYREE and R. L. STRAUBE: Proc. Soc. exp. Biol. (N. Y.) **73**, 18 (1950).
— E. B. TYREE, R. L. STRAUBE and D. E. SMITH: Science **110**, 213 (1949).
QUASTLER, H.: Amer. J. Roentgenol. **54**, 499 (1945); **54**, 457 (1945); **54**, 723 (1945); **61**, 591 (1949).
— et al.: Amer. J. Roentgenol. **57**, 359 (1947); **62**, 555 (1949); **64**, 963 (1950).

## Diskussionsbemerkungen

**H. Gärtner (Tübingen):**

In Ergänzung zu den Ausführungen von Herrn Hofmann möchte ich mitteilen, daß ich Untersuchungen über die Ultrafraktionierung mit schnellen Elektronen einerseits und Radiostrontium andererseits an Hühnerherzfibroblastenkulturen vorgenommen habe. Dabei zeigte sich, daß die strahlenbiologische Reaktion bezüglich der Mitosehemmung nach *kontinuierlicher* Betastrahlung stärker in Erscheinung tritt als nach der diskontinuierlichen Elektronenbestrahlung.

Die Ergebnisse zeigen weiter, daß *qualitative* Wirkungsunterschiede zwischen schnellen Elektronen und Radiostrontium nicht anzunehmen sind. Wohl aber bestehen solche bezüglich des „Ruhezelleffektes" zwischen den Betastrahlen beider benutzten Strahlenquellen und konventioneller Röntgenstrahlung.

Zusammenfassend ist festzustellen, daß Impulsfrequenz oder Ultrafraktionierung zwar nicht zu vernachlässigen sind, daß aber der differentialen Ionisation die größere Bedeutung zukommt.

**R. Kepp (Gießen):**

Die Frage, ob die energiereichen Strahlen auf bösartiges Gewebe eine besonders elektive Wirkung ausüben, ist experimentell schwierig zu überprüfen, da hierfür gleiche zeitliche und räumliche Dosisbedingungen Voraussetzung sind. Die zeitliche Dosisverteilung läßt sich ohne Schwierigkeiten gleich gestalten, die räumliche Dosisverteilung jedoch nicht in Körpern, die eine gewisse Dicke aufweisen. Eine Scheinelektivität wird auch dadurch hervorgerufen, daß die Dosiseffektkurve bei den energiereichen Strahlen dort relativ weniger wirksam ist, wo die Dosis gering ist. Wenn also das Dosismaximum in den Tumor gelegt wird, entsteht gegenüber der Umgebung eine relativ stärkere Strahlenwirkung. Der klinische Eindruck geht dahin, daß die energiereichen Strahlen eine gesteigerte Elektivität aufweisen, jedoch ist dieser Eindruck durch die vorgenannten Fehlerquellen mitbestimmt. Der bisher einzige Beweis für die stärkere elektive Wirkung energiereicher Strahlen sind die von Fräulein Gärtner vorgetragenen Untersuchungen über die stärkere Wirkung auf Carcinomzellen in Gewebskultur gegenüber Hühnerherzfibroblasten bei der Anwendung schneller Elektronen.

**D. Hofmann (Gießen):**

Zur Ergänzung der vorhergegangenen Diskussionen wird auf Deutungsmöglichkeiten der eigenen Ultrafraktionierungsergebnisse eingegangen. Es erscheint durchaus möglich, daß die der primären Ionisationen folgenden chemischen Vorgänge der Radikalbildung und die Diffusionsvorgänge durch die Länge der zur Verfügung stehenden Pause beeinflußt werden. Bei genügender Pausenkürze vermag es so zu einer Art des Konzentrations- oder Sättigungseffektes kommen. Durch Herrn Prof. Rajewsky wird auf die erweiterte Möglichkeit der Enzym- oder Fermentbeeinflussung durch die Radikale hingewiesen.

# Dosimetrie energiereicher Strahlung

## Dosimetrie energiereicher Strahlungen

Von

**R. Glocker**, Stuttgart

Die ganze Problematik der Dosimetrie in ihrem gegenwärtigen Zustand ist daran zu erkennen, daß die Internationale Kommission für radiologische Einheiten zwei verschiedene Dosisbegriffe, die „Bestrahlungsdosis" (exposure dose) und die „absorbierte Dosis" (absorbed dose) aufgestellt und zwei verschiedene Dosiseinheiten, das „Röntgen" und das „rad" hierfür benannt hat. Ausgelöst wurden diese Schwierigkeiten durch den Übergang zu ultraharten Röntgenstrahlen; die erzeugten Sekundärelektronen haben hier so große Reichweiten, daß die Wirkung der Röntgenstrahlen sich nicht mehr auf die unmittelbare Umgebung des Absorptionsortes beschränkt.

Der Begriff „Bestrahlungsdosis" (exposure dose) ist eng verknüpft mit der Definition der Röntgeneinheit. Man denke sich einen Luftwürfel (Fig. 1) bestrahlt und die Ionisationsmessung so durchgeführt, daß alle Ionen, die von den

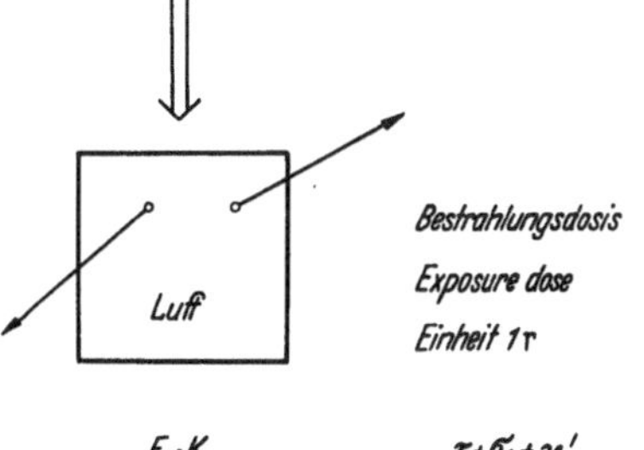


Fig. 1. „Bestrahlungsdosis" (schematisch)

Sekundärelektronen (Photo-, Compton- und Paarelektronen) ausgelöst werden, von der Messung erfaßt werden. Auch die Ionen, die auf den außerhalb des bestrahlten Volumens liegenden Teilen der Elektronenbahnen entstehen, liefern ihren Beitrag. In dieser Weise arbeiten die Standardionisationskammern zur Realisierung der Röntgeneinheit und zur Eichung von Dosimetern in Röntgeneinheiten. Diese so gemessene Luftionisation, ausgedrückt in Röntgeneinheiten, ist die „Bestrahlungsdosis" (exposure dose). Physikalisch bedeutet die „Bestrahlungsdosis" nichts anderes als die in dem bestrahlten Luftvolumen pro Masseneinheit erzeugte Energie an Photo-, Compton- und Paarbildungselektronen. Es ist

$$D_{exp} = \frac{E_0\,K}{88}\,[r], \tag{1}$$

wobei

$$K = \frac{\tau + \sigma_A + \varkappa'}{\varrho}$$

ist.

$E_0$ = auffallende Röntgenstrahlenenergie pro 1 cm² Oberfläche.

$\tau$   = Photoabsorptions-Koeffizient (unter Ausschluß einer charakteristischen Eigenstrahlung).

$\sigma_A$ = der beim Comptoneffekt auf die Bildung von Comptonelektronen entfallende Anteil der Energie.

$\varkappa'$ = Paarbildungskoeffizient nach Abzug der Ruheenergie von Elektron und Positron.

Wie die Fig. 2 zeigt, ist die „absorbierte Dosis" die in 1 g des bestrahlten Luftvolumens wirksame Energie. Sie entspricht im wesentlichen der ursprünglichen Definition der Dosis von R. CHRISTEN vom Jahre 1913. Die wirksame, d. h. in Ionisationen und Anregungen verwandelte Strahlungsenergie rührt her sowohl von den innerhalb des betrachteten Volumens erregten Sekundärelektronen als auch von den von außen her eintretenden. Die Teile der Elektronenbahnen, die Energie in dem betrachteten Volumen abgeben, sind in Fig. 2 mit ausgezogenen Strichen gezeichnet. Die absorbierte Dosis, die für jeden beliebigen Stoff definiert werden kann, wird in einem Energiemaß ausgedrückt, nämlich in rad, wobei 1 rad = 100 erg/g ist. Ergänzend sei bemerkt, daß die „Bestrahlungsdosis" nur für Luft definiert ist. In Form einer Gleichung stellt sich die absorbierte Dosis $D_{abs}$ so dar:

$$D_{abs} = \frac{q E_0 K}{100} + \frac{T}{100} \; [\text{rad}]. \qquad (2)$$

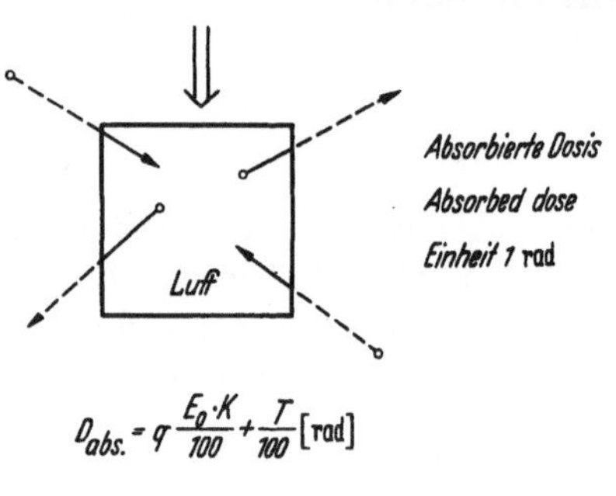


Fig. 2. „Absorbierte Dosis" (schematisch)

Dabei ist $q$ der Anteil der in dem betreffenden Volumen erzeugten Sekundärelektronen-Energie, der dort verbleibt, $T$ ist die Energie, die in dem Volumen von solchen Elektronen abgegeben wird, die in der Umgebung entstanden sind und von außen her eintreten. In zwei Fällen vereinfacht sich dieser Ausdruck:

1. wenn die Reichweite der Sekundärelektronen klein ist gegenüber den Dimensionen des Meßvolumens (z. B. bei weichen Röntgenstrahlen), oder

2. wenn die Meßanordnung so beschaffen ist, daß der Energieverlust durch abwandernde Sekundärelektronen kompensiert wird durch die von außen her einströmende Elektronenenergie (Fall des „Elektronengleichgewichtes").

Ist eine der beiden Bedingungen erfüllt, so vereinfacht sich die Gleichung; es wird

$$D_{abs} = 0{,}88\,D_{exp} . \qquad (3)$$

Der Faktor 0,88 rührt davon her, daß auf der linken Seite das rad, auf der rechten das r als Meßeinheit benützt wird.

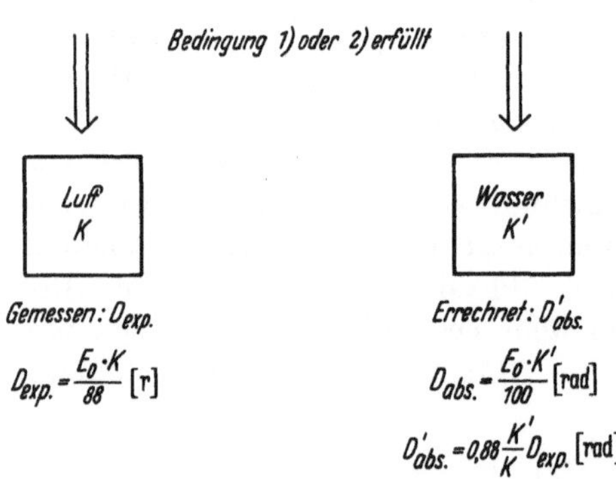


Fig. 3. Ermittlung der absorbierten Dosis von Wasser aus der Ionisationsmessung in r-Einheiten

Wie kann nun aus einer Ionisationsmessung in Röntgeneinheiten die für die Therapie wichtige absorbierte Dosis im Gewebe ermittelt werden? Eine einfache Umrechnung von Luft auf einen anderen Stoff, z. B. Wasser, ist nur möglich, wenn eine der beiden genannten Bedingungen erfüllt ist (Fig. 3). Wird $D_{exp}$

in einer in r geeichten Luft-Ionisationskammer gemessen, so gilt

$$D_{exp} = \frac{E_0 K}{88} \, [\mathrm{r}] \, . \tag{4}$$

Für 1 g Wasser[1] ist die in rad ausgedrückte absorbierte Dosis

$$D'_{abs} = \frac{E_0 K'}{100} \, [\mathrm{rad}] \, . \tag{5}$$

Hieraus folgt unter Benützung der Gl. (3)

$$D'_{abs} = \frac{0,88 \, K'}{K} \cdot D_{exp} \, . \tag{6}$$

Man muß also im Verhältnis der beiden Koeffizienten, deren Zahlenwerte für Wasser, Muskel, Fett, Knochen aus Tabellen entnommen[2] werden können, umrechnen.

Für sehr harte Röntgenstrahlen und für $\gamma$-Strahlen ist von BRAGG und von GRAY das in Fig. 4 dargestellte Meßverfahren angegeben worden. Die absorbierte Dosis in einem Medium, z. B. in Wasser, kann aus einer Ionisationsmessung in einem luftgefüllten Hohlraum indirekt bestimmt werden. Die Dimensionen des Hohlraumes müssen so klein sein, daß der ursprünglich in dem Medium vorhandene Elektronenfluß dadurch nicht gestört wird; die gemessene Ionisation muß praktisch nur von den Sekundärelektronen der Umgebung herrühren; die Ionisation durch die direkte Röntgenstrahlung muß also vernachlässigbar klein sein. In Fig. 4 ergibt sich dann die absorbierte Dosis $D'_{abs}$ für Wasser aus dem im ionisierten Luftraum pro g in elektrostatischen Einheiten gemessenen Ladungstransport $Q$

$$D'_{abs} = 0,88 \cdot Q \cdot s_L^w \, [\mathrm{rad}] \, . \tag{7}$$

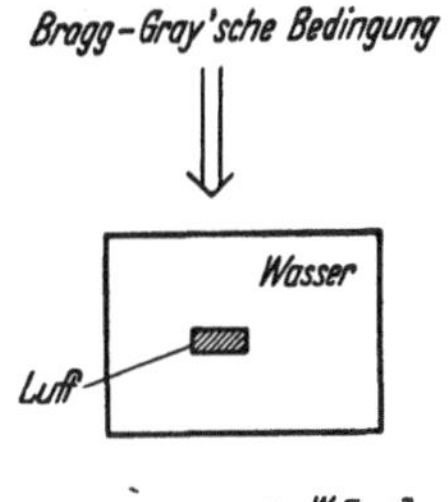

$$D_{abs.} = 0,88 \cdot \varrho \cdot s_L^w \, [\mathrm{rad}]$$

Fig. 4. Bragg-Graysche Bedingung

Dabei bedeutet $s_L^w$ das Verhältnis des Massen-Elektronenbremsvermögens von Wasser zu Luft, je bezogen auf die Masseneinheit.

Bei ultraharten Röntgenstrahlen tritt die Schwierigkeit auf, daß bei gleicher chemischer Zusammensetzung das Elektronenbremsvermögen in der Gasphase wegen des Fermi-Effektes (Polarisationskorrektion) etwas andere Werte hat als im festen oder flüssigen Aggregatzustand. Einige Zahlen sind in Tab. 1 angegeben.

*Tabelle 1*

| MeV | 1 | 5 | 10 | 15 | 30 |
|---|---|---|---|---|---|
| $s_L^w$ | 1,13 | 1,05 | 1,00 | 0,98 | 0,95 |

Das Elektronenbremsvermögen des Wassers ändert sich infolgedessen von 1—30 MeV um rund 20%, verglichen mit dem Bremsvermögen der Luft.

Alle Schwierigkeiten, die durch die großen Reichweiten der hochenergetischen Sekundärelektronen verursacht sind, lassen sich beheben, wenn als Dosierungsgrundlage nicht die Luftionisation gewählt wird, sondern eine Strahlungsreaktion

---

[1] Dabei ist zur Vereinfachung vorausgesetzt, daß die betrachtete Wasserschicht so dünn ist, daß statt $1 - e^{-\mu d}$ näherungsweise $\mu d$ gesetzt werden darf.

[2] Handbook 62 (April 1957) des Nat. Bureau of Standards, Washington (USA).

in einem festen oder flüssigen Stoff von einer dem Gewebe möglichst ähnlichen Zusammensetzung. Ein Weg ist z. B. die Leuchtstoffdosimetrie, ein anderer die chemische Dosimetrie mit hochverdünnten wäßrigen Lösungen[1]. Es konnte von uns[2] gezeigt werden, daß durch Mischung von Leuchtstoffen Substanzen hergestellt werden können, die sich hinsichtlich der Aussendung von Fluorescenzlicht bei Bestrahlung mit Röntgenstrahlen oder schnellen Elektronenstrahlen so verhalten, wie wenn sie die Zusammensetzung der Luft, des Wassers, des Muskels,

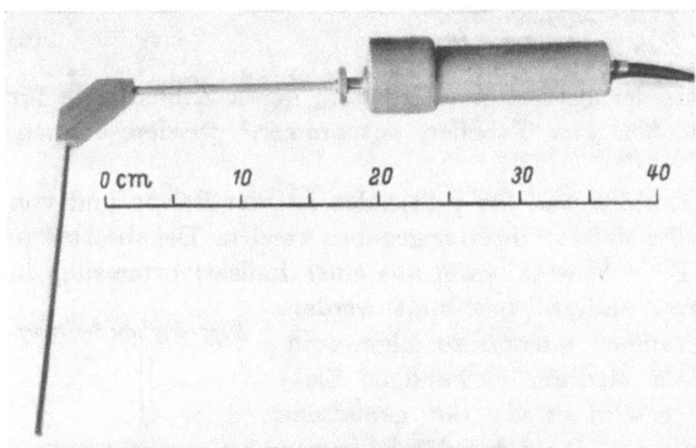



Fig. 5. Ansicht des Leuchtstoffdosimeters

des Knochens usw. hätten. Das Leuchtstoffdosimeter zeigt somit die absorbierte Dosis in diesen Stoffen an, z. B. die Muskeldosis, die Knochendosis usw. Die äußere Form ist in Fig. 5 zu sehen. Am Ende eines Plexiglasstabes, der von einem Aluminiumrohr lichtdicht umgeben ist, befindet sich ein einige Zehntel mm dikkes Scheibchen, das den Leuchtstoff auf Cellophanunterlage enthält. Das Fluorescenzlicht wird durch den Plexiglasstab einem Sekundärelektronenvervielfacher (Photomultiplier) zugeführt, dessen Ausgangsstrom gemessen wird. Ein gewisser Nachteil der Anordnung ist die starre Form des Lichtleiters. Zwei Anwendungsbeispiele seien kurz erwähnt: Es ist möglich,

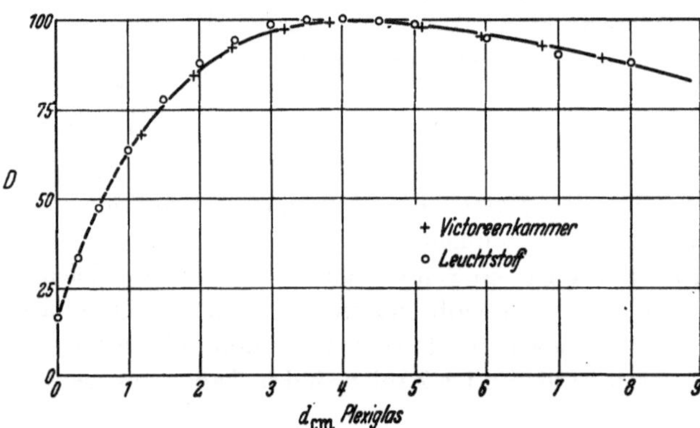



Fig. 6. Tiefendosiskurve von 31 MeV-Röntgenstrahlen

die Tiefendosiskurve eines Betatrons von der Oberfläche an auszumessen. In der Kurve der Fig. 6, die an dem Züricher 31 MeV-Betatron[3] aufgenommen wurde, bezeichnen die Kreuze Ionisationsmessungen[4] mit einem Victoreendosimeter und


[1] Siehe den folgenden Vortrag von S. RÖSINGER.

[2] BREITLING, G., u. R. GLOCKER: Naturwiss. 39, 471 (1952); 42, 483 (1955). — BREITLING, G.: Z. angew. Phys. 4, 401 (1952). — GLOCKER, R., u. G. BREITLING: Strahlenther. 88, 1 (1952). — BREITLING, G., R. GLOCKER u. H. MOHR: Fortschr. Röntgenstr. 84, 561 (1956).

[3] Herrn Prof. Dr. H. SCHINZ danke ich bestens für die Erlaubnis der Benützung des Betatrons.

[4] JOYET, G., u. W. MAUDERLI: Brown-Boveri-Mitt. 38, 284 (1951)

die Kreise die Leuchtstoffmessungen[1]. Die beiden Meßreihen fügen sich gut aneinander. Das zweite Beispiel betrifft den Anschluß der Dosimetrie von schnellen Elektronenstrahlen an die Röntgendosismessung. Mit einem luftäquivalenten Leuchtstoff und einer Luftionisationskammer wurde sowohl bei 200 kV

Röntgenstrahlen als auch bei Elektronenstrahlen[2] von 3—15 MeV gemessen. Das Verhältnis der Ergebnisse beider Methoden ist in Fig. 7 aufgezeichnet, wobei der Wert für 200 kV Röntgenstrahlen = 1 gesetzt ist[3]. Es ist zunächst überraschend, daß das Verhältnis mit der Energie der schnellen Elektronenstrahlen sich ändert. Dieser Gang hat, wie

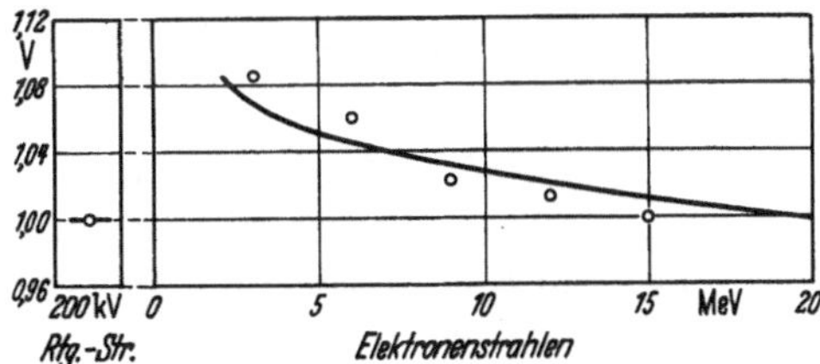


Fig. 7. Ionisationsmessungen und Messungen mit luftäquivalentem Leuchtstoff bei 200 kV Röntgenstrahlen und schnellen Elektronenstrahlen von 3—15 MeV

die Untersuchung im einzelnen ergab, zwei Ursachen: 1. Die Polarisationskorrektion und 2. die verschieden starke Ausnützung der von den primären Elektronen gebildeten energiereichen Sekundärelektronen; diese ist in der Luftschicht infolge der geringeren Massendichte kleiner als in der Leuchtstoffschicht.

Die Grundgleichung der Auswertung lautet:

$$\frac{\left(\dfrac{\text{Fluor. Int.}}{\text{Ionisat.}}\right)_{\text{15 MeV Elstr.}}}{\left(\dfrac{\text{Fluor. Int.}}{\text{Ionisat.}}\right)_{\text{200 kV Rtgstr.}}} = \frac{\eta \cdot p \cdot \left(\dfrac{\Delta E}{\Delta X}\right) \Big/ \eta' \cdot p' \cdot \left(\dfrac{\Delta E}{\Delta X}\right)'}{\sigma_A / \sigma_A'}$$

Dabei bedeutet:

$\dfrac{\Delta E}{\Delta X}$ = Elektronenbremsvermögen.

$p$ = Polarisationskorrektion.

$\eta$ = Ausnützung der energiereichen Sekundärelektronen.

$\sigma_A$ = Comptonabsorptionskoeffizient.

Die theoretischen Werte wurden durch die Leuchtstoffmessungen und parallel dazu vorgenommenen Versuche mit Eisensulfatlösungen gut bestätigt.

Der Korrektionsfaktor $\eta p/\eta p'$ hat im vorliegenden Fall den Wert 0,95 für den luftäquivalenten Leuchtstoff bzw. 0,93 für die Eisensulfatlösung. Was hier für Leuchtstoff und Lösung hinsichtlich der Übertragung der Ionisationsmessung gesagt wurde, gilt sinngemäß auch für das Gewebe.

---

[1] Brown-Boveri, 2. Betatronheft (1953).

[2] Herr Prof. Dr. J. BECKER hat in freundlicher Weise die Heidelberger Elektronenschleuder für die Messungen zur Verfügung gestellt.

[3] BREITLING, G., u. R. GLOCKER: Naturwiss. 42, 11 (1955). — GLOCKER, R.: Z. Physik 143, 191 (1955). — BREITLING, G.: Z. Physik 149, 180 (1957).

# Chemische Dosimetrie*

Von

S. RÖSINGER, Höchst am Main

Unter chemischer Dosimetrie darf ich verstanden wissen die irreversible Veränderung eines chemischen Systems, hervorgerufen durch ionisierende Strahlung. Ein Kennzeichen der chemischen Dosimetrie ist die nur integrierende Messung der in einem bestimmten Volumen absorbierten Energie.

GLOCKER und Mitarbeiter haben vor Jahren das Wirkungsgesetz ionisierender Strahlen festgelegt [10]. In ihm wird ausgesagt, daß nur der Anteil einer Strahlung zu einer Wirkung gelangt, der in Elektronenenergie umgewandelt wird. Auch die chemische Dosimetrie hat dieses Gesetz bestätigt [15].

Ich darf zunächst einige Beispiele verschiedener Methoden chemischer Dosimetrie erwähnen. Sie gliedern sich prinzipiell in Redoxreaktion, wie z. B. die Reduktion von $Ce^{IV} \to Ce^{III}$ [5] oder die Oxydation von $Fe^{2+} \to Fe^{3+}$ [8] und Abspaltung bestimmter chemischer Verbindungen, z. B. die Bestimmung von $CO_2$ und $H_2$ bei einem Ameisensäuredosimeter, wie es HART [14] vorgeschlagen hat, oder der Abspaltung von HCl aus $HCCl_3$. Die Salzsäure stellt einen bestimmten $p_H$ ein, der entweder potentiometrisch oder mit einem Farbindicator gemessen werden kann [22]. Ein weiteres Verfahren ist die Veränderung eines Farbstoffes, (z. B. folgendes Gerüst)

$$\bigcirc\!\!-N\!\!=\!\!N\!-\!\overset{\overset{\displaystyle S}{\|}}{C}\!-\!N\!\!=\!\!N\!-\!\!\bigcirc$$

im Strahlungsfeld [4].

Vom rein physikalischen Standpunkt aus zeigt die chemische Messung weitgehende Ähnlichkeit zur Energieabsorption im Gewebe, denn die Reaktion läuft im wäßrigen Medium ab, dessen Absorptionseigenschaften dem der organischen Substanz weitgehend ähneln. Des weiteren braucht die Polarisationskorrektur — im Gegensatz zur Luftkammer — nicht berücksichtigt zu werden [2]. Für unsere Untersuchungen haben wir das System $Fe^{2+} \to Fe^{3+}$ ausgewählt. Die Messungen wurden an einer belüfteten 0,8 $H_2SO_4$, in der $FeSO_4$ in einer $10^{-3}$ molaren Lösung vorhanden war, durchgeführt. Die Messung der gebildeten $Fe^{3+}$ wurde photometrisch mit Thiocyanat als Komplex bei 480 m$\mu$ ausgeführt. Die Bestrahlung erfolgte in Cuvetten aus Polystyrol mit einem Inhalt von 2 $cm^3$ und einer Schichtdicke von 2 mm. Definierte Zusätze von bestimmten organischen Stoffen zu der zu bestrahlenden Lösung und zu der Meßlösung erhöhen die Empfindlichkeit um den Faktor 10, so daß Dosen von einigen hundert r gemessen werden können [19].

---

* Aus dem Röntgeninstitut der Techn. Hochschule Stuttgart (Direktor: Prof. Dr. R. GLOCKER)

Die Energie wird praktisch allein vom Wasser, zu einem geringen Teil von der Schwefelsäure absorbiert, das $Fe^{2+}$ wirkt nur als Indicator für gebildete Radikale und das $H_2O_2$.

Für Biologie und Medizin ist es letztlich wichtig, die Veränderung in der Substanz zu erkennen. Der Chemie bleibt es vorbehalten, den Reaktionsmechanismus an Modellverbindungen zu erforschen. Als Reaktionsschema darf ich die radiolytische Zersetzung des Wassers anführen, die in ihrer Folgereaktion die Oxydation des $Fe^{2+} \rightarrow Fe^{3+}$ bewirkt.

Wir unterscheiden eine Radikalreaktion und eine Molekularreaktion [1]

$$H_2O \rightarrow H + OH \quad (R)$$
$$2\,H_2O \rightarrow H_2 + H_2O_2 \quad (M),$$

die beide einen Anteil an der gemessenen Wirkung tragen und je nach der Ionisationsdichte der Strahlung in verschiedenem Verhältnis beitragen. Hier setzt bereits die chemische Reaktionskinetik ein, nach der alle Folgereaktionen verlaufen müssen. Nach WEISS, HART, ALLEN [23] legt man heute folgende Reaktionen zugrunde:

$$H + O_2 \rightarrow HO_2$$
$$HO_2 + H^+ + Fe^{2+} \rightarrow Fe^{3+} + H_2O_2$$
$$H_2O_2 + Fe^{2+} \rightarrow Fe^{3+} + OH + OH^-$$
$$OH + Fe^{2+} \rightarrow Fe^{3+} + OH^-.$$

**Ergebnisse.** Vergleicht man zunächst einmal die chemische Wirkung bei 15 MeV Elektronenstrahlen und 200 kV (0,5 mm Cu) Röntgenstrahlen [12], die beide für die Therapie von Bedeutung sind — wir benutzten Elektronen und keine $\gamma$-Strahlen wegen der definierten Energieverhältnisse der Sekundärelektronen —, so finden wir entsprechend der Gleichung [11]

$$Q = \frac{(C/i)\ 15\,\mathrm{MeV}}{(C'/i')\ 200\,\mathrm{kV}} = 0{,}96,$$

wobei $C$ und $C'$ die chemischen Ausbeuten und $i$ und $i'$ die Ionisationsströme bei 15 MeV Elektronen und 200 kV Röntgenstrahlen bedeuten.

Die chemische Wirkung ist bei gleicher Dosis (mit einer Ionisationskammer gemessen) bei 15 MeV Elektronenstrahlen kleiner als bei 200 kV Röntgenstrahlen, so wie es aus der

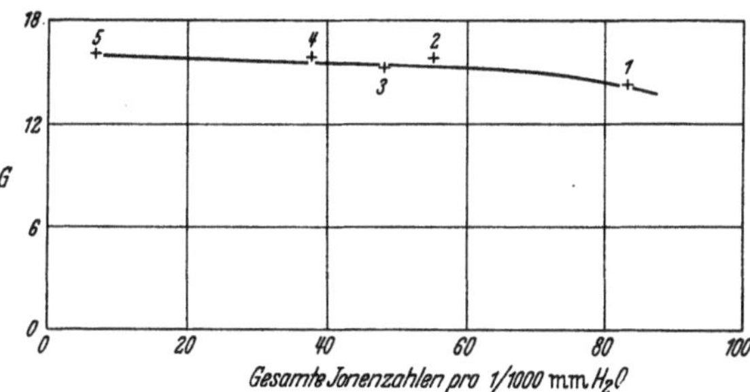


Fig. 1. Abhängigkeit des $G$-Wertes der $Fe^{II}$-Oxydation von der Ionisationsdichte. (1) Cu K $\alpha$ 8 kV (Ni-Filter), (2) 200 kV (0,5 mm Cu), (3) Mo K $\alpha$ 17 kV (Zr-Filter), (4) Ag K $\alpha$ 23 kV (Pd-Filter), (5) 15 MeV-Elektronen

Theorie von GLOCKER zu erwarten war. Berücksichtigt man aber bei dieser Messung die Polarisationskorrektion, so findet man, daß bei 15 MeV pro 100 eV absorbierter Energie mehr $Fe^{3+}$ Ionen gebildet werden als bei einer 200 kV-Röntgenstrahlung.

BURTON hat für die 100 eV-Ausbeute den sog. $G$-Wert eingeführt.

Man erkennt auf der Fig. 1 die Abhängigkeit des $G$-Wertes von der Ionisationsdichte. Durch die Comptonstreuung liegt der Wert der 200 kV-Strahlung zwischen

158                                          S. Rösinger:

dem einer gefilterten $CuK\alpha$ (8 kV) und $MoK\alpha$ (17 kV) Strahlung bei einer
Gesamtionendichte von etwa 55 Ionen, während für die 15 MeV Elektronen sich
eine Ionendichte von 7—8 errechnet. Mit abnehmender Ionisationsdichte ist eine
geringe Steigerung des $G$-Wertes zu beobachten.

**Diskussion.** Vergleicht man die gefundenen $G$-Werte mit bereits in der Literatur publizierten Ergebnissen, so ist wegen der Energieabhängigkeit eine Bestimmung der $G$-Werte bei verschiedener Strahlungsqualität vom gleichen Autor
notwendig. ZSULA, LUIZZI und LAUGHLIN haben ebenfalls bei Elektronenstrahlen
eine leichte Erhöhung gefunden. Auch aus der Arbeit von HAYBITTLE, SAUNDERS
und SWALLOW ist ein Anstieg des $G$ von 220 kV zu 30 MeV zu entnehmen.
ZUPPINGER und MINDER haben wohl eine Zunahme des $G$ von 250 kV zu 31 MeV
Röntgenstrahlen gefunden, doch liegen ihre Absolutwerte mit 16,2 und 16,9
relativ hoch. CHAPIRO, EBERT und GRAY nehmen innerhalb ihrer Meßfehler
keinen Unterschied zwischen 37 kV und 2 MeV Röntgenstrahlen an, während
FRITZ-NIGGLI, ohne einen Absolutwert anzugeben, zu dem gleichen Ergebnis
kommt.

Als Zusammenfassung darf man heute folgendes feststellen: Es scheint so,
als ob mit geringer werdender Ionendichte, also mit steigender Energie der
Elektronen im absorbierenden Medium der $G$-Wert leicht ansteigt, daß man aber,
ohne einen großen Fehler zu begehen, mit einem Mittelwert von 15,5 von etwa
20 kV bis 30 MeV rechnen kann.

*Tabelle 1*

| | Röntgenstrahlen | | $Co^{60}$ | Elektronenstrahlen | | Röntgenstrahlen | |
| --- | --- | --- | --- | --- | --- | --- | --- |
| | kV | $G$ | | MeV | $G$ | MeV | $G$ |
| GLOCKER, MESSNER, RÖSINGER | 200 (0,5 Cu) | 15,7 | | 15 | 16,0 | | |
| ZSULA, LUIZZI, LAUGHLIN | | | 15,7 | 6,3 | 15,5 | | |
| | | | | 16,0 | 15,7 | | |
| HAYBITTLE, SAUNDERS, SWALLOW | 220 (0,65 Al) | 15,0 | 15,5 | | | 30 | 16,3 |
| ZUPPINGER, MINDER | 250 (0,25 Cu) | 16,2 | | | | 31 | 16,9 |
| FARMER, RIGG, WEISS | 200 | 16,0 | | | | | |
| GORMACK, HUMMEL, JOHNS | | | 16,9 | | | 23 | 16,2 |
| WEISS, BERNSTEIN, KUPPER | | | | | | | |
| SCHULER, ALLEN | | | | | | 2 | 16,0 |
| SALDICK, ALLEN | | | | 1 | 15,6 | 2 | 15,5 |
| EHRENBERG, SAELAND | 175 (0,5 Cu, 1 Al) | 16,6 | | | | | |
| HOCHANADEL, GHORMLEY | | | 15,7 | | | | |
| LAZO, DEWHURST, BURTON | | | 15,8 | | | | |
| CHAPIRO, BOAG, EBERT, GRAY | 37 (Mo) | 20,3 | Ra 20,9 | | | 1,2 | 20,8 |
| FRITZ-NIGGLI | 180 | ⟶ | | gleicher Umsatz ⟵ | | 31 | |

Bisherige $G$-Wert-Bestimmungen der $Fe^{II}$-Oxydation im Gebiet von 200 kV-Röntgenstrahlen, von $Co^{60}$ und von hochenergetischen Bestrahlungsanlagen.

Bemerkung: Alle $G$-Werte wurden auf 34,0 eV umgerechnet.

## Literatur


*1.* ALLEN, A. O.: Rad. Res. **1**, 86 (1953).

*2.* BREITLING, G., R. GLOCKER u. S. RÖSINGER: Naturwiss. **42**, 507 (1955).

*3.* CHAPIRO, A., I. W. BOAG, M. EBERT et L. H. J. GRAY: J. Chim. physique **50**, 468 (1953).

*4.* CLARK, G. L., and P. E. BIERSTEDT JR.: Rad. Res. **2**, 295 (1955).

*5.* — and W. S. COE: J. Chem. Phys. **5**, 97 (1957).

*6.* EHRENBERG, L., u. E. SAELAND: Jener-Publ. Nr. 8 (1954).

*7.* FARMER, F. T., T. RIGG and J. WEISS: J. Chem. Soc. (London) **1954**, 3248.

*8.* FRICKE, H., and S. MORSE: Amer. J. Roentgenol. 18, 420 (1927). — FRICKE, H.: Physic. Rev. **31**, 17 (1928).

*9.* FRITZ-NIGGLI, H., u. N. SCHMIDLIN-MESZAROS: Strahlenther. **95**, 551 (1954).

*10.* GLOCKER, R.: Z. Physik **43**, 827 (1927); **46**, 764 (1928); **136**, 352 (1953).

*11.* — Z. Physik **143**, 191 (1955).

*12.* — D. MESSNER u. S. RÖSINGER: Z. phys. Chem. (im Druck).

*13.* CORMACK, D. V., R. W. HUMMEL, H. E. JOHNS and I. W. SPINKS: J. Chem. Phys. **22**, 6 (1954); **23**, 162 (1955).

*14.* HART, E. J.: J. Phys. Chem. **56**, 594 (1952).

*15.* HAYBITTLE, J. L., R. D. SAUNDERS and A. J. SWALLOW: J. Chem. Phys. **25**, 1213 (1956).

*16.* HOCHANADEL, C. J., and J. A. GHORMLEY: J. Chem. Phys. **21**, 880 (1953).

*17.* LAZO, R., H. P. DEWHURST and M. BURTON: J. Chem. Phys. **22**, 1370 (1954).

*18.* RÖSINGER, S.: Z. phys. Chem. N. F. **10**, 310 (1957).

*19.* — R. GLOCKER u. J. GOUBEAU: Z. phys. Chem. (im Druck).

*20.* SALDICK, J., and A. O. ALLEN: J. Chem. Phys. **22**, 438 (1954).

*21.* SCHULER, G. H., and A. O. ALLEN: J. Chem. Phys. **24**, 54 (1956).

*22.* SHULTE, J. W., and S. F. SUTTLE: J. Amer. Chem. Soc. **75**, 2222 (1953).

*23.* WEISS, J.: Proc. Roy. Soc. A **211**, 375 (1952). — ALLEN, A. O.: J. Phys. Coll. Chem. **52**, 479 (1948).

*24.* — W. BERNSTEIN and J. B. H. KUPER: J. Chem. Phys. **22**, 1593 (1954).

*25.* ZSULA, J., A. LUIZZI u. J. S. LAUGHLIN: Priv. Mitt. (1957).

*26.* ZUPPINGER, A., u. N. MINDER: Radiol. clin. (Basel) **24**, 347 (1955).

# Mesure par films de la distribution en profondeur de la dose pour les électrons de haute énergie*

Par

**J. M. Dutreix**, Paris

La dosimétrie par films du faisceau d'électrons a été déjà largement étudiée dans le domaine d'énergie correspondant aux rayons beta émis par les éléments radioactifs [*7*, *10*], et nous sommes proposés d'utiliser cette méthode pour le faisceau d'électrons de grande énergie produit par le bétatron.

L'avantage le plus évident du film est la haute résolution spatiale offerte par son extrême minceur (environ 0,02 cm), et par la technique habituelle de densitométrie qui permet une exploration du film par un pinceau lumineux étroit (de diamètre environ 1 mm.); cette haute résolution spatiale est particulièrement utile dans les régions où la variation de dose est rapide.

Mais le film présente un inconvénient majeur: le processus complexe de formation et du développement de l'image ne permet pas d'utiliser une relation théorique entre la densité optique et l'énergie absorbée, ou la dose; il est nécessaire d'étaloner la densité du film à l'aide d'un dosimètre plus fidèle, tel qu'une chambre d'ionisation.

Le but de la dosimétrie par film étant de substituer le film à la chambre d'ionisation, la première étape est d'établir la relation entre la densité optique et la dose, c'est-à-dire de définir les *courbes d'étalonnage* du film. Nous avons utilisé le film Eastman Kodak, type M, et la densité optique a été mesurée à l'aide d'un densitomètre ANSCO qui permet une évaluation précise de la densité de 0,1 à 3, ce qui constitue un domaine suffisamment vaste pour les besoins pratiques de la radiothérapie.

**Courbes d'étalonnage.** Les courbes representant la variation de la densité en fonction de l'exposition sont des lignes droites; l'ordonnée, pour l'exposition 0, représente le «voile de fond» (ou «fog»), c'est-à-dire la densité pour un film non exposé et développé; cette densité est petite (0,1) et pour des densités entre 1 et 3, la courbe d'étalonnage peut être aussi bien considérée comme une ligne droite passant par l'origine, ce qui simplifie les mesures dans ce domaine en permettant de négliger le «voile de fond».

La pente des courbes d'étalonnage varie en fonction des conditions de développement. Les Fig. 1 et 1a représentent la variation due à la température $\Theta$, quand le temps de développement $t$ est fixé, et les Fig. 2 et 2a celle qui est due au temps de développement à une température définie.

---

* Physics Department, Memorial Hospital — New York. Adresse actuelle: Institut Gustave Roussy à Villejuif (Seine)

Les familles de courbes obtenues pour différentes valeurs de $\Theta$ et de $t$ permettant de calculer la correction correspondant à une variation donnée dans les conditions de développement, mais un procédé plus sûr consiste à utiliser des conditions définies; nous avons adopté:

$\Theta = 22$ degrés centigrades
$t = 5$ minutes.

Les autres étapes du développement, bien que leur constance soit moins impérieuse, sont cependant standardisées:

bain d'arrêt: 1 minute
bain de fixage: 15 minutes
bain de lavage: 20 minutes.

L'agitation est également standardisée de la façon suivante:

chaque minute, le film est retiré du bain pendant quelques secondes et replongé sans autre agitation.

L'âge du révélateur parait également avoir quelque importance: dans un bain neuf, la densité est augmentée et la courbe caractéristique présente une petite incurvation. Ceci disparaît après le développement d'environ 0,5 m² de film, et la qualité du révélateur demeure constante jusqu'à 5 et 10 m².

**Precision attendue dans la mesure de la densité.** La précision des mesures dépend de plusieurs facteurs que nous pouvons diviser en 2 groupes:

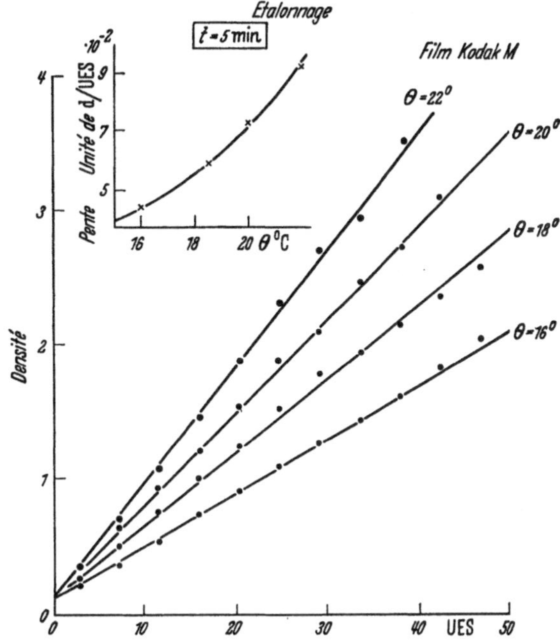


Fig. 1 et 1a. Courbes d'étalonnage des films pour une durée de développement constante (5 minutes). La densité optique est proportionnelle à l'exposition. Le coefficient de proportionnalité croît avec la température du bain; sa variation est représentée sur la figure 1a

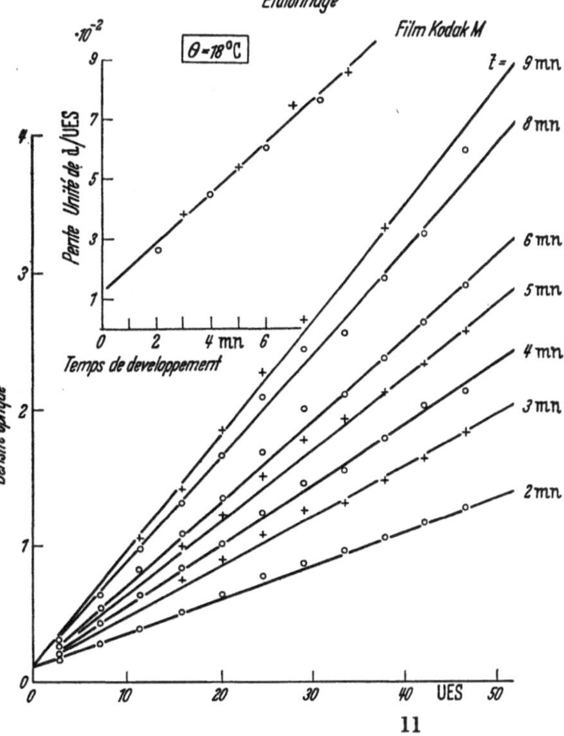


Fig. 2 et 2a. Courbe d'étalonnage des films pour une température constante du bain (18° C). La densité optique est proportionnelle à l'exposition. Le coefficient de proportionnalité croît avec la durée du développement, sa variation est représentée sur la figure 2a

162　　　　　　　　　　　　　　　　J. M. DUTREIX:

*a) Conditions de développement:* La pente $s$ de la droite d'étalonnage au point $t = 5$ mn, $\Theta = 20°$ C est $7 \cdot 10^{-2}$ unités de densité par unité d'exposition (VES/cm³). La variation de $s$ en fonction des paramètres $t$ et $\Theta$ est en ce point (voir Fig. 1 a, 2 a):

$$\frac{ds}{d\Theta} = 6 \cdot 10^{-3}$$ (unité de densité par unité d'exposition et par degré centigrade)

et $\frac{ds}{dt} = 1{,}55 \cdot 10^{-2}$ (unité de densité par unité d'exposition et par minute).

Une variation de $1/10°$ C sur $\Theta$ introduit une variation de $1\%$ sur la densité, et une variation de $1/10$ mn sur $t$, une variation d'environ $2\%$. Ces variations représentent approximativement la précision effective des conditions de développement et l'erreur correspondante sur la densité est de l'ordre $\pm3\%$. (Le facteur «agitation» qui est inaccessible à une mesure, sera considéré dans le second groupe).

*b) Conditions de mesure.* L'erreur précédente peut être éliminée si l'on a seulement l'intention de déterminer la densité relative de certains points sur le même film, ou sur des films développés simultanément. En fait, même dans de telles conditions, les points expérimentaux ne se situent pas rigoureusement sur une ligne droite, mais sont dispersés autour d'elle avec un écart d'environ $\pm2{,}5\%$.

Cet écart contient les imprécisions:

1. sur la quantité de dose délivrée au film (erreur du moniteur),

2. sur la lecture au densitomètre,

3. erreur due au film lui-même, telle que l'irrégularité de l'émulsion, ou l'irrégularité de l'imprégnation dans le bain (facteur d'agitation).

En résumé:

En dosimétrie «directe», quand une lecture de *densité optique* doit être directement traduite en termes de *dose*, l'erreur attendue est d'environ $\pm5\%$. Mais il est nécessaire d'avoir un étalonnage pour chaque livraison car la sensibilité du film peut varier de plus de $10\%$ d'une livraison à une autre.

En dosimétrie «relative» quand le but est seulement de déterminer le rapport entre les doses en différents points du même film ou sur des films développés simultanément, le premier groupe d'erreurs peut être négligé, et la précisi on attendue est environ $2{,}5\%$. Quand l'information est donnée par la mesure d'un grand nimbre de points, comme par exemple dans le tracé d'une courbe de distribution de la dose en profondeur, la précision attendue est encore plus grande.

Cette précision a été vérifiée expérimentalement par la mesure d'un grand nombre de films: la dispersion dans les mesures directes, par exemple les densités mesurées sur les films exposés à la même dose au cours de deux mois est de l'ordre de $\pm5\%$, tandis que les courbes qui représentent par exemple la distribution de la dose en profondeur dans le même faisceau coïncident à environ $1\%$ près.

**Influence de l'énergie.** La réponse du film ne dépend pas de l'énergie des électrons ni du diffuseur placé à la fenêtre du tube pour égaliser la dose dans le faisceau. La densite du film conserve une proportionnalité constante avec la mesure fournie par une chambre d'ionisation dans le domaine d'énergie de 5 à 24 MeV, et pour les 4 conditions différentes de diffusion (pas de diffuseur, $^1/_2$ mm d'aluminium, 0,125 mm de plomb, 0,250 mm de plomb). Ce résultat était attendu après les conclusions rapportées par plusieurs auteurs qui ont souligné qu'il n'y avait pas de dépendance de l'énergie au-dessus de 2 MeV [*10, 11*], et même 0,5 MeV [*6*].

**Influence de l'angle d'incidence.** Cette influence fut étudiee de 2 façons différentes :

1. Le film est roulé en forme d'un tube cylindrique de 2 cm de diamètre dont l'axe est perpendiculaire au faisceau; le noircissement est homogène sauf le long de 2 bandes étroites plus claires, qui correspondent aux 2 génératrices où les électrons passent tangentiellement au cylindre.

2. Le film est placé en sandwich dans un cylindre massif de plexiglas (diamètre 5 cm, longueur 20 cm) coupé en 2 moitiés le long d'un plan diamétral. L'incidence du faisceau est modifiée en faisant tourner le cylindre autour de son axe qui est perpendiculaire au faisceau; il n'y a pas de variation significative dans la densité mesurée sur l'axe de rotation.

Ce résultat était également attendu après les expériences de DUDLEY qui n'ont pas montré d'influence de l'angle d'incidence aux énergies supérieures à 1,5 MeV, sauf à de très grands angles. MARKUS et PAUL [9] ont souligné que la diffusion dans le film doit être supérieure à celle dans le fantôme et ils ont recommandé d'utiliser le film perpendiculairement au faisceau plutôt que parallèlement; dans le domaine d'énergie que nous avons utilisé ce phénomène n'est pas apparu très significatif.

**Utilisation en radiothérapie.** *A. Distribution de la dose en profondeur.* Les mesures de la densité optique des films irradiés entre des plaques de plexiglas, soit parallèles, soit perpendiculaires à l'axe du faisceau d'électrons, sont en bon accord avec les résultats obtenus à l'aide d'une chambre d'ionisation.

Les mesures par films dans le plexiglas ont été comparées aux mesures d'ionisation dans l'eau, la profondeur étant exprimée en densité superficielle (1 cm de plexiglas = 1,18 g/cm² [1]. La Fig. 3 montre le bon accord obtenu à différentes énergies.

Ce résultat est d'environ 10% différent des mesures effectuées pour des électrons de 15 MeV par VON DER DECKEN [2] qui a trouvé que les distributions des doses en profondeur dans l'eau et le plexiglas étaient très voisines (environ 5% d'écart) quand les profondeurs sont exprimées en unité de longueur.

Les mesures par films furent répétées dans d'autres milieux: le Mix D, l'Aluminium et le Cuivre. Pour les 2 premiers milieux, les courbes de densité en profondeur s'accordent assez exactement aux mesures effectuées dans le Plexiglas, les profondeurs étant exprimées en termes de densités superficielles (Fig. 4) et le parcours extrapolé dans les 4 matériaux est le même: 11,3 g/cm² à 22,5 MeV [2].

Il est possible d'admettre que:

1. Le Mix D et le Plexiglas sont des milieux très satisfaisants pour confectionner un fantôme équivalent à l'Eau: le Mix D a l'avantage d'une densité égale à 1, tandis que le Plexiglas a des propriétés mécaniques bien meilleures.

2. La mesure de la densité optique du film est une méthode fidèle qui donne des résultats comparables à ceux obtenus par une chambre d'ionisation.

---

[1] Il serait aussi logique d'exprimer la profondeur en terme de densité électronique superficielle; le rapport entre ces densités pour l'eau et le plexiglas étant 1/1,15; ceci revient a reculer les courbes relatives au plexiglas de 3%

[2] La formule proposée par KATZ et PENFOLD [8], $R$ (g/cm²) = 0,530 E (MeV) — 0,116 donne à cette énergie $R = 11,8$ g/cm²

Il est nécessaire de placer le film dans une enveloppe de papier, ou d'opacifier les surfaces des plaques quand le film est utilisé dans un milieu transparent tel que le plexiglas, pour éviter un noircissement supplémentaire dû à la lumière émise par les électrons de haute énergie (Cf. chapitre III).

*Discussion.* Les courbes de distribution de la densité en profondeur dans le plexiglas, à 22,5 MeV, obtenues sur 10 films, correspondent à moins de 1%. La même précision peut être attendue des mesures par une chambre d'ionisation.

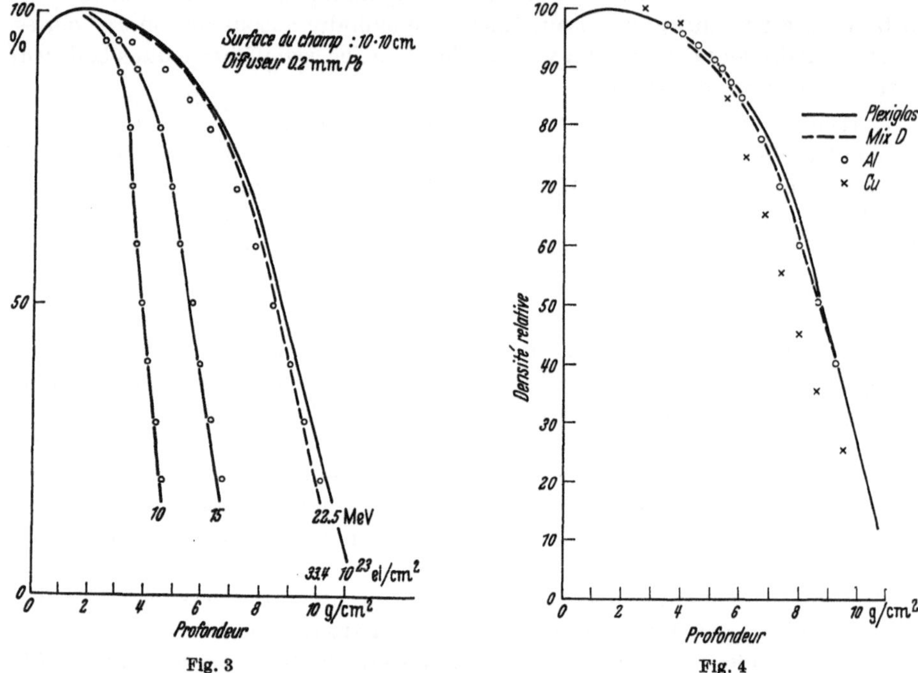


Fig. 3                          Fig. 4

Fig. 3. Distribution de la dose en profondeur. Les courbes en trait plein représentent la variation de la dose mesurée par densitométrie des films dans le plexiglas. Les croix représentent les mesures à la chambre d'ionisation. Pour les 2 groupes de mesures, la profondeur est exprimée en masse superficielle (en g/cm²). La courbe en tirets représente la distribution en profondeur mesurée par film dans le plexiglas à 22,5 MeV, lorsque la profondeur est exprimée en densité électronique superficielle (1 cm de plexiglas → 1,18 g/cm² → 3,34 10²³ électrons/cm²)

Fig. 4. Distribution de la densité optique en profondeur dans différents milieux à 22,5 MeV. On observe un excellent accord entre les courbes du Plexiglas, Mix D et Aluminium. La courbe correspondant au Cuivre se montre par contre très différente, bien que le parcours extrapolé soir sensiblement le même

Cependant, il apparaît, entre les 2 techniques de mesure, une petite différence qui n'est pas significative pour les besoins pratiques de la dosimétrie, mais qui est cependant trop élevée pour être attribuée au hasard.

La chambre d'ionisation, dont le diamètre est de 4 mm; peut être considéré comme respectant de façon satisfaisante la condition de Bragg Gray; l'émulsion, bien que extrèmement mince, a une masse superficielle de 6 mg/cm², ce qui correspond à une épaisseur de 5 cm d'air, et au parcours d'électrons de 60 keV. Dans ces conditions, une certaine fraction de la densité optique du film est due aux électrons secondaires provenant de l'émulsion elle-même.

Mais, d'autre part, l'émulsion n'est pas assez épaisse pour que la densité soit due exclusivement à l'énergie absorbée dans l'émulsion elle-même. La participation du milieu environnant est mise en évidence en plaçant des écrans très fins et de nature différente, au contact de l'émulsion; ceci ne change pratiquement pas e flux des électrons primaires, mais provoque cependant une modification sensible de la densité.

La densité optique du film est donc simultanément fonction de l'énergie absorbée dans l'émul-leonielle-même, et dans le milieu sinvronnant. Le rapport entre les pertes d'énergie dans les 2 milieux est sensiblement constante au-dessus de 100 keV (et environ égal à 1,5), mais il augmente progressivement quand l'énergie des électrons diminue (il est égal à 2 à 10 keV). On peut donc considérer que la signification physique de la densité optique est moins précise que celle de la mesure d'ionisation dans une chambre de petit volume.

*B. Courbes isodose.* Les courbes isodose obtenues dans le plexiglas par mesure de la densité optique du film sont en bon accord avec celles obtenues dans l'eau par des mesures d'ionisation, la profondeur étant exprimée en densité superficielle. Les isodoses pour une incidence oblique, Fig. 5 et 6, sont obtenues à l'aide de 3 séries de films:

1. parallèles à la surface,

2. perpendiculaires à la surface et parallèles à l'axe du faisceau,

3. à 90° sur les 2 directions précédentes.

Les isodoses courent parallèlement à la surface d'entrée du fantôme, et les courbes de distribution de la dose en profondeur au long de chaque pinceau élémentaire sont très voisines de la courbe de distribution de la dose en profondeur pour un

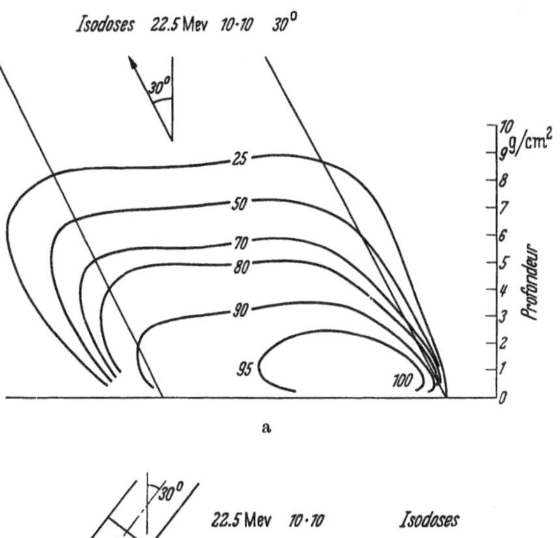


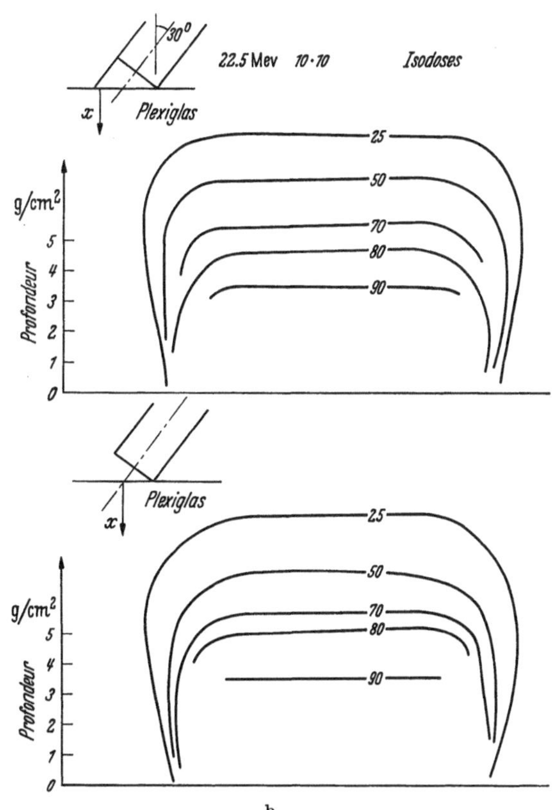


Fig. 5 a et b. Isodoses mesurées par film dans le Plexiglas pour un champ oblique (30°). a) Mesures dans le plan médian du faisceau perpendiculaire à la surface d'entrée b) Mesures dans 2 plans transversaux indiqués sur les schémas

                J. M. Dutreix:

faisceau perpendiculaire. La dosimétrie par film est particulièrement commode quand la surface d'entrée est irrégulière. Les Fig. 7 à 9 montrent quelques exemples de courbes isodoses obtenues dans des fantômes de cire de formes irrégulières.

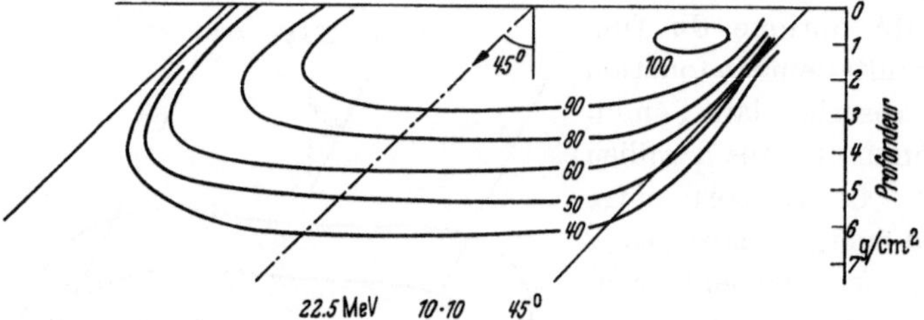


Fig. 6. Isodoses mesurées par film dans le Plexiglas pour un champ oblique à 45°

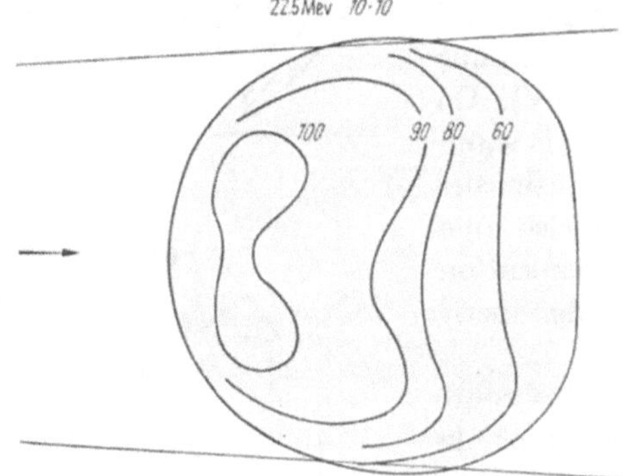


Fig. 7. Isodoses dans un cou fantôme de densité homogène (cire)

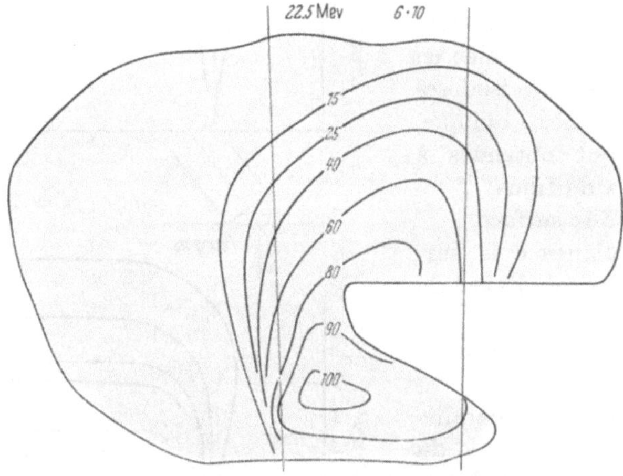


Fig. 8. Isodoses pour une surface et un milieu de forme irrégulière (mesurée dans un fantôme en presswood). Il s'agit d'un champ latéral droit chez un malade ayant subi un évidement large du sinus maxillaire

*Courbes d'isodoses en radiothérapie sous grille.* Le film est particulièrement commode pour l'étude de la distribution de la dose dans des régions où sa variation est rapide, ce qui est le cas sous une grille.

La Fig. 10 correspond à une grille à ouvertures carrées, de rapport 40%. Elle montre que les faisceaux individuels restent parfaitement définis jusqu'à une profondeur de 7 g/cm² à 22,5 MeV; au-delà, le défaut d'homogénéité disparaît en raison de la diffusion, mais la dose est alors inférieure à 20% de la dose à la surface d'entrée. Ce résultat est en bon accord avec les mesures rapportées par BECKER et Coll. [1].

**Lumière émise par le Plexiglas irradié.** Quand le film est placé directement dans le fantôme de plexiglas, sans aucune enveloppe opaque, un noircissement supplémentaire est observé qui est dû aux photons lumineux qui prennent naissance dans le volume irradié. Cette lumière est arrêtée par une simple feuille de papier jaune ordinaire, dans lequel le film est enveloppé. Une petite fraction de cette lumière semble être distribuée de façon isotropique (fluorescence), mais la partie la plus importante est fortement dirigée vers l'avant (rayonnement CERENKOV).

La Fig. 11 montre l'accroissement de la densité dû à ce rayonnement en profondeur; les ordonnées représentent le pourcentage de cette densité

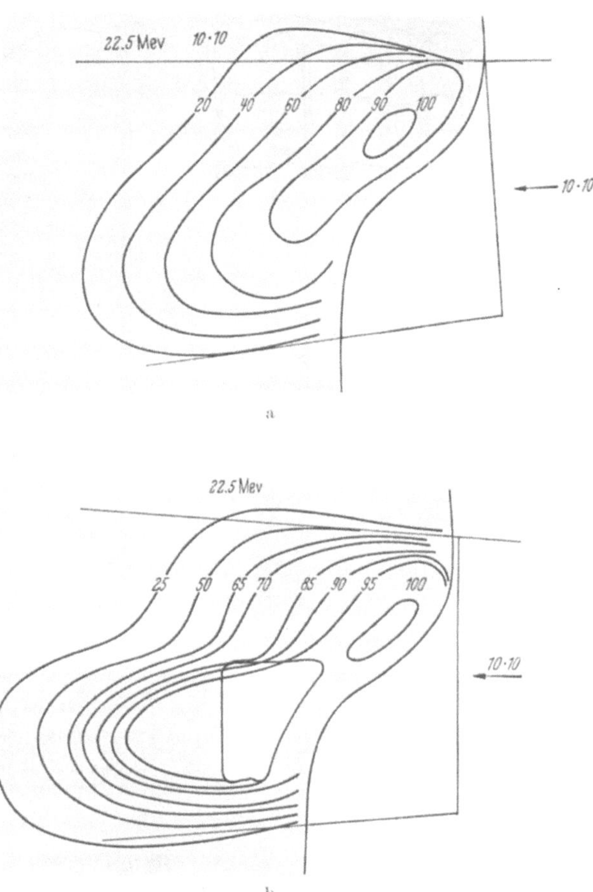


Fig. 9a et b. Isodoses dans le plan médian pour la région cervicale. a) Fantôme homogène. b) Une cavité a été creusée pour représenter la cavité laryngée. Sa présence entraîne une perturbation importante des isodoses

par rapport à la densité due effectivement aux électrons au même point: la production de ce rayonnement semble cesser à une profondeur de 8 g/cm²; la ligne doite correspondant à la première partie de la courbe est probablement due à la conjonction fortuite de différents facteurs. La présence de ces radiations oblige, dans ce milieu transparent, à envelopper le film dans une feuille de papier opaque ou à le placer entre deux plaques opacifiées (cette opacification fut obtenue en vaporisant de l'aquadag et une mince couche protectrice de vernis plastique)[1].

---

[1] Nous sommes heureux d'exprimer ici notre reconnaissance au Docteur J. J. LAUGHLIN, Chef du Physics Department, du Memorial Center, New York, qui nous a offert toutes les facilités pour effectuer ce travail dans son Service, ainsi qu'a Monsieur J. W. BEATTIE dont l'aide et les conseils nous ont été précieux.

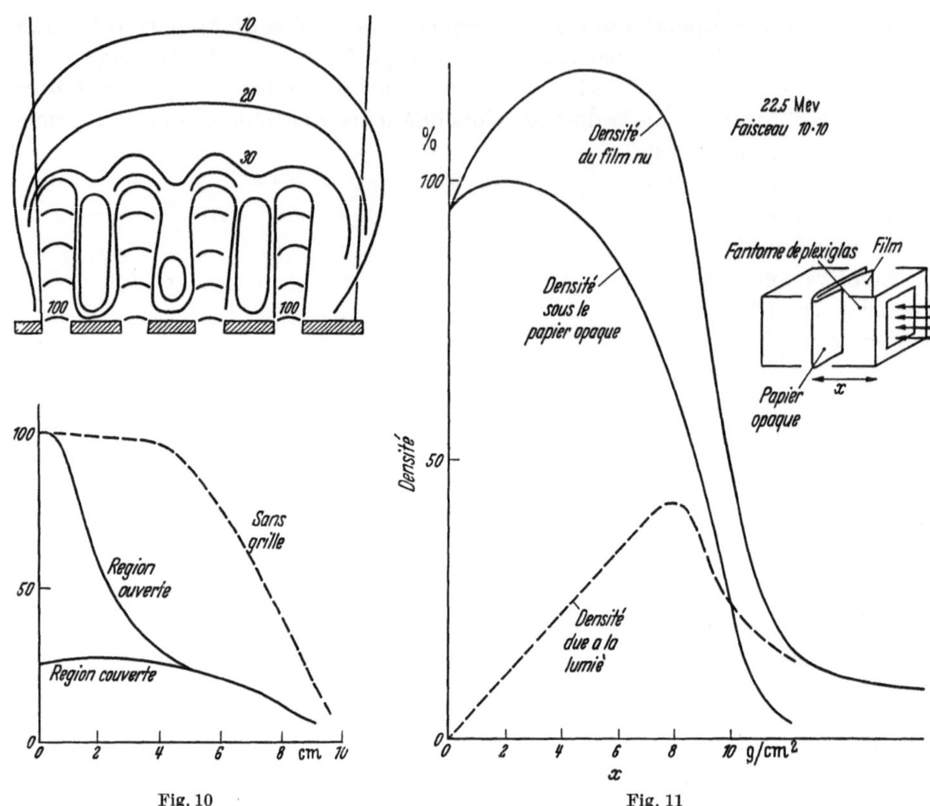


Fig. 10          Fig. 11

Fig. 10. Courbes isodoses sous une grille de plomb. La dose décroît plus vite sous les ouvertures que dans le cas du faisceau total (sans grille). Elle devient homogène sur toute la surface du champ à une profondeur où elle est réduite à 20% de la dose à la surface

Fig. 11. Mise en évidence de l'effet Cerenkov dans le Plexiglas. Une moitié du film est nue; l'autre moitié est recouverte de papier opaque. La variation en profondeur est exprimée en % de la dose maximum sous le papier opaque (densité due aux électrons). La différence, courbe en pointillés, représente la densité due à la lumière (essentiellement effet Cerenkov)

## Références


1. Becker, J., G. Weitzel et C. B. von der Decken: Strahlenther. **99**, 213 (1956).
2. Decken, C. B. von der: Strahlenther. **99**, 227 (1956).
3. Dudley, R. A.: Nucleonics **12**, 26 (1956).
4. — Radiation Dosimetry. New York: Academic Press 1956.
5. — et B. M. Dobyns: Science (Lancaster, Pa.) **109**, 321 (1949).
6. Fleeman, J., et F. S. Frantz: J. of Res., N. B. S. **48**, 117 (1952).
7. Jetter, Evelyn S., et H. Blatz: Nucleonics **10**, 63 (1952).
8. Katz, L., et A. S. Penfield: Royal Soc. Canad. Abstracts **119**, 111 (1951).
9. Markus, B., et W. Paul: Strahlenther. **14**, 612 (1955).
10. Storm, E.: LA **1951**, 1284.
11. Tochilin, E., et R. Golden: Nucleonics **11**, 26 (1953).

# Rückstreuung von Elektronen in verschiedenen Medien*

Von

G. BREITLING, Tübingen

Für die Energieabgabe in einem mit Elektronen bestrahlten Körper ist die Gestaltung des Strahlungsfeldes und damit die elastische Streuung von großer Bedeutung. Hier soll auf einen Sonderfall der Streuung, nämlich die Rückstreuung, näher eingegangen werden und ihre Bedeutung für die therapeutische Anwendung schneller Elektronen anhand einiger Beispiele gezeigt werden. Die Rückstreuung kommt praktisch allein durch allmähliche Änderung der Bahnrichtung, also durch eine große Zahl aufeinanderfolgender Einzelstreuakte mit geringer Ablenkung zustande. Für die auf die Masseneinheit des Streuers bezogene Zahl der in einen bestimmten Winkelbereich gestreuten Elektronen gilt nach BOTHE [2] und MOLIÈRE [8]

$$n \sim \frac{Z}{E^2} \, .$$

Der Einfluß der Streuung wird also um so stärker in Erscheinung treten, je höher die effektive Ordnungszahl des bestrahlten Mediums und je energieärmer die Elektronen sind.

Im Gegensatz dazu ist das Massenbremsvermögen, das die Reichweite der Elektronen bestimmt, nach BETHE-BLOCH [1] bzw. MÖLLER [7] weitgehend unabhängig von der Ordnungszahl des bremsenden Mediums und für Energien größer als 1 MeV praktisch konstant, wenn der Einfluß der Polarisation in Festkörpern [6] berücksichtigt wird.

Wegen der verschiedenen Abhängigkeit des Streu- und Bremsvermögens von der Ordnungszahl $Z$ und der Energie $E$ werden sich heterogene $\beta$-Strahlen beim Eindringen in Materie verschiedener Ordnungszahl $Z$ recht unterschiedlich verhalten. In Medien mit niedriger Ordnungszahl, wie z. B. im menschlichen Gewebe, werden Elektronen mit einer Energie von einigen MeV, wie sie in der Therapie Verwendung finden, zunächst ohne wesentliche Streuung abgebremst. Erst mit der Erniedrigung der Energie tritt eine stärkere Aufstreuung ein. Die Elektronen besitzen aber dann keine ausreichende Energie mehr, um in oberflächennahe Bereiche zurückdiffundieren zu können. Die Rückstreuung ist daher in der Nähe der Oberfläche sehr gering. Sie steigt mit zunehmender Tiefe an, erreicht aber auch in größeren Tiefen keine nennenswerten Ausmaße. Im Gegensatz dazu ist die Streuung in Medien hoher effektiver Ordnungszahl der dominierende Prozeß. Hier wird schon in geringen Tiefenlagen, wo noch keine wesentliche Verminderung der mittleren Energie der Elektronen eingetreten ist, eine starke Aufstreuung hervor-

---

* Aus dem Medizinischen Strahleninstitut der Universität Tübingen (Direktor: Prof. Dr. R. BAUER)

gerufen und damit der Zustand der vollständigen Diffusion erreicht. Die Rückstreuung ist daher in Medien hoher effektiver Ordnungszahl bereits an der Oberfläche beachtlich.

Die Überlegungen werden durch Messungen an Elektronen von 6—15 MeV
bestätigt. Die Untersuchungen wurden an Platten des betreffenden Materials
durchgeführt, die senkrecht zum Strahlengang angeordnet waren. Der Durch-

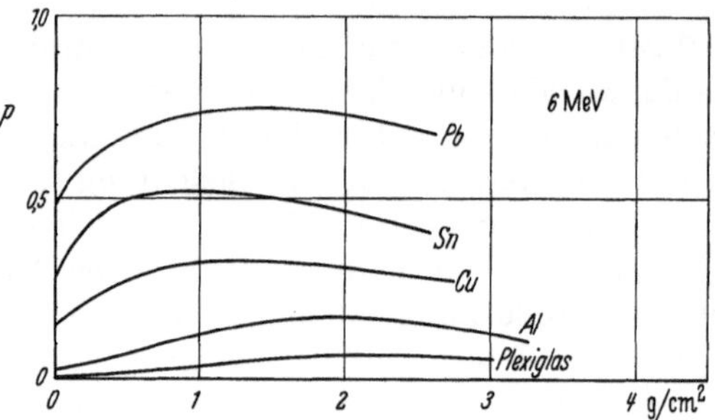


Fig. 1. Abhängigkeit des Rückstreukoeffizienten von der Massenbelegung für verschiedene Materialien

messer des Strahlenfeldes war dabei jeweils größer als die Reichweite der Elektronen in dem Medium. Dadurch war gewährleistet, daß die Elektronen des gesamten
Winkelbereiches erfaßt wurden. Als Strahlungsempfänger diente eine Ionisationskammer; gemessen wurde also die in einem in das Medium eingebrachten Luftraum abgegebene Energie, und zwar mit und ohne Rückstreuung. Das Verhältnis

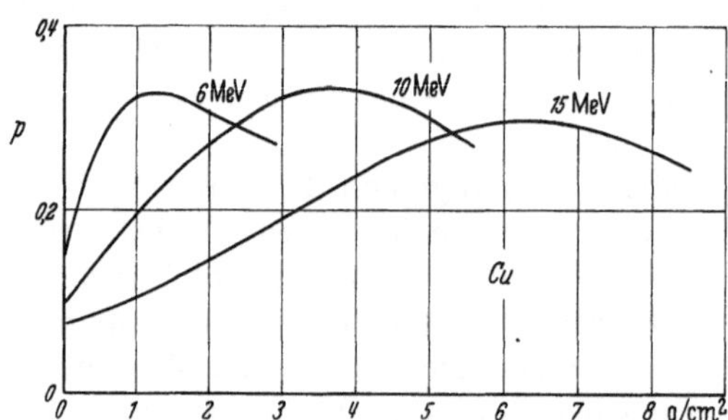


Fig. 2. Abhängigkeit des Rückstreukoeffizienten in Kupfer von der Massenbelegung für verschiedene Energien

der aus dem Winkelbereich 2 π rückgestreuten Energie zu dem von vorn abgegebenen Energiebetrag wird im folgenden als Rückstreukoeffizient $p$ bezeichnet. In
Fig. 1 sind die Meßergebnisse für Elektronen mit einer Primärenergie von
6 MeV wiedergegeben. Man sieht, daß der Rückstreukoeffizient $p$, entsprechend
den Erwartungen, bei Plexiglas an der Oberfläche nur etwa 1% beträgt. Er steigt
mit der Eindringtiefe der Elektronen an und erreicht nahe dem Ende der Reichweite der Elektronen einen Maximalwert von etwa 8%. Der darauffolgende Abfall ist auf die starke Abnahme der Elektronenzahl bei Annäherung an die Reichweite zu erklären. Die geringe Rückstreuung in Plexiglas, das hinsichtlich Bremsung und Streuung dem menschlichen Gewebe recht nahe kommt, hat zur Folge,
daß die Isodosenbilder nur unwesentlich geändert werden, wenn der bestrahlte

Körperteil eine geringere Tiefenausdehnung als die Reichweite der Elektronen aufweist. Mit zunehmender Ordnungszahl tritt die Rückstreuung immer stärker in
Erscheinung und erreicht bei Blei für 6 MeV-Elektronen einen Wert von $p = 46\%$
an der Oberfläche bzw. von $p = 75\%$ im Maximum. Gleichzeitig rückt das Maximum des Rückstreukoeffizienten, entsprechend der mit $Z$ wachsenden Streuung
und dem damit verbundenen früheren Eintreten der vollständigen Diffusion,
immer weiter nach geringen Tiefenlagen hin.

Die Abhängigkeit der Rückstreuung von der Primärenergie der Elektronen
geht aus Fig. 2 hervor. Hier ist der Rückstreukoeffizient von Cu für Elektronenenergien von 6, 10 und 15,8 MeV als Funk-
tion der Eindringtiefe aufgetragen. Man
sieht, daß der Wert von $p$ an der Oberfläche,
wie erwartet, mit steigender Energie ab-
nimmt. Der bei vollständiger Diffusion
erreichte Maximalwert ist dagegen in Übereinstimmung mit Messungen von BRAND [4],
BOTHE [3], BURTT [5] und SELIGER [9]
praktisch konstant.

Die Abhängigkeit der Rückstreuung
von der Ordnungszahl des bestrahlten
Mediums ist auch für die Therapie von
Bedeutung. Der menschliche Körper ist ja
keineswegs homogen aufgebaut. In einer
rohen Einteilung kann zwischen Muskel-
gewebe, Fett und Knochen unterschieden
werden. Die größten Unterschiede in der
effektiven Ordnungszahl treten dabei zwischen Fett und Knochen auf, während
Muskel und Fett nur wenig voneinander
abweichen. Es erhebt sich daher die Frage,

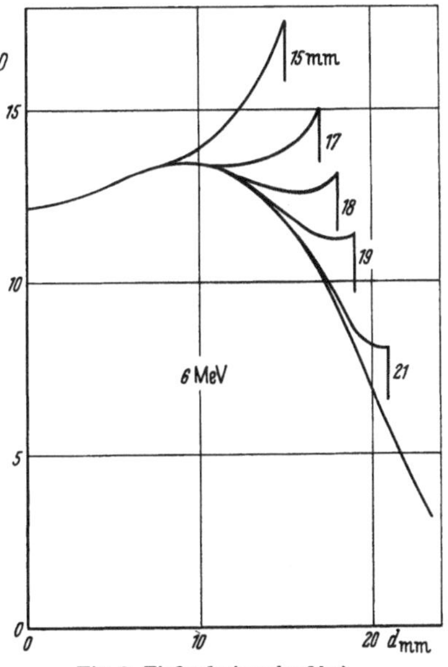



Fig. 3. Tiefendosisverlauf bei
Rückstreuung an Blei

inwieweit die Dosisverteilung in bestrahltem Gewebe durch die Rückstreuung
von dahinterliegenden Knochenpartien beeinflußt wird. Wie die Messungen
zeigten, ist die durch die Rückstreuung bedingte Dosiserhöhung von der Tiefenlage
abhängig, in der sich der Knochen befindet. Bei einer Primärenergie von 15 MeV
beträgt die Dosiserhöhung an der Grenzfläche Gewebe—Knochen in einer Tiefe von
1 cm z. B. 3%. Sie steigt auf etwa 12% an, wenn sich der Knochen in einer
Gewebetiefe von 5 cm befindet. Stärkere Erhöhungen treten auf, wenn die Elektronen auf Materialien höherer Ordnungszahl auftreffen, wie es z. B. bei Zahnkronen oder Knochennägeln vorkommt. Tiefendosiskurven mit Blei als rück-
streuendem Medium, das hinsichtlich Streuung und Bremsung etwa dem für
Zahnkronen verwendeten Gold entspricht, sind in Fig. 3 wiedergegeben. Man
erkennt, wie der Tiefendosisverlauf durch die vom Blei rückgestreuten Elektronen angehoben wird. Die Dosiserhöhung beträgt jeweils etwa 50%. Die Messungen erklären die starken Hautreaktionen, wie sie bei den an Goldkronen anliegen-
den Hautpartien beobachtet wurden. Zu ihrer Verhütung müssen die Kronen mit
gewebeäquivalentem Material umgeben werden. Wie aus Fig. 3 weiter hervorgeht, kann die Rückstreuung aber auch herangezogen werden, um eine homogene

Energieabgabe in einem bestrahlten Gewebestück bei vollständiger Schonung der dahinterliegenden Partien zu erzielen. So würde in dem dargestellten Fall ein 18 mm dickes Gewebestück, z. B. eine Wange, mit Elektronen von 6 MeV homogen durchstrahlt, wenn ein Bleiblech von 1 mm Dicke als Rückstreufolie verwendet wird. Die Dosis hinter dem Bleiblech beträgt weniger als 1% der im Gewebe abgegebenen. Das wiedergegebene Beispiel zeigt, welche Vorsicht bei der Elektronen-Bestrahlung geboten ist, wenn Materialien hoher Ordnungszahl getroffen werden, wie andererseits aber die Rückstreuung für die Gestaltung des Strahlungsfeldes und damit für die Dosisverteilung herangezogen werden kann.

## Literatur

1. Bethe, H. A.: Ann. Physik 5, 325 (1930).
2. Bothe, W.: Handbuch der Physik. Band XXII. Berlin: Julius Springer 1933.
3. — Z. Naturforsch. 4a, 542 (1949).
4. Brand, J. O.: Ann. Physik 5, 609 (1936).
5. Burtt, B. P.: Nucleonics 5, Nr. 2, 28 (1949).
6. Fermi, E.: Physic. Rev. 57, 485 (1940).
7. Möller, C.: Ann. Physik 14, 531 (1932).
8. Molière, G.: Z. Naturforsch. 2a, 133 (1947); 3a, 78 (1948).
9. Seliger, H. H.: Physic. Rev. 78, 491 (1950).

# Absolutdosimetrie der Elektronen eines Betatrons*

Von

B. MARKUS, Göttingen

Vom Praktiker wird man, wenn man eine Elektronenschleuder betreibt, immer wieder gefragt: „Wie messen *Sie* die Dosis? In ‚Röntgen' oder in ‚rad'?" Vor Beantwortung der Frage müßte man im Grunde immer die Gegenfrage stellen: „Was verstehen Sie unter ‚Dosis'?"

Dem heutigen Wissen über die physikalische, chemische und biologische Wirkung ionisierender Strahlen entsprechend ist es allein sinnvoll, einer Dosisdefinition die wirklich absorbierte Energie, und zwar kinetische Energie der primären und sekundären Elektronen im betrachteten Volumen, zugrunde zu legen. Das gilt auch für die schnellen Elektronen eines Betatrons.

Die heute im Bereich der „klassischen" Strahlen noch allgemein angewandte Dosimetrie in Luft mit der Einheit „Röntgen" (r) benutzt in ihrem Gültigkeitsbereich indirekt ebenfalls diese „energetische" Fundierung ihres Dosisbegriffs.

Es sei betont, daß das „r" stets nur die Dosis in Luft mißt, und zwar sowohl der Definition nach als auch dem Wesen nach. Diese entscheidende Tatsache bringt das Scheitern aller Versuche mit sich, das „r", allgemein die Gasionisation, zur Dosismessung im höheren Energiebereich der Quantenstrahlung, aber auch der Elektronen, zu benutzen.

Die Situation ist folgende: Von einer brauchbaren Dosimetrie hat man zu fordern, daß sie 1. ein genaues Bild der räumlichen Verteilung der spezifischen Dosis im bestrahlten Körper zu liefern vermag, wenn dieser unter ganz bestimmten physikalischen Bedingungen bestrahlt wird; 2. eine Strahlenquelle in absoluten Einheiten der spezifischen Dosisleistung zu eichen gestattet und damit für die relative Dosisverteilung im Körper in einem ihrer Punkte den Anschluß an den Absolutwert leistet, der durch die jeweilige Bestrahlung gegeben wird.

Zur Bestimmung der relativen Dosisverteilung im Körper hat man einen Detektor solcher Art zu verwenden, daß seine Anzeige proportional der in seiner unmittelbaren Umgebung verabfolgten spezifischen Dosis ist, also der spezifischen Energieabsorption im angegebenen Sinne. Als solcher Detektor hat sich die „Bragg-Graysche Luftblase" im Bereich der klassischen Strahlen bestens bewährt, in praktischer Ausführung: die kleine Luft-Ionisationskammer mit luftäquivalenten Wänden.

Nun hat man versucht, und tut es auch heute noch, mit derartigen Fingerhutkammern auch die relative Dosisverteilung ultraharter Röntgenstrahlung oder schneller Elektronen im Körper punktweise auszumessen. Da man durch diesen

---

* Aus der Universitäts-Hautklinik Göttingen (Direktor: Prof. Dr. H.-G. BODE), Abteilung Elektronenschleuder. — Auszugsweise vorgetragen

Vorgang eine spezifische Luftionisation mißt, ist man versucht, die Werte als in „Röntgen" gemessen anzugeben. Leider ist das nicht allgemein richtig, da die „Röntgen"-Definition das Bestehen des Elektronengleichgewichts gerade *in Luft* bei der Messung fordert. Daher die Schwierigkeit der Dosismessung im Bereich höherer Quanten- und auch Elektronenenergien: In Röntgeneinheiten kann man auf diese Weise nur Dosismessungen bis zu einigen Hundert KeV Quanten- oder Elektronenenergie durchführen.

Oftmals kann man der Lösung eines Problems näher kommen, wenn man diesem, etwas drastisch ausgedrückt, einen Namen gibt. Da den in genannter Weise im Körper gemessenen Ionisationswerten unter Berufung auf das Bragg-Graysche Prinzip große Bedeutung zuzukommen scheint, wurde dafür von Fränz und Hübner (1957) der Name „Gewebe-Ionendosis", verbunden mit der Einführung eines speziellen zusätzlichen Dosisbegriffs, vorgeschlagen. Die Einheit sei dabei durch die Ionisation der Kammerluft in Höhe einer el.-stat. Einheit pro $cm^3$ Normalluft in einem beliebigen Punkte des Körpers realisiert, ohne Rücksicht auf Elektronengleichgewicht.

Nun besteht jedoch im Bereich ultraharter Strahlen eine weitere Schwierigkeit. Hat man nämlich dort auf diese Weise Dosisverteilungen ermittelt und gibt an, ihre Werte seien in „relativen Einheiten" (z. B. Ionisationen pro $cm^3$ Normalluft) gemessen, so muß der unbefangene Betrachter dies so verstehen, daß ein fester, konstanter Proportionalitätsfaktor existiert, über welchen man sich für jeden Raumpunkt der Verteilung die jeweils verabfolgte spezifische Dosis im Energiemaß ausrechnen kann. Das ist nun leider nicht der Fall, wenn man sich oberhalb von etwa $^1/_2$ oder 1 MeV, je nach verlangter Genauigkeit einer anderen Grenze, befindet. Der Proportionalitätsfaktor, gegeben im wesentlichen durch das Verhältnis des differentiellen Energieverlustes von Elektronen in Luft einerseits, Körpersubstanz andererseits, ist nämlich von dieser Energie an nach oben nicht mehr annähernd eine Konstante, also energieunabhängig, sondern nimmt merklich ab. Die Abweichung hat mehr als nur akademisches Interesse, sie beträgt für Wasser von $^1/_2$—15 MeV bis zu etwa 15% [Glocker (1955)] und wächst weiter mit der Energie. Die Ursache ist eine prinzipielle Erscheinung: Der sog. „Polarisationseffekt" jeder kondensierten, gegenüber gasförmiger Materie. Auf diese entscheidende Tatsache im Zusammenhang mit der Dosimetrie wurde in den letzten Jahren durch Glocker wiederholt hingewiesen.

Auf Grund dieses Sachverhaltes erscheint also die relative oder absolute Ionisation eines Gases im beschriebenen Sinne als neuer Dosisbegriff und Dosismaß, weil wieder nur in einem beschränkten Energiebereich verwendbar, nicht sehr vorteilhaft. Wohl aber kann sie auch im höheren Energiebereich ihre Bedeutung behalten, speziell in Form von Luftionisationskammern, wenn diese lediglich als Sekundärstandard zur Eichung für ganz bestimmte Strahlenquellen benutzt werden.

Von Glocker, Breitling und Mitarbeitern [Ref. Glocker (1956)] wurde in den letzten Jahren konsequent eine Festkörperdosimetrie entwickelt, und zwar auch im niederen Energiebereich, um die Gasdosimetrie zu ersetzen: Leuchtstoffe in Verbindung mit Sekundärelektronenvervielfachern.

Die bei uns in Göttingen in der Dosimetrie ultraharter Strahlen viel angewandte Filmmethode [Markus und Paul (1953)] ist ebenfalls eine Festkörperdosimetrie.

Nach dem Vorangehenden, das sich in gleicher Weise auf Quanten, als auch auf Elektronen bezieht, sollen für schnelle Elektronen in der Arbeit folgende Definitionen verwendet werden, wie sie schon früher vorgeschlagen wurden [MARKUS und PAUL (1953)]:

1. Die *Dosis D*, die einem Körper bei der Einwirkung von schnellen Elektronen zugeführt wird, ist definiert durch den in ihm insgesamt absorbierten Betrag an kinetischer Energie der primären und sekundären Elektronen, zusammengesetzt aus den in den einzelnen Volumelementen $\Delta V$ des Körpers absorbierten Energiebeträgen $\Delta E$:

$$D = \Sigma \Delta E \quad [\text{erg}].$$

Diese Größe wird oftmals als „Volum"- oder „Raumdosis" bezeichnet.

2. Die *spezifische Dosis* $\delta$ ist gegeben durch die Definitionsgleichung

$$\delta = \frac{\Delta E}{\Delta V} \quad [\text{erg/cm}^3],$$

also gleich der pro cm³ Körpersubstanz absorbierten Energie im obigen Sinne. Da diese Größe differentiell definiert ist, kann man von der spezifischen Dosis „in einem Punkte" bzw. von der räumlichen Verteilung der spezifischen Dosis sprechen.

Der Begriff der „Spezifischen Dosis" ist identisch mit demjenigen, welcher landläufig, aber undeutlich, mit „Dosis" bezeichnet wird, bei welcher Bezeichnung man sich immer aufs neue wundern müßte, daß bei einem Feld von 1 cm Durchmesser die verabfolgte „Dosis" die gleiche ist wie bei einem Feld von 10 cm Durchmesser!

3. Die *spezifische Oberflächendosis* $\delta_0$ sei der Wert der spezifischen Dosis in einem Volumelement unmittelbar unter der Oberfläche des bestrahlten Körpers.

4. Die *spezifische Dosisleistung w* ist gleich der in der stofflichen Bezugseinheit pro Sekunde absorbierten Energie:

$$w = \frac{d\delta}{dt} \quad [\text{erg/cm}^3 \cdot \text{sec}].$$

Bei so eindeutigen physikalischen Definitionsmöglichkeiten erscheint es unvorteilhaft, noch neue Dosisbegriffe wie: absorbierte Dosis, Energiedosis, Ionendosis u. ä. einzuführen. Man sollte vielleicht auch den erzieherischen Wert einer einfachen und klaren Dosisdefinition nicht unterschätzen; ist doch der damit Arbeitende gezwungen, seine Meß- und Versuchsanordnung genau zu überprüfen, ob sie ihm die Dosismessung oder -berechnung in diesen eindeutigen Einheiten gestattet. Ist nämlich erst einmal ein Dosisbegriff geschaffen, der nur unter ganz speziellen Bedingungen anwendbar ist, so wird doch versucht, diesen auch außerhalb seiner Gültigkeitsgrenzen zu verwenden, wenn er scheinbar bequemer anzuwenden ist.

Hat man sich nun für schnelle Elektronen zur energetischen Konzeption des Dosisbegriffs entschieden, so erhebt sich die Frage, wie kann die so definierte spezifische Dosis wirklich gemessen werden? — Am besten, man betrachtet wieder die beiden Teilaufgaben der Dosimetrie getrennt: 1. Messung der relativen Dosisverteilung, 2. Bestimmung eines Absolutwertes dieser Verteilung. Über die räumliche Dosisverteilung soll jetzt nicht mehr gesprochen, sondern diese nach dem Vorangehenden als fertig vorliegend angesehen werden. Zur Gewinnung eines Absolutwertes dieser Verteilung unter bestimmten Bestrahlungsbedingungen haben nun Glocker und Mitarbeiter [Glocker (1956)] den Anschluß ihrer Leuchtstoff- oder auch chemischen Dosimeter an konventionelle Röntgenstrahlen gewählt, indem erst der Dosismesser in r-Einheiten mit Röntgenstrahlen geeicht wird und dann rechnerisch erst der Übergang zu Energieeinheiten für die spezielle Röntgenstrahlung, dann weiter zu absoluten Einheiten für Elektronen, durchgeführt wurde. Durch die Übereinstimmung der relativen Energieabhängigkeit der benutzten Anordnung in Theorie und Experiment wurde die Richtigkeit der Umeichung bestätigt.

Ein anderer Weg ist der direkte: Man mißt die Zahl der pro cm² eingefallenen Elektronen und multipliziert sie mit dem Energieverlust des einzelnen Elektrons in einer dünnen Schicht des betreffenden Materials. Das Ergebnis ist die spezifische Dosis in der Schicht. Daraus erhält man unmittelbar die spezifische Oberflächendosis für den Körper durch Multiplikation mit dem Rückstreufaktor des betreffenden Materials. Für die Praxis schließt man sich die spezifische Dosisleistung der Elektronenschleuder, jetzt bezogen auf die Körperoberfläche, z. B. an eine beliebige Ionisationskammer an, die damit in absoluten Einheiten geeicht ist (Elektronen/cm² · sec bzw. erg/cm³ · sec).

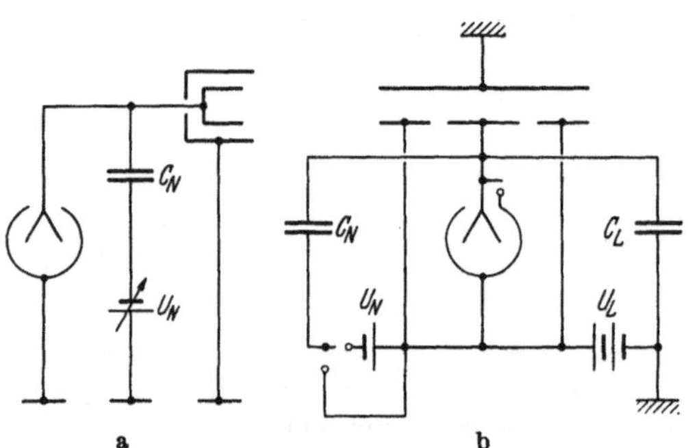


Fig. 1 a u. b. a Prinzipschaltbild des Faraday-Kastens mit Elektrometer. $C_N$ Normalkondensator, $U_N$ Kompensationsspannung. b Prinzipschaltbild des Ionisationskammer-Monitors. $C_L$ Ladekondensator, $U_L$ Lade-Kammer- Spannung

Die Größe des differentiellen Energieverlustes (Bremsvermögens) für Elektronen wird z. B. mit einem Betaspektrometer gemessen, es liegen bereits für viele Stoffe die Werte in der Literatur vor. Der Rückstreufaktor, der in unserem Energiebereich von 3—15 MeV etwa 4% beträgt, kann mit jedem Festkörperdosimeter gemessen werden; da es sich um eine Relativmessung handelt, mit meist ausreichender Genauigkeit auch mit einer flachen, dünnwandigen Ionisationskammer.

In der Arbeit soll speziell über die Messung der Elektronenstromstärke und Eichung einer handelsüblichen Fingerhutkammer am 15 MeV-Siemens-Betatron berichtet werden.

Zur Messung der Elektronenstromstärke wurde der direkte Weg gewählt: Bestimmung der Elektronenladungen durch Auffangen in einem Faraday-Käfig (F. K.). Fig. 1 a zeigt das Prinzip-Schaltbild: Die aufgefangene Elektronenladung erzeugt zwischen dem Auffänger des F. K. und seiner Abschirmung bzw. Erde eine Spannung, die mit einem geeigneten Voltmeter gemessen wird. Praktisch wurde die Elektronenladung durch Kompensation mit einer influenzierten Gegenladung über einen geeichten Kondensator (Kapazität $C_N$) gemessen, an welchen eine sehr genau gemessene Spannung $U_N$ angelegt wurde. Die aufgefangene Elektronenladung beträgt dann $C_N \cdot U_N$. Zur Null-Messung der zu kompensierenden Spannung am F. K. wurde ein Gleichspannungsverstärker (Schwingkondensator-System) benutzt (Fig. 2), an welchen ein Kompensationsschreiber angepaßt wurde, welcher automatisch immer die richtige Gegenspannung an den Eingang des Meßkreises gab und diese gleichzeitig fortlaufend registrierte. Die Eichung erfolgte über den Normalkondensator $C_N$ in Verbindung mit einem technischen Kompensator und Normalelement.

Die Apparatur besteht weiterhin aus einer Parallelplatten-Ionisationskammer als Monitor, durch welche der zu messende Elektronenstrahl unbehindert hin-

durchgeht. Die gemessene Ionisation pro Sekunde ist proportional der Elektronenstromstärke. Das Prinzipschaltbild der Ionisationsmessung ist in Fig. 1b gezeigt. Die Ionisation wurde integrierend gemessen.

Die zu eichende Fingerhutkammer wurde mit einer Zentriervorrichtung zwischen F. K. und Monitor gebracht.

Die Messung des Elektronenstroms im Fara-

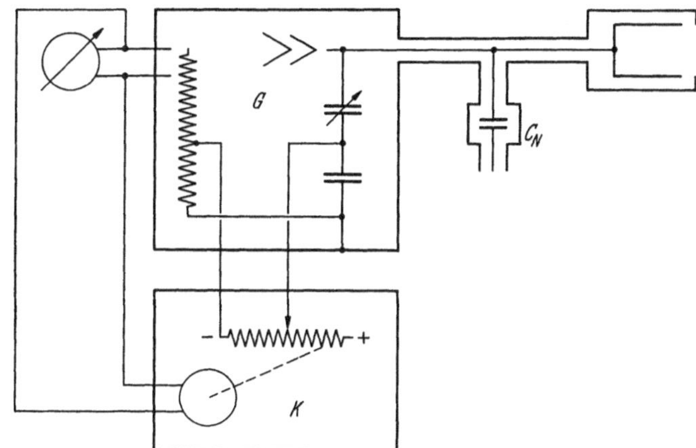

Fig. 2. Schaltbild des Faraday-Kastens. *G* Gleichspannungsverstärker, *K* Kompensationsschreiber, $C_N$ Normalkondensator

day-Käfig erfordert für die hohen Elektronenenergien eines Betatrons gewisse Vorsichtsmaßnahmen. Es genügt leider nicht, die Auffängerwände so dick zu machen, daß die Elektronen vollständig abgebremst werden. Durch die im

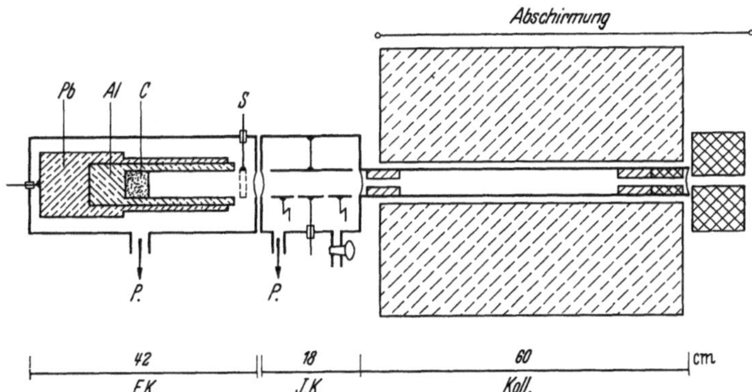


Fig. 3. Schema der Gesamtapparatur. *Koll.* Kollimator, *J.K.* Ionisationskammer, *F.K.* Faraday-Kasten, *S* Gegenfeldanschluß, *P* Pumpenanschluß

Auffänger entstehende Bremsstrahlung werden nämlich unvermeidlich Sekundärelektronen sowie Paarpositronen erzeugt, die einesteils, soweit sie den Auffänger verlassen, die zu messende Ladung verfälschen, zum anderen selbst wieder Bremsstrahlung erzeugen und damit diesen Effekt verstärken. Der Auffänger muß also so massiv gebaut werden, daß diese Störeffekte vernachlässigbar bleiben, indem die Bremsstrahlung wesentlich bereits im Auffänger absorbiert wird. Fig. 3 zeigt

genauer den Aufbau der gesamten Apparatur. Man erkennt den Aufbau des Auffängers aus verschiedenem Material: erst Graphit und Aluminium zur Abbremsung
der primären Elektronen mit möglichst geringer Bremsstrahlerzeugung, darum
einen Bleimantel zur Quantenabsorption. Die Abmessungen sind aus der Figur zu
entnehmen. Der ganze F. K. mußte weiterhin als Hochvakuum-Apparatur gebaut
werden, um Störeffekte durch Ionisation auszuschalten. Ferner mußte der Ladungsverlust durch Rückstreuung aus dem Auffänger vernachlässigbar klein gehalten
und kontrolliert werden. Genauere Einzelheiten sollen a. a. O. gesondert folgen.

Die Strahldefinition erfolgte durch Ausblenden in einem evakuierten Kollimator. In Fig. 3 ist die Anordnung der Blenden, wiederum erst aus nieder-

Fig. 4. Apparatur mit teilweise aufgebautem Strahlenschutz

atomigem Material, dann aus Blei bestehend, angegeben. Der Strahldurchmesser
betrug 15 mm beim Verlassen des Kollimators, 17 mm Halbwert-Durchmesser
beim Eintritt in den Auffänger infolge Streuung in den Fensterfolien (30 $\mu$ Al)
und in der Kammerluft. Zwischen Betatron und Kollimator fand noch eine Vorausblendung statt (dicke Kunststoffblende).

Ein ziemlich ernstes Problem bildete die unvermeidliche äußere Abschirmung
der kritischen Apparateteile gegen Streustrahlung. Die Abschirmung wurde
um den Kollimator herum angeordnet und bestand aus Blei- und Baryt-Strahlenschutzblöcken. Der Gesamtabschirmwert betrug mit 6 t/m² mehr als 20 HWS in
Strahlrichtung, das Gesamtgewicht der Abschirmung etwa 180 kg. Die ganze
Apparatur mußte trotzdem zum wiederholten Auf- und Abbau beweglich gehalten
werden, da die Messungen an der klinischen Elektronenschleuder durchgeführt
wurden.

Fig. 4 zeigt die Apparatur selbst, und zwar mit teilweise aufgebautem
Strahlenschutz. Rechts befindet sich die Elektronenschleuder.

Eine gewisse Komplikation der Messung verursachte die Ausleuchtung der
zwischen Monitor und F. K. zentrierten Fingerhutkammer. Da eine gleichmäßige

Ausstrahlung einen so großen Strahldurchmesser erfordert, daß dadurch die Elektronenstrommessung erheblich erschwert wird, wurde auf gleichmäßige Ausstrahlung verzichtet. Dadurch war es aber nötig, das genaue Verhältnis der auf das empfindliche Volumen der Fingerhutkammer treffenden Elektronen zur insgesamt aufgefangenen Elektronenzahl zu messen. Fig. 5 zeigt die Fingerhutkammer („Siemens-Kleinkammer") in Durchstrahlung mit 15 MeV-Elektronen

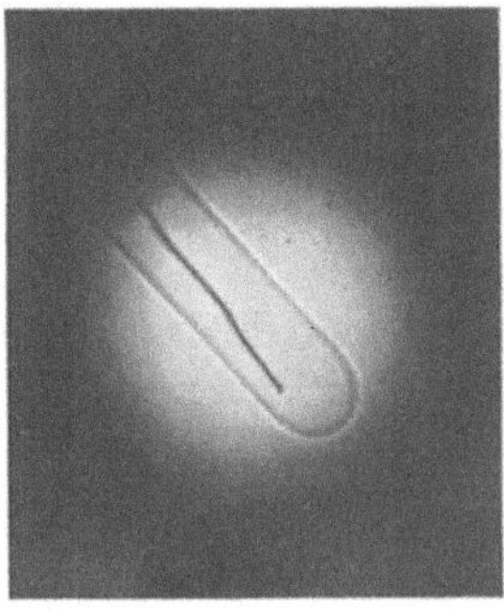

Fig. 5. Fingerhutkammer, vom Elektronenstrahl durchstrahlt (15 MeV)

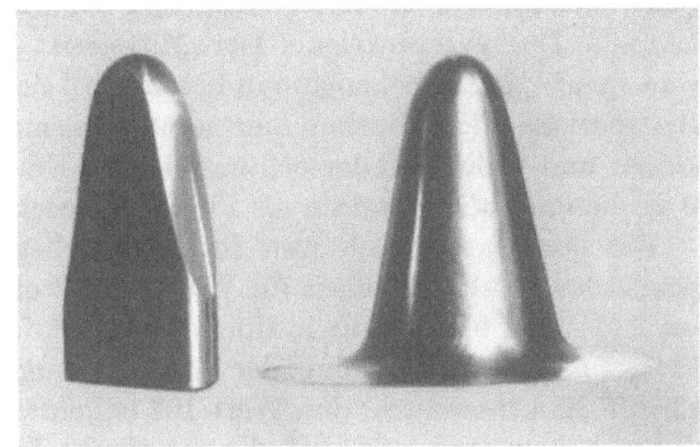

Fig. 6. Räumliche Modelle der Intensitätsverteilung im freien Elektronenstrahl und über dem wirksamen Kammerquerschnitt

und ihre Lage im Elektronenstrahl. Die Messung des ausgenutzten Elektronenbruchteils geschah mit einem 3-dimensionalen Modell der Intensitätsverteilung im Strahlquerschnitt, welche nach der Filmmethode bestimmt worden war, und einem ebensolchen Modell der dabei auf die Ionisationskammerfläche entfallenden Intensität (der Durchdringungskörper einer angenäherten Gauß-Verteilung mit einem Zylinder, errichtet über dem Meßvolumen der Kammer). Die beiden Modellkörper wurden aus Stahl gefertigt, ihr Gewichtsverhältnis und damit das gesuchte Verhältnis der Elektronenzahlen betrug 38% (Fig. 6). Fig. 7a zeigt nochmals die Verteilungsfunktion der Intensität in Abhängigkeit vom Radius des freien Strahles, Fig. 7b die Lage der drei äußersten Kammerpunkte auf einem Kreis, parallel zur Grundebene,

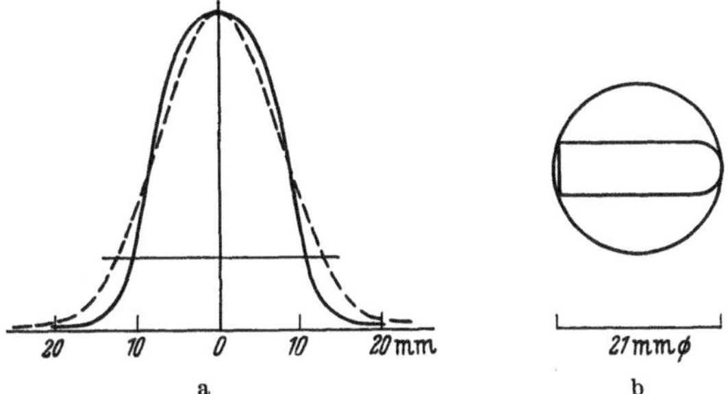


Fig. 7a u. b. a Intensitätsverteilung im Elektronenstrahl am Ort der Fingerhutkammer (ausgezogene Kurve: gemessen; strichliert: angenähert berechnete Kurve). b Lage der wirksamen Kammerfläche im Strahlquerschnitt

also einer „Isodose". (Die strichliert eingezeichnete Kurve ist die für den vorgegebenen, endlichen Strahldurchmesser nach der Theorie der Vielfachstreuung schneller Elektronen angenähert berechnete Verteilung, normiert auf gleiche Intensität auf der Strahlachse.)

Die eigentliche Messung erfolgte nun so, daß die Monitor-Ionisation in Elektronen/cm² geeicht, dann die Fingerhutkammer an die festgelegte Stelle zwischen F. K. und Monitor gebracht wurde und mit bestimmten Elektronenzahlen bestrahlt wurde. Die dabei erhaltene Ionisationsladung der Fingerhutkammer wurde mit einem in „r" geeichten, zugehörigen Röntgendosismesser (Siemens-Momentan-Dosimeter) gemessen und gab damit die letztlich gesuchte Anzahl „r" (Skalenteile) pro Elektron und $cm^2$ Kammerfläche. Die Elektronenstromstärke mußte dabei so klein gehalten werden, daß Sättigung in der Fingerhutkammer und in der Monitorkammer erreicht blieb; sie betrug etwa $10^{-13}$ A.

Eine wirkliche Messung ergab beispielsweise folgende Werte: $6,35 \cdot 10^5$ Elektronen pro $cm^2$ Kammerfläche und Sekunde, dazugehörig $1,94 \cdot 10^{-2}$ „r" pro Sekunde Dosimeteranzeige. Der Mittelwert betrug $3,22 \ (\pm 0,15) \cdot 10^7$ Elektronen/$cm^2 \cdot$ „r". Es sei nochmals betont, daß sich die Angabe „r" ausschließlich auf eine spezielle Fingerhutkammer samt Dosismesser und auf 15 MeV-Elektronen bezieht und nichts mit der echten Röntgen-Einheit zu tun hat; die Größe des „r" ist in diesem Falle geradezu als Firmeneigenschaft zu bezeichnen[1].

Wie groß ergibt sich nun für das vorliegende Beispiel einer Eichung die spezifische Oberflächendosis für Wasser? Mit einem differentiellen Energieverlust von 2 MeV/cm für Wasser in dünner Schicht für Elektronen von 15 MeV, einem Rückstreufaktor 1,04 und der Umrechnung $1 \text{ MeV} = 1,6 \cdot 10^{-6}$ erg, ergeben $3,2 \cdot 10^7$ Elektronen/$cm^2$ den Wert 107 erg/$cm^3$.

Sieht man nach, wie groß die spezifische Ionisation in der Fingerhutkammer ist unter der Annahme, daß bei 1 el.-stat. Ladungseinheit ($2,09 \cdot 10^9$ Ionenpaare pro $cm^3$ Kammerluft) 1 r angezeigt wird, so erhält man mit obigen Meßwerten $2,09 \cdot 10^9 : 3,22 \cdot 10^7 = 65 \pm 3$ Ionenpaare pro Elektron pro $m^3$ Kammerluft. Dieser Wert steht recht gut im Einklang mit der heutigen Kenntnis der spezifischen Ionisation schneller Elektronen in Luft, hat aber wegen der willkürlichen Kammergeometrie wiederum keine allgemeinere Bedeutung.

Zur Genauigkeit des erhaltenen Eichwertes ist zu sagen, daß sich der mittlere Gesamtfehler von $\pm 10\%$ wie folgt zusammensetzt: Streuung der Elektronenzahlwerte: $\pm 3\%$; Streuung der Ionisations-(r)-Werte: $\pm 5\%$; Elektronenverluste: 1 bis 2%; Sättigungsverluste an Ionisation: 0,1%; Genauigkeit der Kapazitäts- und Spannungsangabe: 0,2%.

Zur Reproduzierbarkeit ist zu sagen, daß diese dadurch mehrfach geprüft wurde, daß die Apparatur wiederholt weitgehend abgebaut und nach jedem Neuaufbau vor der Elektronenschleuder neu justiert werden mußte; die Meßwerte der absoluten Elektronenzahl, bezogen auf eine bestimmte, absolute Ionisation des Monitors, lagen bei der nötigen Sorgfalt stets innerhalb der angegebenen Grenzen von $\pm 3\%$.

Ein Ausbau dieses absoluten Eichverfahrens, verbunden mit der Einbeziehung kleinerer Elektronenenergien, ist vorgesehen.

Die Arbeit wurde mit großzügiger finanzieller Unterstützung der Deutschen Forschungsgemeinschaft durchgeführt; ich möchte an dieser Stelle allen verantwortlichen Herren meinen Dank aussprechen.

**Zusammenfassung.** Die Schwierigkeiten bei der Dosimetrie energiereicher Strahlungen in konventionellen Dosiseinheiten werden rekapituliert.

Zur Begründung der in der Arbeit verwendeten Dosisdefinitionen wird gezeigt, daß es im Energiebereich schneller Elektronen eines Betatrons — ebenso wie in

---

[1] Insbesondere ist eine unterschiedliche Richtungs- und Flächenempfindlichkeit der Fingerhutkammer unberücksichtigt

dem der ultraharten Röntgenstrahlung — sinnvoll ist, die pro Volumelement des
bestrahlten Körpers (cm³) absorbierte kinetische Energie der primären und
sekundären Elektronen (erg/cm³) dem Dosisbegriff, insbesondere dem der „spezifischen Dosis", zugrunde zu legen.

Meßmöglichkeiten der spezifischen Elektronendosis werden diskutiert und
eine Apparatur beschrieben, welche gestattet, Sekundärstandards (z. B. Fingerhutkammern) in Absolutwerten sowohl der Elektronenzahl/cm² Kammerfläche, als
auch der spezifischen Oberflächendosis in erg/cm³ eines bestrahlten Körpers,
wenn Elektronenbremsvermögen und Rückstreufaktor des betreffenden Materials
bekannt sind, am Betatron zu eichen. Die Absolutmessung des Elektronenstroms
außerhalb des Beschleunigungsgefäßes erfolgt durch Auffangen der Elektronen
in einem speziell konstruierten Faraday-Kasten. Ein Beispiel einer durchgeführten Absoluteichung wird mitgeteilt.

## Literatur

FRÄNZ, H., u. W. HÜBNER: Strahlenther. **102**, 590 (1957).
GLOCKER, R.: Z. Physik **143**, 191 (1955).
— Strahlenther. **35**, 276 (1956).
MARKUS, B., u. W. PAUL: Strahlenther. **92**, 599 u. 612 (1953).

## Diskussionsbemerkungen

H. BERGER (Erlangen):

Es ist zu überlegen, ob die logischen Bedenken von Herrn MARKUS gegen die genormte
Dosisdefinition Dosis = absorbierte Energie pro Massenelement, schwerwiegend genug sind, um
die neue, von Herrn MARKUS benutzte Definition Dosis = absorbierte Energie integriert über
den ganzen Körper, entgegen dem üblichen Sprachgebrauch zu rechtfertigen. Beide Begriffe
nebeneinander erschweren die Verständigung, und der übliche Dosisbegriff ist durchaus im
Einklang mit der Vorstellung, daß die „Dosis" diejenige Größe ist, die am meisten Einfluß
auf den biologischen Effekt hat.

B. MARKUS (Göttingen):

Ich möchte Herrn BERGER in Kürze folgendes antworten:

1. Die von ihm zitierten, genormten Dosisdefinitionen der DIN-Blätter beziehen sich
nur auf konventionelle Strahlen, legen dem Dosisbegriff die r-Einheit zugrunde und enthalten dementsprechend die ultraharten Strahlen nicht inbegriffen. Dies ist auch der Grund,
daß fast jedes Jahr von irgendeiner Seite neue Vorschläge zu Dosisdefinitionen für den hohen
Energiebereich gemacht werden.

2. Die in meiner Arbeit verwendeten Dosisbegriffe und -definitionen, insbesondere die der
„spezifischen Dosis", stammen nicht von mir, sondern wurden z. B. von Herrn Prof. POHL,
Göttingen [Naturwissenschaften **38**, 147 (1951)], sowie von Herrn Prof. PAUL (1953) dringend
zur Benutzung empfohlen; sie werden gelegentlich auch z. B. von GLOCKER [Z. Physik **136**,
367 (1953)] benutzt.

3. Schließlich sollte man den erzieherischen Wert einer klaren und einfachen Dosisdefinition
nicht unterschätzen. Die bestehenden physikalischen Grundeinheiten, wie das „erg", reichen
völlig aus, um Dosiseinheiten — *rein physikalische Größen* — zweckmäßig zu definieren. Es
war übrigens in der Arbeit nicht beabsichtigt, sonstige bestehende Dosisdefinitionen refor-
mieren zu wollen; die benutzten Definitionen wurden nur deshalb vorgezogen, weil sie sich als
ganz besonders klar und praktisch erweisen.

# Untersuchungen über Dosimetrie und Ausblendung von 30 MeV-Elektronenstrahlen*

Von

M. SEMPERT und R. WIDERÖE, Baden/Schweiz

Wir haben mit großem Interesse von Herrn Prof. BECKER erfahren, daß die therapeutischen Frühresultate mit Elektronenbestrahlung als gut bezeichnet werden können. Es sind gewisse Anzeichen vorhanden, daß man mit Elektronen mehr als mit Röntgenstrahlen erreichen kann, auf alle Fälle verdient dieses neue Hilfsmittel der Strahlentherapie die allergrößte Beachtung.

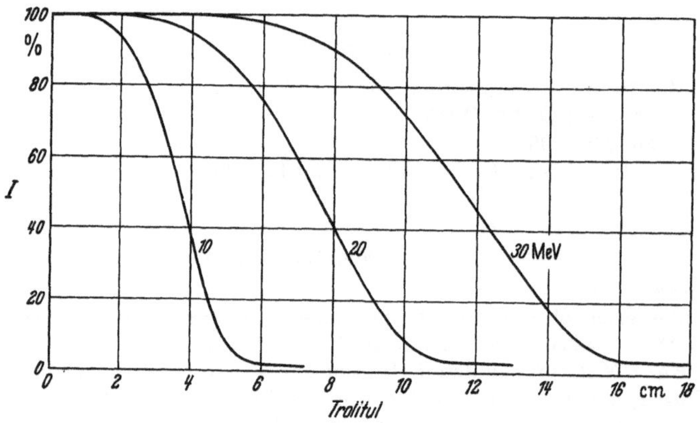


Fig. 1. $\beta$-Tiefendosiskurven in Trolitul hinter Kreisblende (Messing) von 100 mm $\varnothing$ . Abstand vom Austritt $= 1,6$ m

In einer Hinsicht müssen aber die Arbeiten der Heidelberger Schule noch als recht vorläufig betrachtet werden. Die Reichweite der verwendeten 15 MeV-Elektronen ist klein, therapeutisch reicht man kaum über 4 cm Wassertiefe hinaus (bei Gitterbestrahlung nur 2—3 cm Tiefe), und von einer Tiefentherapie kann man nicht gut reden.

Wir haben bei Brown-Boveri in Baden mit 30 MeV-Elektronenstrahlen gearbeitet, und ich möchte Ihnen hier einige unserer Resultate schildern und auch etwas über unsere im Gang befindliche Weiterentwicklung erzählen. Bevor ich auf nähere Einzelheiten eingehe, bringe ich zunächst einen kurzen Überblick über die wichtigsten Eigenschaften der 30 MeV-Elektronenstrahlen.

Fig. 1 zeigt Tiefendosiskurven für 10, 20 und 30 MeV-Elektronen. Wie Sie sehen, erreicht man mit 30 MeV-Elektronen in 8 cm Wassertiefe noch eine Intensität von etwa 90%. Ein photographischer Längsschnitt eines Elektronenfeldes von 100 mm Durchmesser (30 MeV) in Trolitul ist in Fig. 2 dargestellt.

---

* Aus dem Strahlenlaboratorium der AG. Brown-Boveri u. Cie., Baden (Schweiz)

An der Oberfläche des Phantoms liegt die Intensität meist etwas unterhalb des in 2—3 cm auftretenden Höchstwertes (= 100%), da aber dieser Teil der Kurve leicht durch gestreute Elektronen kleinerer Energien beeinflußt werden

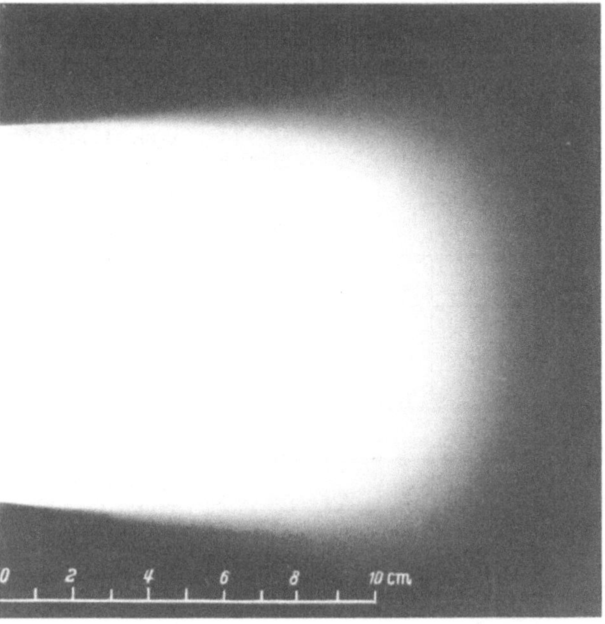


Fig. 2. Photographischer Längsschnitt durch ein 30 MeV-Elektronenfeld in Trolitul (spez. Gewicht 1,05). Feldgröße 100 mm Durchmesser

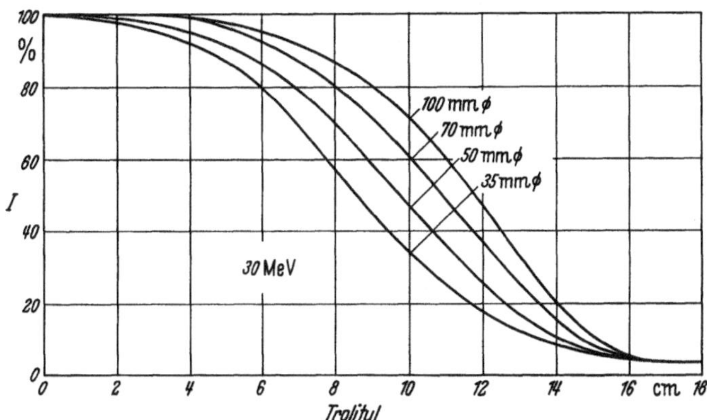


Fig. 3. $\beta$-Tiefendosiskurven bei verschiedenen Feldgrößen. Abstand von der Austrittstelle = 1,6 m, ohne Streuplatten

kann, haben wir es vorgezogen, hier eine konstante Intensität anzunehmen, was wohl auch häufig dem praktischen Fall entsprechen wird. Die von uns gemessene Tiefendosiskurve für 20 MeV-Elektronen entspricht sehr gut der von LAUGHLIN bei 21,6 MeV-Elektronen gemessenen Tiefendosiskurve; das gleiche gilt auch für die 10 MeV-Kurve[1].

---

[1] LAUGHLIN, J. S., et al.: Some physical aspects of electronbeam therapy. Radiology 60, No. 2, 165—184 (1953)

Fig. 3 stellt dar, wie sich die Tiefendosiskurven mit der Feldgröße ändern. Die Ursache dieser Änderung ist natürlich die Seitenstreuung der Elektronen im Phantom, die sich bei den kleineren Feldern stärker bemerkbar macht. Sie sehen aus diesem Bild, daß wir bei einer Feldgröße von 100 mm Durchmesser eine recht gute Tiefendosiskurve erreichen, während die Tiefendosiskurven bei den kleineren Feldern eher besser sein könnten.

Fig. 4 zeigt einen Vergleich von Tiefendosiskurven in Wasser für verschiedene Strahlenarten. Die Kurve $A$ gilt für 200 kV Röntgenstrahlen. Man ersieht hieraus das Übel der hohen Hautbelastung. Die Kurve $B$ gilt für die 1,15 MeV-$\gamma$-Strahlen von Cobalt[60]. Die Hautbelastung ist auch in diesem Falle noch sehr störend (etwa 150% der Tumordosis), und erst bei der Kurve $D$, die für 31 MeV-$\gamma$-Strahlen gilt, verschwindet die Eintrittsdosis fast vollständig. Die Austrittsdosis von etwa 60 bis 70% stört in diesem Falle nur selten. Die Kurve $C$ zeigt die Tiefendosiskurven für 30 MeV-Elektronenstrahlen. Die Eintrittsdosis beträgt etwa 110%, ist also wesentlich günstiger als bei der Cobalt[60]-Strahlung. Hinter dem Tumor fällt die Intensität der Elektronenstrahlen steil auf wenige Prozente herunter. Man erkennt aus diesem Vergleich, daß die 30 MeV-Elektronenstrahlen neben den 31 MeV-$\gamma$-Strahlen wohl eines der wichtigsten Hilfsmittel der Tiefentherapie werden können.

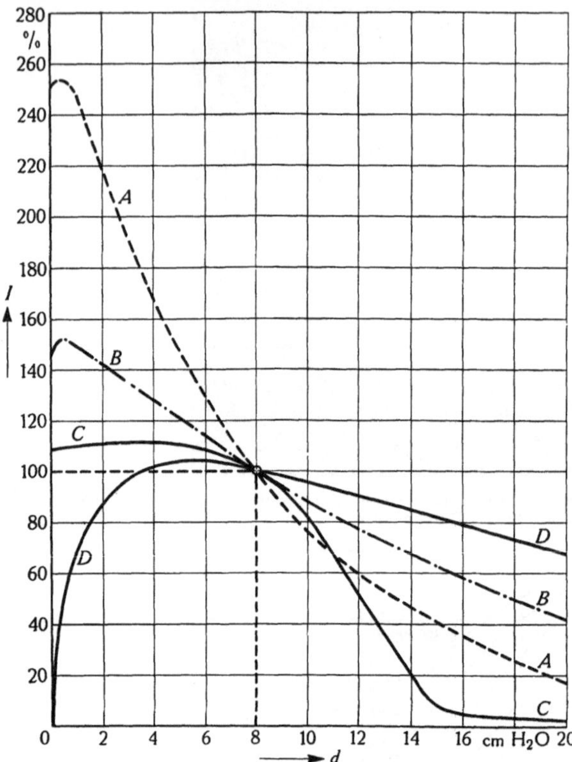

Fig. 4. Vergleich verschiedener Tiefendosiskurven in Wasser.
$A$ 200 kV-Röntgenstrahlen. Feldgröße 200 cm² FHA = 50 cm.
$B$ Cobalt-60 $\gamma$-Strahlung. Feldgröße 200 cm², FHA = 80 cm.
$C$ 30 MeV-Elektronenstrahlen, Feldgröße 78,5 cm² (100 mm Durchmesser), FHA = 100 cm. $D$ 31 MeV-Betatronstrahlen ($\gamma$), Feldgröße beliebig, FHA = 100 cm

Nach dieser ersten Übersicht über die Eigenschaften der Elektronenstrahlen werde ich jetzt näher auf unsere Arbeiten eingehen und vorerst etwas rein Praktisches über Meßverfahren und Dosimetrie berichten.

Bei der Dosimetrie von 30 MeV-Elektronenstrahlen haben wir zunächst einige bemerkenswerte Erfahrungen gewonnen. Es wurden drei Victoreen-Kammern (25, 100 und 250 r Meßbereich) vor einem Trolitulphantom aufgestellt und mit Filmen die Intensitätsverteilung an der Oberfläche und in der Tiefe des Phantoms gemessen (Fig. 5.) Es zeigte sich, daß sogar die dünnen, aus Isolierstoff angefertigten Kammerwände, durch Streueffekte (die Kammern waren 25 cm vor dem Phantom angebracht) einen erheblichen, bis zu 40% betragenden Intensitätsabfall der Elektronenstrahlen erzeugten. Dieser Intensitätsabfall ließ sich sehr deutlich bis tief in das Phantom hinein verfolgen. Es hat sich somit gezeigt, daß es unstatthaft ist, die Meßkammer vor oder hinter dem zu bestrahlenden Objekt anzubringen. Wir verwenden für unsere Messungen (beispielsweise bei der Bestrahlung biologischer Objekte) im allgemeinen flache, das gesamte Feld erfassende Ionisationskammern, bestehend aus drei nur 0,035 mm dicken Aluminiumfolien als Elektroden. Bei der Aufnahme von Tiefendosiskurven haben wir im allgemeinen im Phantom eingelegte Victoreen-Kammern benutzt. Die Tiefendosiskurven wurden auch mit Filmen aufgenommen; wir verwenden dazu entweder Filme quer zur Strahlrichtung oder auch einen Film, der schwach zur Strahlrichtung geneigt ist. Als Film wurde der Röntgenfilm Kodak AA verwendet, der erst bei einer Schwärzung von $S = 2,5$ Sättigungseffekte aufzuweisen beginnt.

Die mit Filmen gewonnenen Tiefendosiskurven liegen im allgemeinen etwas günstiger als die Messungen mit der Victoreen-Kammer. Wir sind uns aber bewußt, daß diese Kammer das Feld etwas verändert und bauen jetzt eine kleine, dünne Flachkammer, mit der wir auch in der Tiefe des Phantoms Messungen ausführen können, ohne das Feld nennenswert zu stören.

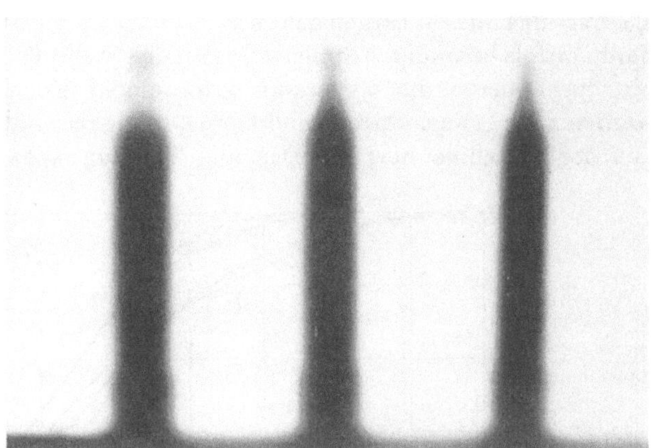

Fig. 5. Elektronenradiographie von drei Victoreen-Kammern. Von links nach rechts: 25, 100, 250 r Meßbereich. Kammern 25 cm vor dem Film

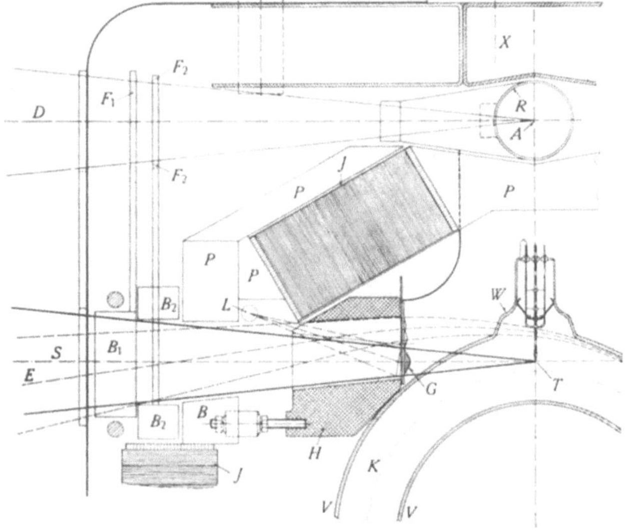


Fig. 6. Schnitt durch den Strahlentransformator in der Kreisbahnebene mit Röntgen- und Elektronenstrahlen. $V$ Vakuumröhre; $K$ Elektronenkreisbahn; $T$ Antikathode (Target); $A$ Antikathode der 125 kV-Röntgenröhre $(R)$; $G$ Ausgleichskörper; $H$ Absorberblock aus Isoblei; $B$ Blende: $B_1$, $B_2$ bewegliche Blenden; $F_1$, $F_2$ Feldanzeige; $D$ Diagnostikfeld; $P$ Bleiabschirmung; $L$ Lichtquelle für die optische Feldanzeige; $J$ Joche des Betatrons; $X$ Drehachse. Die Röntgenstrahlen gehen geradlinig von der Antikathode aus und haben ihre größte Intensität in der Strahlachse $(S)$. Die durch das Fenster $W$ heraustretenden Elektronen durchlaufen wegen der magnetischen Streufelder gekrümmte Bahnkurven $E$, die an der Austrittsstelle aus der Maschine einen Winkel von etwa 8° mit der Strahlachse $S$ bilden

Fig. 6 zeigt im Prinzip unsere Versuchsanordnung. In der Kreisröhre V befindet sich ein Target $T$ für die Erzeugung von Röntgenstrahlen, die in dem Kegel mit der Achse $S$ emittiert werden.

Wenn die Elektronen aus der Röhre herausgebracht werden sollen, werden sie erst über dieses Target gehoben, so daß sie hier nicht auftreffen können, und dann mittels besonderer Spulen etwa an der Stelle $W$ gegen die Glaswand gebracht. Sie durchqueren die Glaswand unter einem Energieverlust von etwa 1 MeV, werden auch etwas gestreut und treten auf gekrümmten Bahnkurven („Achse" $E$) aus der Maschine heraus. Das aus Trolitul (spez. Gewicht 1,05) bestehende

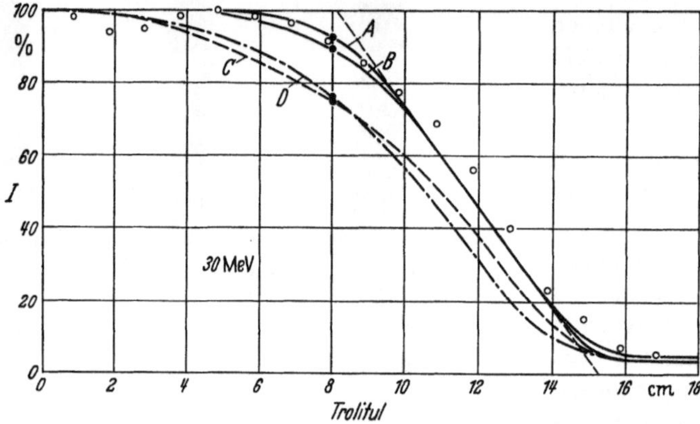


Fig. 7. $\beta$-Tiefendosiskurven. Feldgröße 100 mm $\varnothing$. *A* Abstand von der Austrittstelle 1 m. Ohne Streuplatte. Gemessen in Bern 19. 6. 57. *B* Abstand 1,6 m. Ohne Streuplatte. Gemessen in Baden 28. 2. 57. *C* Abstand 1 m. Dünne Streuplatte. Messingtrichter vor der Blende. Gemessen in Baden 26. 6. 57. *D* Abstand 1 m. 4 mm Al als Streuplatte. Gemessen in Baden 28. 6. 57. Meßpunkte: Radiographie. Anordnung wie *C*, ohne Messingtrichter. Gemessen in Baden 28. 6. 57

Phantom befand sich meist in 1 bzw. 1,6 m Abstand von der Strahlaustrittsstelle $W$, während das Feld durch eine 20 mm dicke Messingplatte dicht vor dem Phantom abgegrenzt wurde. Fig. 7 zeigt nun verschiedene Tiefendosiskurven, die wir in dieser Weise gemessen haben. Die Kurven $A$ und $B$ entsprechen „guten" Tiefendosiskurven, die 8 cm-Werte von etwa 90% oder mehr aufweisen. Die Kurve $A$ wurde in Bern bei Prof. Zuppinger gemessen; die dortige 31 MeV-Maschine wird jetzt gerade für Elektronentherapie eingerichtet.

Wir haben vor der Messingblende einige Messingplatten in der Form eines Trichters aufgestellt, um eine früher von der Heidelberger Schule angegebene Anordnung zu prüfen. Unser Ergebnis war aber nicht günstig, die Tiefendosiskurve sinkt auf einen 8 cm-Wert von etwa 75% herunter, wie die Kurve $C$ zeigt. Dies rührt daher, daß in den Messingblechen Streuelektronen kleinerer Energien entstehen, die die Tiefendosiskurve verschlechtern. Dieser Versuch zeigt deutlich, daß man Streuelektronen vermeiden muß; die Primärelektronen sollen nach Möglichkeit nur von der Blende abgefangen werden. Irgendwelche Vorblenden haben sich einstweilen nicht als vorteilhaft erwiesen, höchstens eine kleine Strahlbegrenzung direkt an der Kreisröhre.

Unsere weiteren Arbeiten betrafen den Intensitätsausgleich. In der Kreisröhre sind die Elektronen in der Kreisbahnebene sehr gut gebündelt; beim Austritt durch die Glaswand erfolgt eine gewisse Streuung in der Richtung senkrecht zur

Bahnebene; aber der Intensitätsabfall ist, wie die Kurve *A* in Fig. 8 zeigt, immer noch zu groß. Wir können diesen Intensitätsabfall durch eine vorgeschaltete Aluminiumplatte von 4 mm Stärke genügend vermindern (Kurve *C*), aber leider erzeugt diese Streuplatte recht viele Streuelektronen und die Tiefendosiskurve wird sehr verschlechtert (Kurve *D* in Fig. 7).

Wir haben diese Schwierigkeit dadurch überwunden, daß wir eine profilierte Streuplatte (beispielsweise aus Plexiglas) verwendeten, die in der Mitte die Intensität stärker vermindert als an den beiden Rändern. Die Kurve *B* (Fig. 8) zeigt die mit einer derartigen Streuplatte gewonnene Intensitätskurve. Die profilierte Streuplatte wird etwa 7 mal dünner als eine äquivalente Streuplatte konstanter Dicke, sie verringert die Elektronenintensität nur wenig und beeinflußt die Tiefendosiskurve praktisch nicht.

Fig. 9 zeigt Isodosekurven in einem Trolitulphantom, die mit einer profilierten Streuplatte aufgenommen wurden. Die Isodosekurven wurden photographisch gemessen (Film Kodak AA). Man sieht, daß die Streuplatte nicht genau in der Mitte des Feldes, sondern etwas zu hoch montiert war. Dies ist ein an sich unwichtiges Detail, welches leicht zu korrigieren wäre und nur geringen Einfluß auf den Verlauf der Isodosekurven in der Tiefe hat.

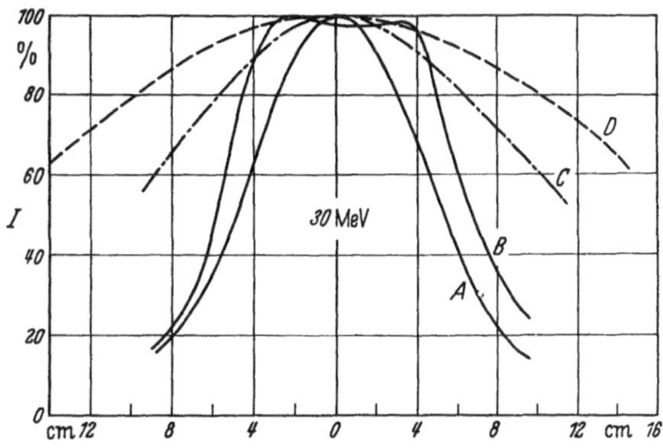


Fig. 8. Intensitätsverteilung der Elektronenstrahlen. *A* ohne Streuer, *B* Streuplatte variabler Dicke, *C* Streuplatte 4 mm Al, *D* Verteilung in der Bahnebene ohne Streuplatte (*A—C*. Verteilung senkrecht zur Bahnebene)

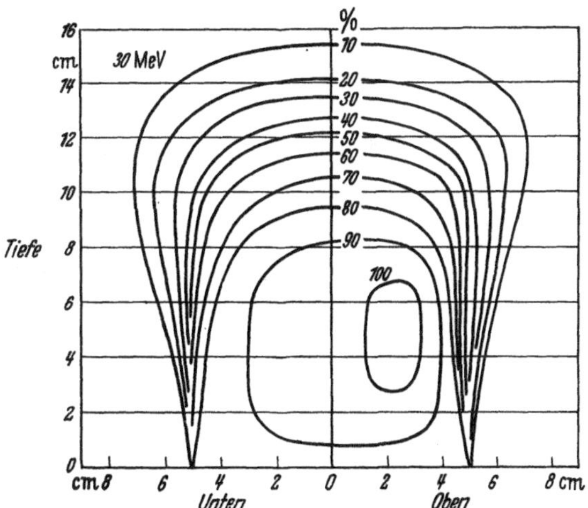


Fig. 9. *β*-Isodosekurven in Trolitul. Abstand von der Austrittstelle = 1 m, Feldgröße = 100 mm $\varnothing$, dünne Streuplatte

Die Isodosekurven bestätigen die Tiefendosiskurven sehr gut, und sie zeigen eine etwas kleinere Seitenstreuung, als man eigentlich aus früheren Messungen erwarten sollte. Dies rührt vielleicht daher, daß wir nur wenige Streuelektronen im Feld hatten.

Wir haben auch Siebfelder (Gittertherapie) untersucht. In Fig. 10 ist der photographische Längsschnitt eines derartigen Feldes dargestellt. Fig. 11 zeigt, wie die Intensität der durch eine Messingplatte abgedeckten Stellen durch Streuung ansteigt, wenn die Elektronen tiefer in das Phantom eindringen. Wenn man im

Sieb Löcher von 20 mm Durchmesser verwendet, die 4 mm voneinander entfernt sind (Abstand der Lochzentren ist 24 mm), so haben wir in 6 cm Tiefe eine Feldschwankung von etwa 83% erreicht (Kurve $A$ in Fig. 8). Fig. 12 zeigt die Tiefen-

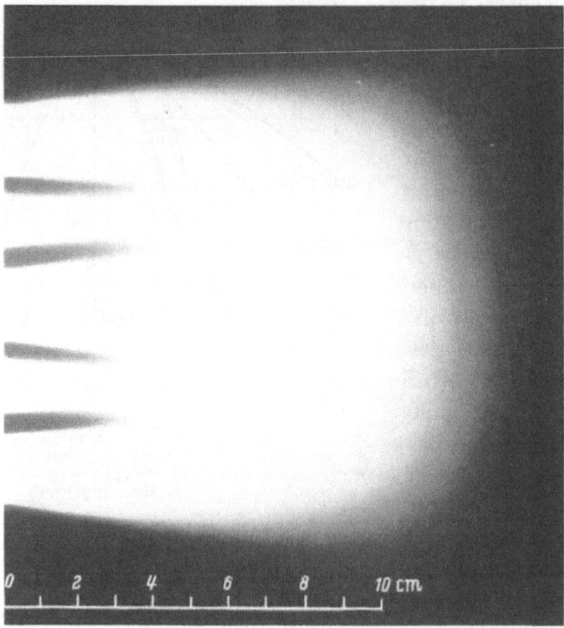


Fig. 10. Photographischer Längsschnitt durch ein 30 MeV-Elektronen-Siebfeld in Trolitul. Lochdurchmesser 20 mm, Abstand zweier Lochzentren 24 mm. Feldgröße etwa 100 mm Durchmesser

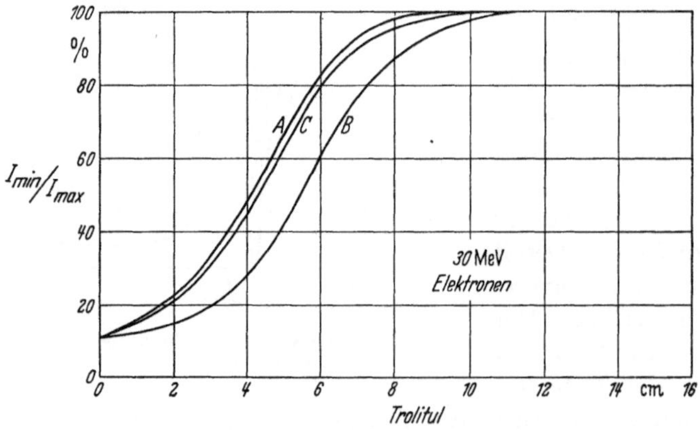


Fig. 11 Intensitätsverteilung in einem Trolitulkörper bei Siebbestrahlung. Abstand von der Austrittstelle = 1 m. $A$ 20 mm ∅ Löcher in 24 mm Abstand, 63% der Fläche belichtet, $B$ 20 mm ∅ Löcher in 28 mm Abstand, 46,4% der Fläche belichtet,   $C$ 14 mm ∅ Löcher in 20 mm Abstand, 44,3% der Fläche belichtet

dosiskurven, gemessen an den Stellen, wo die Intensitäten hoch sind. Wie zu erwarten, werden die Tiefendosiskurven ungünstig beeinflußt, wenn nur ein kleiner Teil des Feldes ausgeleuchtet wird (z. B. bei den Löchern von 20 mm Durchmesser in 8 mm Abstand), und noch schlechter, wenn gleichzeitig auch die Löcher kleiner gemacht werden (Kurve *C* für Löcher von 14 mm Durchmesser). Auf Grund dieser Messungen möchten wir empfehlen, quadratische Löcher von 20 mm Seitenlänge mit einem Abstand von 4 mm zu verwenden. Man muß sich aber darüber im klaren

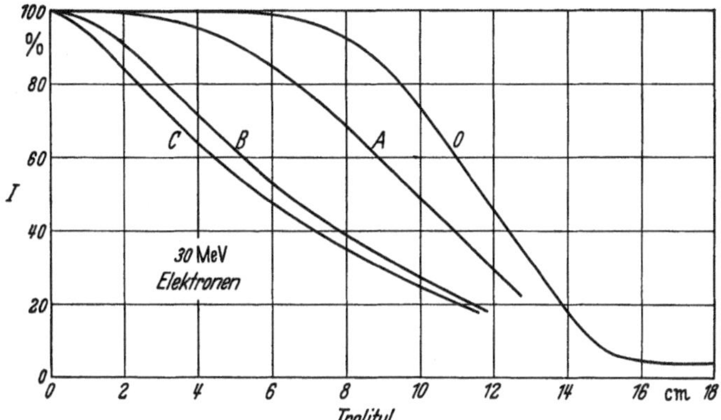


Fig. 12. β-Tiefendosiskurven bei Siebbestrahlung in 1 m Abstand vom Austritt. Dünne Streuplatte, Tiefendosiskurve ohne Sieb, großes Feld (Bern). *A* 20 mm ⌀ Löcher in 24 mm Abstand, 63% der Fläche belichtet. *B* 20 mm Löcher ⌀ in 28 mm Abstand, 46,4% der Fläche belichtet. *C* 14 mm ⌀ Löcher in 20 mm Abstand. 44,3% der Fläche belichtet

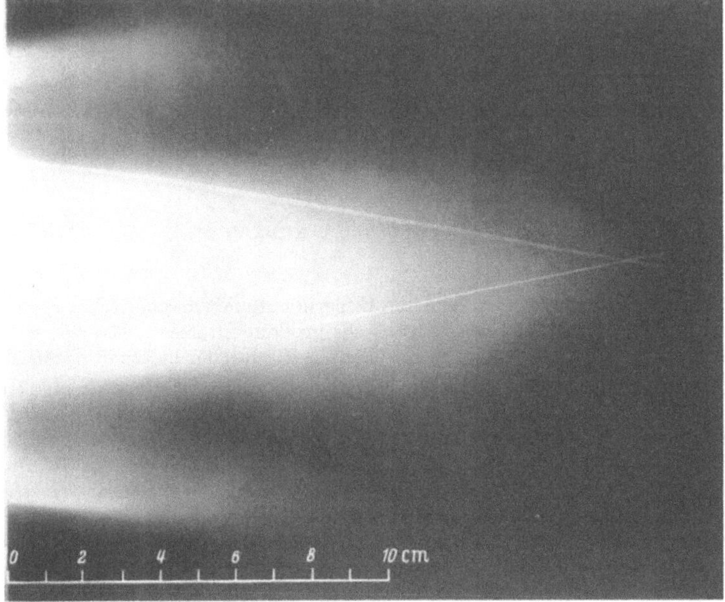


Fig. 13. Photographischer Längsschnitt eines mit magnetischen Linsen konzentrierten 30 MeV-Elektronenfeldes in Plexiglas (spez. Gewicht 1,18)

sein, daß die Siebbestrahlung immer eine gewaltige Verschlechterung der Tiefendosiskurve zur Folge haben wird, und das Verfahren stellt deshalb wohl kaum die endgültige Lösung der Elektronentherapie dar.

Zum Schluß möchte ich kurz eine bei uns im Gang befindliche Entwicklung berühren, die schon gewisse Erfolge gezeitigt hat.

Wir vermuten, daß die Tiefendosiskurve für das Feld von 100 mm Durchmesser schon weitgehend den Wünschen der Radiologen entspricht, sind aber der Ansicht, daß insbesondere die kleineren Felder noch verbesserungsbedürftig sind. Es liegt nahe, hier die Anwendung von Elektronensammellinsen zu versuchen, um die Einflüsse der Streuung zu vermindern und die Intensität in der Tiefe zu steigern.

An der Technischen Hochschule in Karlsruhe hat Dr. ZIEGLER auf unsere Veranlassung mit dem dortigen 31 MeV-Betatron einige Versuche über die Konzentrierung von 30 MeV-Elektronenstrahlen ausgeführt. Er verwendete zwei gekreuzte magnetische Vierpollinsen (Zylinderlinsen), die synchron mit dem Elektronenstrahl kurzzeitig erregt wurden. Die Linsenöffnung betrug etwa 5 cm und die Brennweite etwa 20—25 cm. Es handelt sich hier um einen rohen Vorversuch mit einer Anordnung, die für den vorgesehenen Zweck nicht sehr gut geeignet war (u. a. wirkte sich die Dicke der beiden Linsen recht störend aus). Trotzdem ließ sich aber mittels photographischer Aufnahmen in einem Plexiglasphantom sehr schön die konzentrierende Linsenwirkung auf die Elektronenstrahlen zeigen Fig. 13. Die Tiefendosiskurven zeigen einen deutlichen Anstieg von etwa 30% von der Oberfläche bis zu einer Tiefe von etwa 3 cm.

Diese Versuche waren so ermutigend, daß wir auf diesem Wege weitergehen wollen. Das Problem ist nicht leicht zu lösen, denn eine Linse, die es gestattet, die 30 MeV-Elektronen in 8 cm Tiefe zu sammeln, erfordert eine magnetische Induktion von etwa 4000 Gauß, und wir kommen unter bestimmten Voraussetzungen auf eine maximale Linsenscheinleistung von etwa 10 MVA. Mit einer derartigen Anordnung sollte es möglich sein, die Oberflächendosis auf weniger als die Hälfte zu reduzieren.

## Diskussionsbemerkungen

J. BECKER (Heidelberg):

Die von Herrn Dr. WIDERÖE angeführten Experimente mit magnetischer Fokusierung von Elektronenstrahlbündeln sind zweifellos von sehr großem Interesse. Es wird sich allerdings erst in der Praxis zeigen, ob der nicht unerhebliche Aufwand, der hierzu erforderlich ist, lohnt.

Die begrenzte Eindringtiefe von 15 MeV Elektronen gestattet immerhin eine so vielseitige Indikation, daß wir täglich etwa 40 Patienten mit Elektronen bestrahlen. Die charakteristische Tiefendosiskurve für Elektronen läßt diese zur Behandlung solcher Tumoren geeignet erscheinen, die sich von der Körperoberfläche nach der Tiefe hin erstrecken. In den nicht sehr häufigen Fällen, in denen dabei eine Tiefenausdehnung von 6 cm überschritten wird, konnten wir zum Teil durch Anwendung eines kleinen Kunstgriffes dennoch eine Elektronentherapie durchführen. Wir gehen dabei so vor, daß wir im Anschluß an die Elektronentherapie mit 15 MeV, die in 6 cm Tiefe etwa 50% und in 8 cm Tiefe annähernd 0% der Maximaldosis ergibt, eine Pendelbestrahlung mit 15 MeV Röntgen-Strahlen durchführen und dabei das Maximum in eine Tiefe von 8 cm legen. Dieses Maximum schließt sich sehr gut an den abfallenden Teil der Elektronenkurve an, so daß sich aus der Summierung eine nahezu homogene Dosis bis zu einer Tiefe von 10 cm ergibt, der dann, entsprechend dem steilen Gradienten bei der Pendelung mit ultraharten Röntgenstrahlen, ein scharfer Dosisabfall nach der Tiefe hin folgt. Ein solches Vorgehen setzt zwar voraus, daß an der betreffenden Körperstelle die Pendelbestrahlung durchführbar ist und erfordert auch einen zusätzlichen Zeitaufwand, doch gestattet es in einigen der wenigen Fälle, in denen die wirksame Eindringtiefe der 15 MeV-Elektronen nicht ausreichend ist, wenigstens den räumlich größten Teil des Tumorgebietes mit Elektronen zu bestrahlen.

R. WidERöE (Baden/Schweiz):

Es ist richtig, wie Herr Prof. BECKER es erwähnte, daß man in gewissen Fällen durch Kombination verschiedener Bestrahlungsanordnungen Vorteile schaffen kann. Indessen erscheint es mir von primärer Bedeutung, daß man zunächst die Eigenschaften der einzelnen Bestrahlungskomponenten so günstig wie möglich gestaltet. Die hier von uns beschriebenen Arbeiten müssen von diesem Gesichtspunkt aus beurteilt werden: Wir haben uns bemüht, möglichst günstige Bedingungen für die Tiefenbestrahlung mit Elektronen zu schaffen. Die Angabe von Prof. BECKER, daß die schon heute erreichten Tiefendosiskurven mit 30 MeV-Elektronen für die Therapie von großer Bedeutung sein werden, hat uns sehr erfreut.

Ich möchte noch mit wenigen Worten auf die von Prof. BECKER vorgeschlagene Kombination von 15 MeV $\gamma$-Strahlen und 15 MeV Elektronenstrahlen (ohne Gitterfeld) zurückkommen. Eine derartige Kombination dürfte gerade die weniger günstigen Eigenschaften der beiden Strahlenarten zur Wirkung bringen und wird deswegen wohl nur in ganz seltenen Fällen von besonderem Vorteil sein.

Bei den $\gamma$-Strahlen ist die hohe Austrittsdosis weniger günstig, bei den Elektronenstrahlen stört uns die hohe Eintrittsdosis. Bei der Kombination beider Strahlarten ergänzen sich die ungünstigen Eigenschaften, ohne daß wir im Tumorgebiet etwas gewinnen. Dies ist ja auch der Grund, warum wir die $\gamma$-Strahlen immer peinlich von Elektronenstrahlen säubern, während wir bei der Elektronenbestrahlung die Erzeugung von $\gamma$-Quanten zu vermeiden suchen. In dem Kurvenblatt (Fig. 1) habe ich Beispiele für die mit dieser Kombination möglichen Tiefendosiskurven aufgezeichnet. Sie ersehen aus diesen Kurven folgendes:

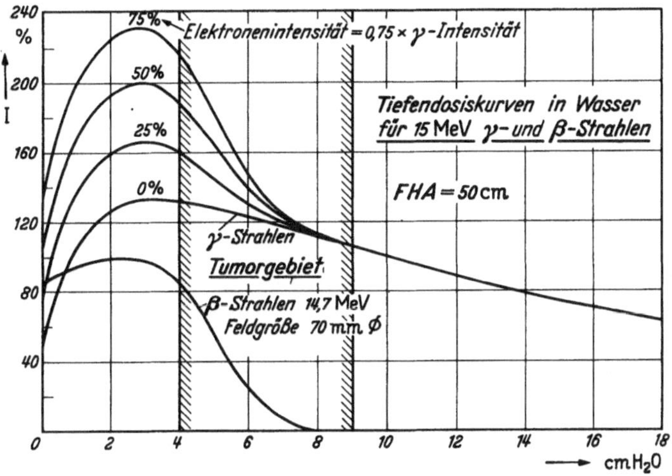


Fig. 1. Tiefendosiskurven für kombinierte $\gamma$- und $\beta$-Bestrahlung

1. Im Tumorgebiet, beispielsweise von 4—8 cm Tiefe, ist die Intensität sehr ungleichmäßig und muß nach den bisherigen Ansichten der Therapeuten im allgemeinen als sehr unvorteilhaft bezeichnet werden.

2. Die Intensität in der Hautoberfläche ist von der gleichen Größenordnung wie im Tumorgebiet.

3. Wir haben im Tumorgebiet fast ausschließlich eine $\gamma$-Strahlung. Die Bestrahlungszeiten werden wegen der kleinen $\gamma$-Intensität der 15 MeV-Maschine somit lang sein.

Diese Untersuchung bestätigt somit die vorhin geäußerte Ansicht, daß eine Kombination von Elektronen- und $\gamma$-Strahlentherapie keine Vorteile, sondern nur Nachteile bringt.

An dieser Tatsache läßt sich auch durch eine Bewegungstherapie nichts ändern; man kann höchstens die einfache Tatsache verschleiern und eine sehr unerwünschte Unklarheit schaffen.

# Strahlenschutzmessungen an schnellen Neutronen*

Von

W. Pohlit, Frankfurt a. M.

Bei dem Betrieb von Bestrahlungsanlagen mit energiereichen Strahlungen
treten eine Reihe von Schwierigkeiten für den Strahlenschutz auf. Erstens ist die
von der Strahlung betroffene Umgebung und damit der in Mitleidenschaft
gezogene Personenkreis sehr groß. Zum anderen benötigt man nicht wie bei den
Röntgenstrahlen Strahlenschutzstoffe in Mengen von Kilogramm, sondern immer
gleich in Tonnen, so daß man sich mit diesen Fragen eingehend beschäftigen muß,
um den Strahlenschutz möglichst rationell
durchzuführen. Zum dritten tritt aber bei
Energien über 10 MeV noch eine weitere unerwünschte Strahlungskomponente dazu, die
Neutronen, auf die hier etwas näher eingegangen werden soll. Diese Neutronen
entstehen durch Kernphotoeffekte in den
Blenden, in der Wand des Beschleunigungsgefäßes und in den von der Gammastrahlung
getroffenen Raumwänden. Sie erfüllen damit
die vom Bedienungspersonal bzw. von den
Patienten benutzten Räume und aus diesem
Grund sollten Geräte vorhanden sein, damit
man sie genau wie die Gammastreustrahlung
messen und ständig kontrollieren kann, um

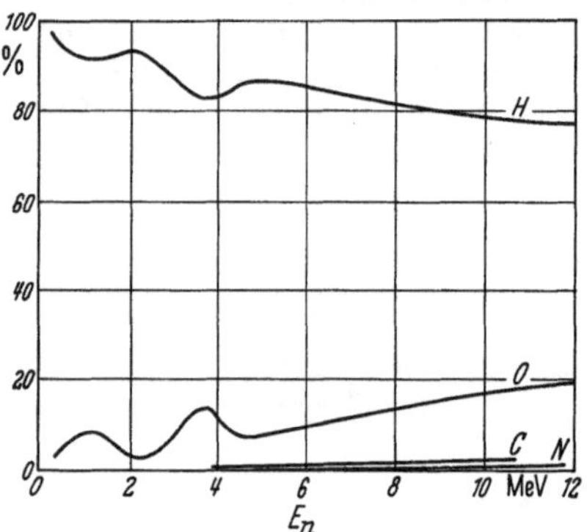

Fig. 1. Energieübertragung auf das Gewebe

daraus die schädigende Wirkung auf die anwesenden Personen abschätzen
zu können. Die Neutronen selbst ionisieren nicht, sie können aber im Gewebe
mit den Atomkernen in Wechselwirkung treten und auf diese Weise schnelle
geladene Teilchen bilden, die dann die Energieübertragung auf das Gewebe
übernehmen. Die im Körper vorhandenen Elemente sind an dieser Energieübertragung in ganz verschiedenem Maße beteiligt, wie Fig. 1 zeigt. Man
erkennt, daß der Wasserstoff dabei die Hauptrolle spielt, da er einen Atom-
kern besitzt, der etwa die gleiche Masse wie das Neutron hat, also bei Zusammen-
stößen eine große Energie übernehmen kann. Außerdem sind die Wasserstoffkerne
im Gewebe am häufigsten vertreten ($C_5H_{40}O_{18}N$).

Man braucht also bei der Neutronenabsorption im Gewebe in erster Linie
nur den Wasserstoff berücksichtigen und kann statt der Gewebsdosis die Dosis
in einem wasserstoffhaltigen Material bestimmen. Zum Beispiel ist in Fig. 2
die Gewebsdosis und die Äthylendosis als Funktion der Neutronenenergie aufgetragen.

---

* Aus dem Max-Planck-Institut für Biophysik, Frankfurt a. M. (Direktor: Prof. Dr. B.
Rajewsky)

Eine Ionisationskammer mit Polyäthylenwänden und mit einer Äthylen-Gasfüllung liefert demnach einen Strom, der der Gewebsdosis proportional ist und unter Berücksichtigung des in Fig. 2 angegebenen Faktors 1,45 und der für ein Ionenpaar in Äthylen verbrauchten Energie kann der Absolutwert der Gewebsdosis für schnelle Neutronen angegeben werden.

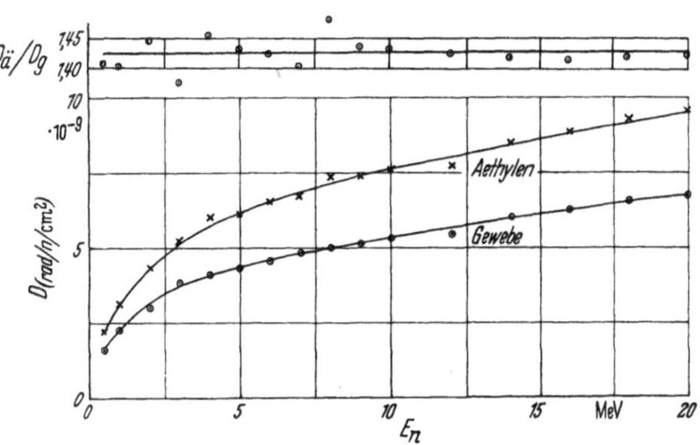

Fig. 2. Gewebsdosis und Äthylendosis als Funktion der Neutronenenergie

Eine besondere Schwierigkeit bereitet aber bei einer solchen Messung die Gammastreustrahlung, die im allgemeinen räumlich und zeitlich sehr inhomogen sein kann. Fig. 3 zeigt z. B. ein solches Streufeld in der Umgebung der Frankfurter 35 MeV-Schleuder.

Die im Mittel vorhandene Neutronendosis ist nur etwa $^1/_{10}$ der vorhandenen Gammadosis, wegen der aber 10 mal größeren biologischen Wirksamkeit der schnellen Neutronen sind dann die schädigenden Wirkungen der beiden Strahlungskomponenten etwa gleich. Die Neutronen dürfen demnach in diesem gemischten Strahlungsfeld durchaus nicht vernachlässigt werden.

Andererseits zeigt dieses Dosisverhältnis $D_n/D_\gamma = 0,1$, daß der Ionisationsstrom in einer Äthylenkammer in der Hauptsache durch die Gammastrahlung bewirkt wird und nicht durch die Neutronen. Aus diesem Grund muß man die Gammastrahlung gesondert bestimmen, z. B. mit einer nur auf Gammastrahlen ansprechenden Kammer aus Graphit und einer $CO_2$-Füllung. Wegen der in Fig. 3 gezeigten Inhomogenität der Gammastreustrahlung kommt ein Nebeneinanderstellen der beiden Kammern nicht

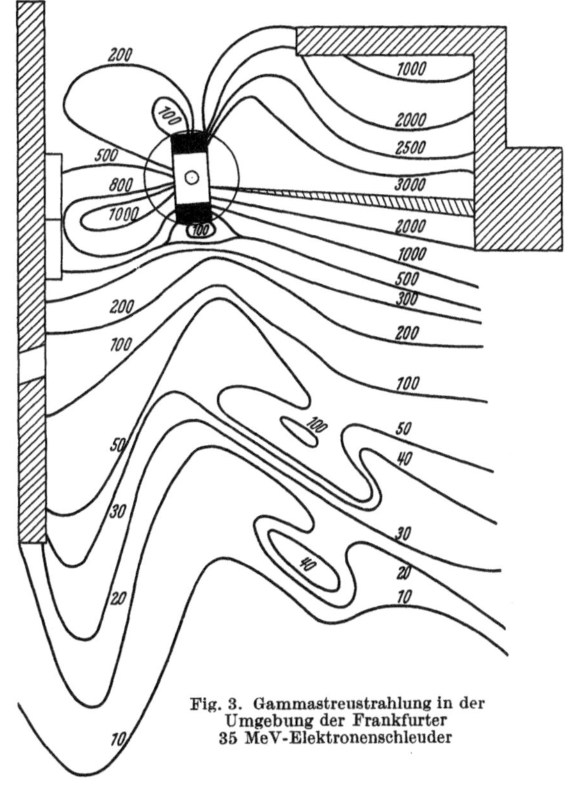

Fig. 3. Gammastreustrahlung in der Umgebung der Frankfurter 35 MeV-Elektronenschleuder

in Frage. Aus diesem Grund haben wir die in Fig. 4 schematisch gezeigte Anordnung der beiden verschieden empfindlichen Kammern gewählt. Infolge der Kugelgestalt aller Kammern und der konzentrischen Anordnung ist eine weit-

gehende Richtungsunabhängigkeit garantiert. Ganz innen befindet sich die
für Neutronen und für Gammastrahlung empfindliche Äthylenkammer, wobei
die Außenwand als Meßelektrode dient. Diese ist gleichzeitig Innenwand und

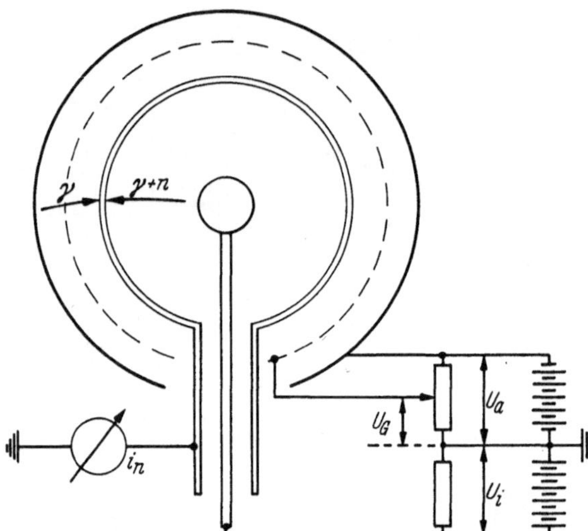


Fig. 4. Meßprinzip der Doppelkammer

Meßelektrode der äußeren, nur
auf Gammastrahlen anspre-
chenden Kammer. Da die
beiden Kammerspannungen
entgegengesetzte Vorzeichen
haben, können sich die Gam-
maionisationsströme bei geeig-
neter Einstellung auf der Meßelektrode gerade aufheben, so
daß nur der Neutronenstrom
übrig bleibt, der dann mit einem
Meßverstärker gemessen und
registriert werden kann. Die
einsatzbereite Anordnung ist
in Fig. 5 wiedergegeben.

Da die Ausschaltung des
Gammastromes auf der Meß-

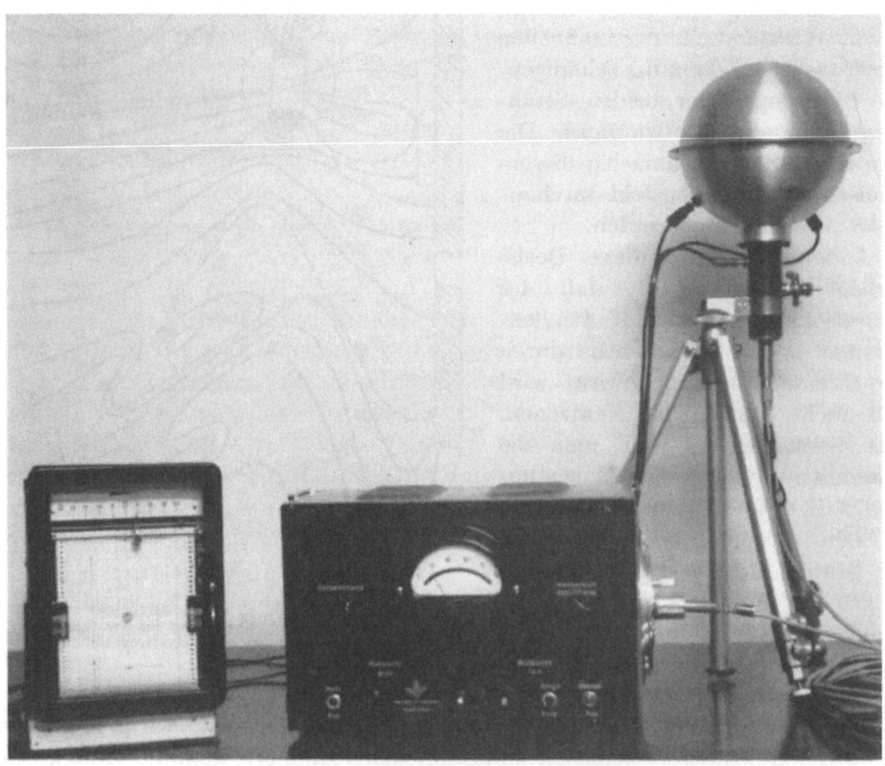

Fig. 5. Neutronen-Ionisationskammer mit Meßverstärker und Schreiber

elektrode sehr genau erfolgen muß, ist in die äußere Gammaionisationskammer ein
elektrisches Gitter eingebaut, das je nach seiner Potentiallage einen Teil des Ionisa-
tionsstromes abfängt. Diese Einstellung ist empfindlich genug, so daß auf diese Weise
die beiden Gammaströme genau kompensiert werden können, und zwar unabhängig
von der vorhandenen Intensität der Strahlung. Man geht also bei einer Messung
so vor, daß man die Doppelkammer mit einem Radiumpräparat aus einer gewissen
Entfernung bestrahlt und das Gitter der Außenkammer solange verändert, bis
die ganze Anordnung auf Gammastrahlung gar nicht mehr anspricht. Geht man
dann damit in ein gemischtes Feld aus Gammastrahlung und Neutronen, dann

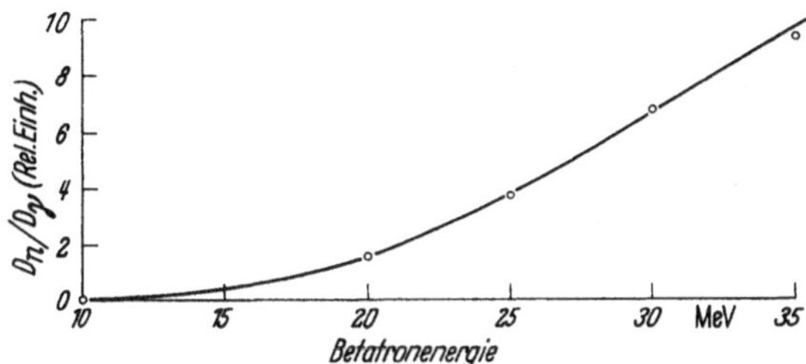


Fig. 6. Neutronenausbeute mit steigender Energie

ist der gemessene Strom ein Maß für die vorhandene Neutronendosisleistung. Bei
einem Betatron mit variabler Energie kann man auch so vorgehen, daß man
zunächst bei kleiner Energie (10 MeV), bei der noch keine Neutronen zu erwarten
sind, diese Gammakompensation durchführt und dann die Energie steigert. Wie
Fig. 6 zeigt, erhält man auf diese Weise die an diesem Ort bei steigender Energie
vorhandenen Neutronendosisleistungen.

Da sich auch andere Institute, teils mit gleichen, teils mit anderen Be-
schleunigern mit diesem Problem beschäftigen, wird man aus den Vergleichen
wertvolle Hinweise für einen rationellen Neutronen-Strahlenschutz erwarten
dürfen.

# Dosisverteilungen bei der Elektronen-Gittertherapie*

Von

F. GUDDEN und A. EHRLY, Heidelberg

Da sich die Elektronen-Gitterbestrahlung bei der Behandlung von Haut- und oberflächennahen Tumoren sehr bewährt hat, war es wünschenswert, ein für die praktischen Bedürfnisse optimales Gitter zu finden. Die Hauptforderungen, die bei der Gittertherapie gestellt werden, sind:

1. Möglichst hohe Dosis in der Tiefe, bezogen auf die Oberflächendosis, damit die großen Vorteile der Gitterbestrahlung auch bei tieferliegenden Herden ausgenützt werden können.

2. Es sollen die Lochdurchmesser nicht größer als 7 mm gewählt werden, weil bei kleinen Gitterlöchern die Hauttoleranz wesentlich erhöht wird, wie Untersuchungen von JOYET und HOHL [2], KEREIAKES [3] und anderen bei 250 kV-Röntgenstrahlung zeigten. Die Narbenbildung bei hoher Dosierung ist schwächer und in kosmetischer Hinsicht befriedigender.

3. Möglichst kleiner Streuzusatz, da die an den Lochrändern gestreuten Elektronen nur die Oberflächendosis erhöhen, ohne nennenswert zur Tiefendosis beizutragen.

Der Streuzusatz ist definiert als das Verhältnis

$$\frac{D^L - D^0}{D^0} \; 100\%$$

Dabei ist $D^L$ die spezifische Dosis hinter einem Gitterloch und $D^0$ die spezifische Dosis an derselben Stelle ohne Gitter, gleicher Elektronenfluß vorausgesetzt. Der Streuzusatz hängt von verschiedenen Parametern ab: 1. Vom Material, aus dem das Gitter hergestellt wurde, 2. von der Lochlänge, 3. vom Lochdurchmesser und 4. von der Elektronenenergie. — Da der mittlere Streuwinkel in erster Näherung proportional der Ordnungszahl des streuenden Mediums ist, wird bei gegebener Lochlänge und festem Lochdurchmesser der Streuzusatz um so kleiner sein, je höher die Ordnungszahl des Gittermaterials ist, worauf schon VON DER DECKEN [1] ausdrücklich hingewiesen hat, weil dann ein hoher Prozentsatz der an den Lochrändern gestreuten Elektronen um so große Winkel abgelenkt wird, daß er in die gegenüberliegende Wand hineinläuft und dort stecken bleibt. Aus dem gleichen Grund wird bei gegebenem Material und Lochdurchmesser der Streuzusatz um so kleiner sein, je länger das Loch ist. Einen Überblick über die Abhängigkeit des Streuzusatzes von den verschiedenen Parametern geben die Tabellen 1—3.

Tab. 1 entnimmt man, daß auf Grund des relativ kleinen Streuzusatzes nur Blei als Gittermaterial in Frage kommt. Aus Tab. 2 würde man schließen, daß große Lochlängen am günstigsten sind. Dem steht aber entgegen, daß die Elektronen des Betatrons von einer Linienquelle ausgehen, so daß bei großen Feldern,

---

* Aus dem Czerny-Krankenhaus für Strahlenbehandlung der Universität Heidelberg (Direktor: Prof. Dr. J. BECKER).

bei denen die Gitterbestrahlung besonders vorteilhaft ist, an den Randlöchern ein starker Halbschatteneffekt entsteht. Man muß also einen Kompromiß schließen. Wir haben uns zu 7 mm dicken Bleigittern mit Lochdurchmessern von 5 und 7 mm entschlossen, bei denen sich die Höhe des Streuzusatzes und die Stärke der Halbschatten in erträglichen Grenzen halten, die, wie aus Tab. 3 folgt, hinsichtlich des Streuzusatzes günstigen Lochdurchmesser von 2 und 3 mm sind wegen des starken Halbschatteneffektes bei größeren Feldern und wegen der schwierigen Reproduzierbarkeit der Einstellung am Patienten in der Praxis unbrauchbar.

Tabelle 1. *Einfluß des Gittermaterials auf den Streuzusatz. Lochlänge 10 mm, Lochdurchmesser 5 mm*

| Gitter aus . . . . . . | Al | Fe | Cu | Pb |
|---|---|---|---|---|
| Streuzusatz in % . . . | 92 | 53 | 40 | 10 |

Tabelle 2. *Abhängigkeit des Streuzusatzes von der Lochlänge. Gitter aus Blei, Lochdurchmesser 5 mm*

| Lochlänge in mm . . . | 5 | 7 | 10 | 12 |
|---|---|---|---|---|
| Streuzusatz in % . . . | 42 | 27 | 17 | 10 |

Tabelle 3. *Abhängigkeit des Streuzusatzes vom Lochdurchmesser. Gitter aus Blei, Lochlänge 5 mm*

| Lochdurchmesser in mm | 2 | 3 | 5 | 7 | 10 |
|---|---|---|---|---|---|
| Streuzusatz in % . . | 29 | 38 | 46 | 46 | 25 |

Zu der 1. Forderung nach möglichst hoher Tiefendosis ist folgendes zu sagen: Wenn man von dem Beitrag der an den Lochrändern in das Loch hineingestreuten Elektronen absieht, ist der Tiefendosisverlauf bei Feldern, deren Durchmesser größer oder gleich der mittleren Elektronenreichweite ist, nach vollständiger Homogenisierung allein durch das Öffnungsverhältnis, d. h. durch das Verhältnis der gesamten Lochfläche zur gesamten abgedeckten Fläche bestimmt. Das folgt einfach daraus, daß die spezifische Dosis der Zahl der Elektronen pro Volumeneinheit proportional ist. Wenn man also bei gegebenem Elektronenfluß durch die Oberfläche einen Bruchteil $x$ der Elektronen durch das Gitter absorbiert, so wird man in einer bestimmten Tiefe, völlige Homogenisierung vorausgesetzt, auch nur den Bruchteil $x$ der Tiefendosis ohne Gitter haben.

Die ersten hier verwendeten Gitter hatten das aus der Röntgen-Gittertherapie übernommene Öffnungsverhältnis von 40:60 (40% offen, 60% abgedeckt). Erhöht man das Verhältnis auf 55:45, was sich noch gut durchführen läßt, ohne daß die Stege zwischen den einzelnen Löchern zu schmal werden, so verschiebt sich die 50%-Isodose

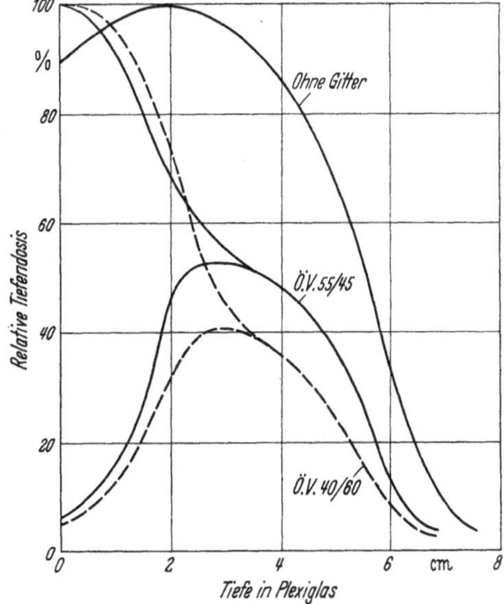


Fig. 1. Relative Dosis als Funktion der Tiefe in Plexiglas für 15 MeV-Elektronen hinter einem Bleigitter mit einem Öffnungsverhältnis von 55:45 (Ö.V. 55/45) und hinter einem Bleigitter mit einem Öffnungsverhältnis von 40:60 (Ö.V. 40/60). Beide Gitter 7 mm dick und mit Lochdurchmessern von 7 mm. Die bei 100% beginnenden Kurvenäste geben die Dosis hinter einem Gitterloch an, die bei 5% beginnenden zeigen die Dosis hinter den abgedeckten Stellen. Zum Vergleich die Tiefendosiskurve ohne Gitter bei einer Feldgröße von 4,5 × 7 cm²

bei 15 MeV-Elektronen von 2,7 cm Tiefe nach 3,7 cm. Dies zeigt Fig. 1. Natürlich tritt bei größerem Öffnungsverhältnis der Punkt des Dosisausgleiches, d. h. die Stelle, an der die spezifische Dosis hinter einem Gitterloch gleich derjenigen hinter einer benachbarten abgedeckten Stelle wird, in etwas geringerer Tiefe auf. Der Eintritt des vollständigen Dosisausgleiches hängt aber insbesondere vom Lochdurchmesser ab: Je kleiner der Lochdurchmesser, desto früher der Dosisausgleich, konstantes Öffnungsverhältnis vorausgesetzt.

Den photometrisch bestimmten Tiefendosisverlauf in Plexiglas für ein 7 mm dickes Bleigitter mit einem Öffnungsverhältnis von 55:45 und 5 mm Lochdurchmesser in Abhängigkeit von der Energie zeigt Fig. 2. Die obere Kurve gibt die spezifische Dosis hinter einem Gitterloch, die untere hinter einer abgedeckten Stelle als Funktion der Tiefe wieder. Die 50%-Isodosen liegen für 15 MeV bei 3,7 cm, für 12 MeV bei 3, für 9 MeV bei 2,5 und für 6 MeV bei 1,5 cm.

Es zeigt sich also, daß die Elektronen-Gittertherapie für alle Herde bis zu einer maximalen Tiefe von 4 cm, bedingt durch die maximale Elektronenenergie von 15 MeV, anwendbar ist, unter Ausnützung der guten Oberflächenschonung und des sehr raschen Dosisabfalles hinter dem Herd.

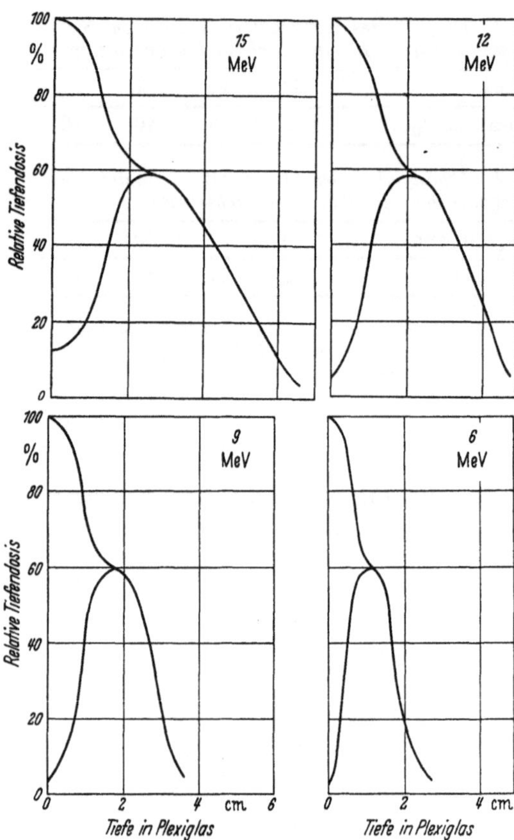


Fig. 2. Relative Tiefendosis in Plexiglas hinter einem Bleigitter von 7 mm Dicke und einem Öffnungsverhältnis von 55:45, Lochdurchmesser 5 mm, für Elektronen-Energien von 15, 12, 9 und 6 MeV

## Literatur


1. BECKER, J., G. WEITZEL u. C.-B. VON DER DECKEN: Strahlenther. **99**, 213 (1956).
2. JOYET, G., u. K. HOHL: Fortschr. Röntgenstr. **82**, 387 (1955).
3. KEREIAKES, J., W. PARR, J. STORER and A. KREBS: Proc. Soc. Exper. Biol. a. Med. **86**, 153 (1954).


## Diskussionsbemerkungen

R. WIDERÖE (Baden/Schweiz):

Wie groß war die Gesamtfläche bei Ihrem Versuch? Wo und an welchen Stellen erfolgten die Messungen, die Sie hier vorgetragen haben?

F. GUDDEN (Heidelberg):

Die Feldgröße betrug 7 × 7 cm und gemessen wurde in der Mitte des Feldes. Wir haben natürlicherweise am Rand des Feldes für die Halbschatteneffekte einen kleinen Abfall. Wir können aber bis zu 8 × 12 cm gehen und die Gitter hierfür sind zum Teil auch schon ausgemessen. Bei kleineren Feldern ist die Gitterbestrahlung nicht mehr von Interesse.

# Calcul de la dose en profondeur dans la thérapeutique de mouvement*

Par

V. Brazzoduro, Rome

Comme on sait dans la thérapeutique de mouvement il n'est pas possible d'effectuer les calculs des doses en profondeur avec une méthode rigoureuse; c'est pourquoi il est nécessaire de recourir à des méthodes approximatives qui sont plus ou moins laborieuses, mais qui introduisent dans la valutation du dosage une certaine erreur.

Même si dans la thérapeutique avec radiations pénétrantes le dosage ne demande pas une précision très poussée il est pourtant nécessaire d'arriver à l'approximation la plus proche de la valeur réelle.

De nombreux systèmes sont employés, beaucoup se ressemblent entre eux, et tous se présentent plus ou moins compliqués, c'est pourquoi nous avons dans notre Centre, élaboré et adopté une méthode particulière; elle dérive de celle publiée par le Groupe Quimby du département radiologique de la Columbia University de New York.

Cette méthode consiste essentiellement à considérer la dose qui vient administrée à un certain tissu pendant une complète rotation de la source, comme si elle provenait de la source même qui au lieu de parcourir sa propre trajectoire d'un mouvement uniforme la parcourût par intermittence en s'arrêtant de brefs intervalles en un certain nombre de points équidistants.

La durée des intervalles de temps est déterminée du temps que la source emploie pour accomplir le cycle tout entier divisé par le nombre de points fixes pris en considération.

La dose qui arrive à un certain point du tissu est déterminée par la somme des doses partielles émises par la source de chacun des points considérés et dans lesquels elle est tenue fixée pour un certain temps. — Notre méthode diffère de celle suivie par le Groupe Quimby dans la façon de calculer les doses partielles.

Les courbes isodoses habituellement disponibles pour le calcul de la dose en profondeur dans la thérapeutique fixe, sont valables pour une distance déterminée source — peau.

Dans la thérapeutique de mouvement au contraire une telle circonstance peut se vérifier seulement en deux points, tandis que pour tous les autres points la distance en est ou majeure ou mineure. Le Groupe Quimby emploie ce même genre de courbes isodoses, même si la distance source — peau est différente de celle à laquelle effectivement elles se rapportent: on retient en effet que l'erreur ainsi introduit, soit insignifiant. Nous avons au contraire préféré d'introduire une

Le présent travail a été exécuté au «Centro di Cure con Cobalto Radioattivo» de Rome, Via Ghirza 15 modification qui, en tenant compte de la distance différente entre la source et la peau dans les divers points, nous consent non seulement une plus grande approximation de résultats, mais aussi une plus grande vitesse de travail.

Il convient à ce propos faire certaines considérations sur le diagramme des courbes isodoses (Fig. 1). Les courbes qui y sont tracées représentent la dose qu'on a aux différentes profondeurs exprimée comme pourcentage de la dose qu'on a à 6 mm. au dessous de la surface de l'entrée de la radiation dans le corps.

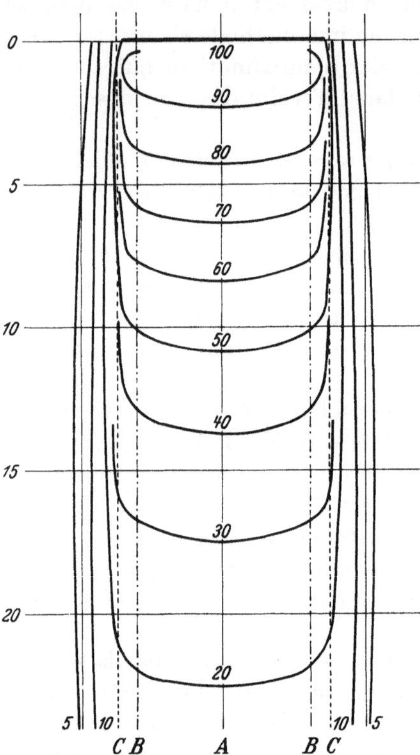

Fig. 1. Courbes isodoses pour le Co 60

La décroissance de la dose avec la profondeur dépend par le produit de deux facteurs, qui sont liés respectivement à la peau et à l'épaisseur du tissu traversé.

Nous pouvons déduire les valeurs de ces facteurs pour les différentes distances du diagramme des courbes isodoses qui reportées sur un second diagramme nous consent un calcul rapide de la dose dans la thérapeutique de mouvement.

Construisons un diagramme (Fig. 2) en papier millimétré semilogarithmique en reportant sur les abscisses (en coordonnées normales) l'épaisseur du tissu traversé et sur les ordonnées (en coordonnées logarithmiques) les valeurs de la dose le long de l'axe central du diagramme des isodoses. La courbe ainsi obtenue aura une allure presque rectiligne et elle nous représentera la diminution de l'intensité le long de l'axe central du faisceau des rayons.

Sur la même feuille nous rapportons maintenant une seconde courbe qui nous représente la diminution de la dose pour le seul effet de l'augmentation de la distance (c'est-à-dire la loi du carré de la distance exprimée par la formule bien connue $\dfrac{(f)^2}{f+d}$ où $f$ est la distance source—peau et $d$ est la profondeur du tissu.

Cette seconde courbe qui aura aussi une allure presque rectiligne, représentera un des deux facteurs dont la première courbe est composée. Ainsi en divisant les valeurs correpondantes de la première courbe pour celles de la seconde, pour ce qu'on vient de dire, on obtiendra une troisième courbe qui représentera le facteur dépendant de l'épaisseur du tissu traversé. Avec ces deux dernières courbes nous pouvons calculer rapidement la dose de la zone centrale de la zone irradiée en n'importe quel endroit. Par exemple pour le calcul de la dose dans le centre de rotation on procède comme suit:

1. on trace (Fig. 3) sur la section du corps des rayons qui passent par le centre de rotation et tournés de 30° l'un envers l'autre;

2. on mesure les longueurs des segments du rayon compris entre le centre considéré et le contour de la section du corps;

3. du diagramme de la Fig. 2 on déduit l'absorption dans chaque direction considérée;

4. on calcule la valeur moyenne de l'absorption;

5. on multiplie cette valeur moyenne pour la valeur de la diminution de l'intensité de la radiation qu'on a pour l'effet du carré de la distance dans le centre de rotation et qu'on obtient aussi du diagramme pour lequel est valable la formule

$$\left(\frac{f}{f+d}\right)^2 \cdot 100,$$

où $f$ est la distance source—peau et $(f + d)$ la distance source—centre de rotation.

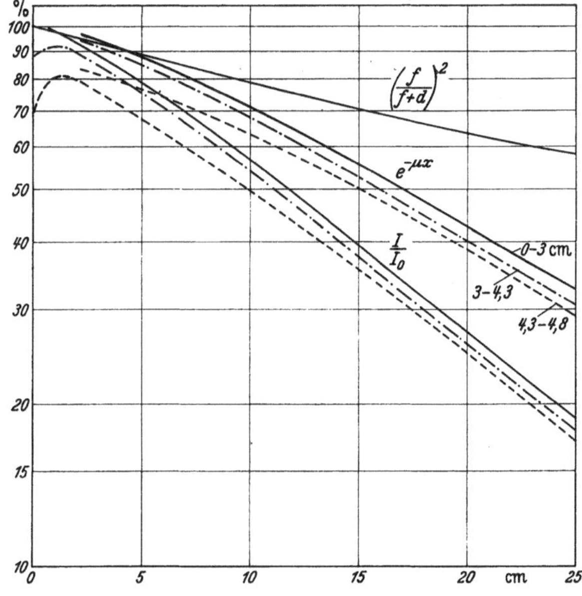


Fig. 2. Courbes de l'absorption

Le nombre ainsi obtenu représente la dose pourcentage qui arrive au centre de rotation rapportée à celle qu'on aurait dans une application fixe en proximité de la peau à la distance standard pour laquelle sont calculées les courbes isodoses, et qu'il ait une durée d'exposition égale au temps employé par la source pour effectuer la rotation complète.

Le calcul de la dose dans les zones éloignées du centre de rotation est un peu plus compliqué car il faut tenir compte que l'intensité de la radiation dans les zones marginales du faisceau des rayons est sensiblement mineure que celle qu'on a dans la zone correspondante centrale, surtout si on entre dans la zone au delà de celle de la pénombre où les courbes isodoses deviennent complètement parallèles à la direction de la bande de rayons parce que en elle agissent seulement les radiations secondaires.

Le diagramme de la Fig. 2 doit être ainsi complété par des courbes obtenues pour des axes parallèles à l'axe central, mais loin de ce dernier. Généralement il suffit de tracer les courbes pour deux de ces axes opportunément choisis. Nous avons appelé la distance de ces deux axes de celui central avec le mot: «Excentricité».

Une fois complété le diagramme, nous pouvons procéder au calcul des points marginaux de la façon suivante (voir Fig. 4):

1. on trace du point considéré les rayons tournés entre eux de 30°, comme pour le centre de rotation;

2. on mesure la longueur de chaque segment de rayon compris entre le point et le contour de la section du corps. Les valeurs ainsi obtenues sont écrites dans une colonne;

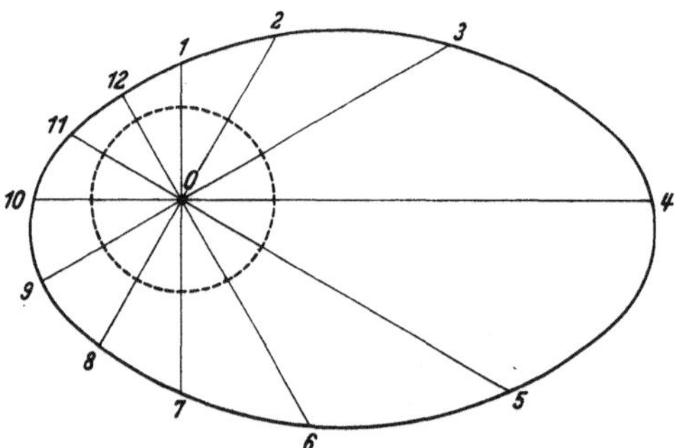

Fig. 3. Diagramme pour le calcul de la dose dans le centre de rotation

3. on écrit dans une deuxième colonne les valeurs de l' «excentricité» et c'està-dire, pour chacune des directions considérées, la distance du rayon qui frappe le point considéré de celui qui frappe le centre de rotation. On relève ces valeurs d'un tableau de facile création;

4. on écrit dans une troisième colonne les valeurs pourcentage de l'absorption pour chaque direction considérée, en les déduisant du diagramme de la Fig. 2 et en choisissant les courbes sur la base de l' «excentricité»;

5. on écrit dans une quatrième colonne les valeurs de pourcentage de la diminution de l'intensité de la radiation qui on a pour l'effet du carré de la distance et qui est différent pour chaque direction considérée. On relève ces valeurs d'un autre tableau de facile création;

6. on marque dans la cinquième colonne les produits des valeurs correspondantes de la troisième et quatrième colonne qui correspondent aux valeurs des doses

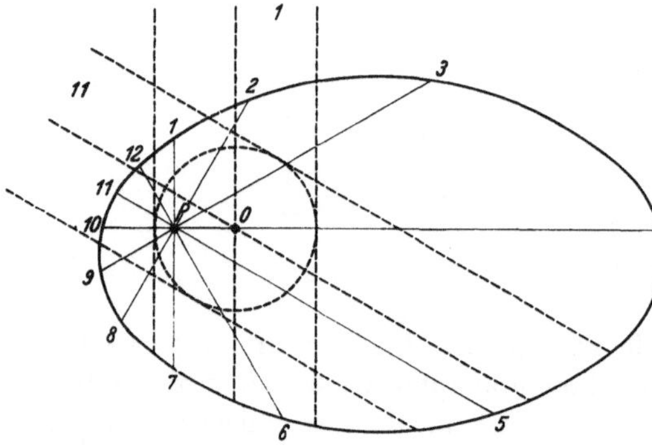

Fig. 4. Diagramme pour le calcul de la dose dans un point quelconque

partielles et qui parviennent au point fixé par les différentes directions considérées;

7. la moyenne des valeurs de la cinquième colonne correspond à la dose qui vient administrée dans le point fixé, exprimée comme pourcentage de la dose qu'on aurait 6 mm. sous la peau en une application fixe, à la distance standard et d'une durée d'exposition égale au temps que la source emploie pour effectuer le cycle complet.

Plus souvent la dose en un point quelconque s'exprime comme valeur de pourcentage de la dose qu'on a au centre de rotation.

Cette méthode est aussi directement applicable aux rotations partielles de la source, c'est-à-dire, pour les «pendulations» comme habituellement on les appelle.

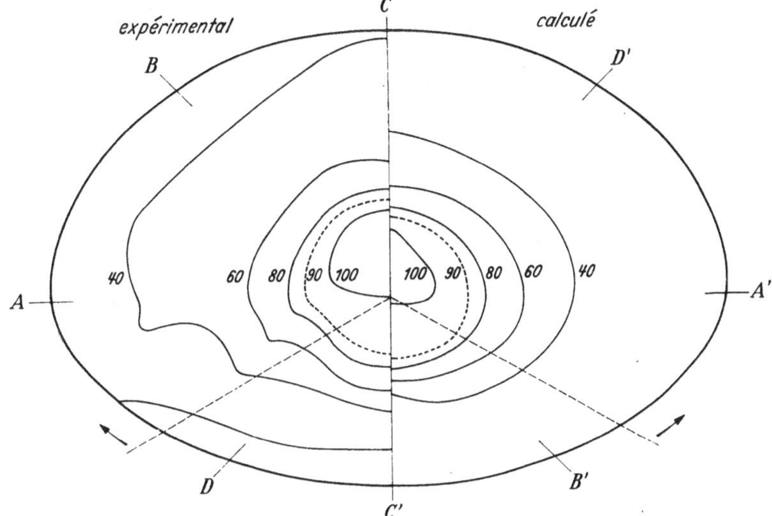


Fig. 5. Courbes isodoses dans une pendulation de 240° (champ 7 × 7)

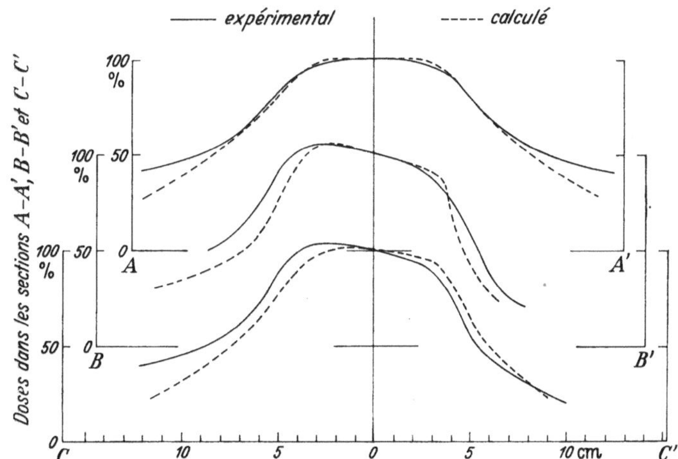


Fig. 6. Courbes des doses dans des sections perpendiculaires au plan considéré

**Contrôle expérimental.** Le contrôle expérimental de la méthode qu'on vient d'exposer a été effectué en se servant d'un fantôme de masonite, à la forme d'un cylindre dont la génératrice court le long d'une ligne correspondante approximativement à la section du tronc du corps humain. Dans le sens de l'axe ont eté creusées des cavités cylindriques à distance régulière aptes à accueillir la chambre de ionisation d'un dosimètre Victoren pour la radiation du Cobalt. Normalement ces cavités sont renfermées avec un petit cylindre en bois.

Le coefficient de l'absorption du fantôme est légèrement inférieur à celui de l'eau et pourtant, moyennant une mesure préliminaire, on a procédé à la construction du diagramme de l'absorption.

La phase préparatoire achevée, on a soumis au «traitement» le fantôme, en répétant les applications toujours dans les mêmes conditions en déplaçant au fur et à mesure les dosimètres dans les différents endroits prédisposés.

Dans la Fig. 5 sont reportées les courbes isodoses pour un cycle de pendulation de 240° avec un champ 7 × 7 cm. Le côté gauche se rapporte à la détermination expérimentale tandis que celui de droit se rapporte aux résultats obtenus avec les calculs.

Pour une comparaison plus facile entre les courbes isodoses obtenues expérimentalement et celles obtenues en se basant sur les calculs ont été construits les diagrammes de la Fig. 6 qui représentent l'allure de la dose le long des quatre sections qui passent par le centre de rotation et déplacées de 45° l'un de l'autre. On voit de tels diagrammes que pour la zone centrale, où la dose est au maximum, on a une correspondance presque parfaite entre le calcul et l'expérience.

Dans les zones marginales au contraire telle correspondence est moins bonne; mais pourtant elle doit être considérée acceptable pour une application courante.

**Conclusions.** Le contrôle expérimental de la méthode proposé pour le calcul de la distribution de la dose en profondeur dans la thérapeutique de mouvement a demontré que cette méthode donne des valeurs assez voisines à celles expérimentales et toutefois acceptables dans le travail courant.

Avec cette méthode il n'est plus nécessaire de faire des contrôles laborieux avec les fantômes.

Je remercie le Professeur Enrico Via pour ses précieux conseils et pour les discussions et le Dr. Orazio Martellacci pour sa collaboration dans l'exécution du travail expérimental.

## Bibliographie

Amer. J. Roentgenol. **73**, N. 5, 815 (1955)